Molecular Response and Genetic Engineering for Stress in Plants, Volume 1

Abiotic stress

Online at: https://doi.org/10.1088/978-0-7503-4921-5

Molecular Response and Genetic Engineering for Stress in Plants, Volume 1

Abiotic stress

Edited by
Pawan Shukla
Seri-Biotech Research Laboratory, Central Silk Board, Bangalore, Karnataka, India

Anirudh Kumar
Department of Botany, Indira Gandhi National Tribal University, Amarkantak, Madhya Pradesh, India

Rakesh Kumar
Department of Life Science, Central University of Karnataka, Kalaburagi, Karnataka, India

Manish K Pandey
Centre of Excellence in Genomics and Systems Biology, International Crops Research Institute for the Semi-Arid Tropics (ICRISAT), Hyderabad, India

IOP Publishing, Bristol, UK

ISBN 978-0-7503-4921-5 (ebook)
ISBN 978-0-7503-4919-2 (print)
ISBN 978-0-7503-4922-2 (myPrint)
ISBN 978-0-7503-4920-8 (mobi)

DOI 10.1088/978-0-7503-4921-5

Version: 20221101

IOP ebooks

British Library Cataloguing-in-Publication Data: A catalogue record for this book is available from the British Library.

Published by IOP Publishing, wholly owned by The Institute of Physics, London

IOP Publishing, No.2 The Distillery, Glassfields, Avon Street, Bristol, BS2 0GR, UK

US Office: IOP Publishing, Inc., 190 North Independence Mall West, Suite 601, Philadelphia, PA 19106, USA

Contents

Preface

The global climate is changing in an unpredictable manner due to anthropogenic activity over the last five to six decades. It is predicted that agricultural crops may face extreme climate change in the form of increases in temperature, drought, flood, and increases in sea level that will lead to intrusion of salt on arable land. These abiotic stress factors are the key growth stressors that adversely affect natural resources and sustainable production of crops through various changes in molecular, biochemical, physiological, and morphological parameters. These stress factors pose a serious threat to food and nutritional security to this world, especially in developing and underdeveloped countries.

In response to abiotic stress, plants have developed specific strategies to survive under different environmental conditions. These strategies are the outcome of evolutionary adaptation in plants, allowing them to grow in various environmental conditions. In addition to this, specific mechanisms help the plant to respond to day-to-day changes in its surrounding conditions. The signal for any change in the environment eventually leads to the expression of the different genes involved in a range of stress-responsive proteins including antioxidant enzymes, ion transporter proteins, chaperones, a protein involved in synthesising osmolytes, hormone biosynthesis, etc. The adoption mechanism through natural as well as human efforts is very slow for even the existing stress factors, while climate change may put further pressure on the plant to adapt to more adverse conditions in a faster manner.

In the last several years, researchers have been trying to understand the mechanisms underlying abiotic stress in order to improve plants' tolerance, particularly for crop production in agriculture. Frequent environmental changes have a major influence on the vegetative development of plants from a range of perspectives, leading to molecular, cellular, physiological, and morphological changes. Recent breakthroughs in this discipline have begun to elucidate on these key processes, which possibly contribute to crop yield enhancement in unfavourable environments. In the recent past, the advent of omics tools has revolutionised the agriculture field. For instance, next-generation sequencing of many agriculturally important crop plants has been accomplished through genomics, transcriptomics, proteomics, and metabolomics during different abiotic stress conditions. The focus of research towards developing climate-smart crops has now been accelerated due to the introduction of marker-assisted selection, genomic selection, genome editing, and metabolic engineering. The deployment of modern science-based solutions may speed up the process of providing the required genetic and management shield in plants from existing and other emerging challenges such as climate change.

Keeping in mind the above facts, this book, *Molecular Response and Genetic Engineering for Stress in Plants, Volume 1: Abiotic stress,* comprehensively and coherently reviews the current research and developments in different abiotic stress-response mechanisms in plants including rapidly growing omics approaches such as genomics, transcriptomics, proteomics, metabolomics, and strategies through genetic engineering and molecular breeding for crop improvement. It provides a detailed examination of how genes and metabolites are modulated in plants in response to various stressors. It covers the various abiotic stress-response mechanisms in plants in view of climate change and global food security. It will help students and professionals to understand plants' response to different abiotic stresses. In addition, this book also provides a way forward to address these upcoming challenges, which will be of interest to readers. Moreover, it will serve as a textbook for students pursuing undergraduate and postgraduate studies in biotechnology and advanced plant molecular biology in postgraduate universities, including in agricultural universities. In addition, the book can also be an asset to modern researchers, plant breeders, and agriculture biotechnologists.

To conclude, understanding how plants respond to abiotic stress factors is crucial for the development of new varieties of crops that are more suited to increasingly harsh climate conditions. This new book, edited by Pawan Shukla, Anirudh Kumar, Rakesh Kumar, and Manish K Pandey, presents a detailed explanation of the molecular basis of plant responses to external stress, such as drought, heat, salinity, cold, heavy metals, oxidative stress, UV radiation, submergence and waterlogging, to aid in the development of stress-resistant crops.

Editors biographies

Pawan Shukla

Dr Pawan Shukla (shklpwn@gmail.com) completed his postgraduate and doctorate degrees from the University of Hyderabad. He was also a Dr D S Kothari postdoctoral fellow at the University of Hyderabad. Currently, he is associated as a Scientist with the Central Silk Board. He has worked at the biotechnology division of Central Sericultural Research and Training Institute (CSR&TI), Central Silk Board, Pampore, Jammu & Kashmir (UT), India and, at present, he is working at the Seri-Biotech Research Laboratory, Central Silk Board, Kodathi, Bangalore. His laboratory is working on the development of mulberry improvements using modern omics tools. He has more than 10 years of research experience in the areas of plant molecular biology and genetic engineering. He has published several research papers in reputed international and national journals. Recently, he has edited two books (*Omics Technologies for Sustainable Agriculture and Global Food Security*, *Vol. I* and *Vol. II*, ISBN 978-981-16-0831-5, ISBN 978-981-16-2956-3), published by Springer Nature.

Anirudh Kumar

Dr Anirudh Kumar, PhD, (anirudh.kumar@igntu.ac.in) is a research faculty member in the Department of Botany, Indira Gandhi National Tribal University (IGNTU), Amarkantak, MP, India. He has more than 10 years research experience in the area of molecular plant pathology. He received his MSc and PhD degrees from the University of Hyderabad, India and postdoc experience from CCCM, Hyderabad, India and ARO, Israel. His current research interests span from antioxidants studies to the antimicrobial properties of plant extracts. He is the author and coauthor of several papers on different aspects of plant biology. He also teaches courses for BSc, MSc, and PhD degrees. For the past few years, his research group has been trying to study the phytochemical profiles of native plants traditionally used by the tribal healers of Amarkantak, MP, India. He has recently edited two books (*Omics Technologies for Sustainable Agriculture and Global Food Security*, (*Vol. I and Vol. II*, ISBN 978-981-16-0831-5, ISBN 978-981-16-2956-3)), published by Springer Nature.

Rakesh Kumar

Dr Rakesh Kumar (rakeshkumar@cuk.ac.in) is a plant biotechnologist, working as assistant professor in the Department of Life Science, Central University of Karnataka, India. He obtained his PhD from the University of Hyderabad, and is an alumnus of the International Crops Research Institute for the Semi-Arid Tropics (ICRISAT). He works in the area of molecular plant developmental biology for crop improvement. His expertise includes genomics and OMICS approaches such as mutagenesis, TILLING, Eco-TILLING, transcriptomics, sequencing, QTL-seq, trait and gene discovery, and LC-MS- and GC-MS-based proteomics and metabolomics. He has published several high-impact research papers, reviews, book chapters, and has obtained research grants from both national and international funding agencies. He has recently edited two books (*Omics Technologies for Sustainable Agriculture and Global Food Security, Vol I* and *Vol II,* ISBN 978-981-16-0831-5, ISBN 978-981-16-2956-3) published by Springer Nature.

Manish K Pandey

Dr Manish K Pandey (M.Pandey@cgiar.org) is working as Principal Scientist-Genomics, Prebreeding & Bioinformatics at International Crops Research Institute for the Semi-Arid Tropics (ICRISAT), Hyderabad-India. He completed his PhD in Genetics and Plant Biotechnology from Osmania University, Hyderabad, India and postdoctoral research experience from the University of Georgia/USDA-ARS, Tifton, USA. His key contributions include reference genome sequence for diploid progenitor and both the subspecies of cultivated tetraploid groundnut, gene expression atlas, low to high density SNP genotyping assays, diagnostic markers for key traits, quality control panel, marker-assisted pyramiding and genomic selection in groundnut. Through genomics-based breeding, his team developed two high oleic groundnut varieties which are released for cultivation in six states of India. He has published more than 170 scientific articles (>7000 citations) in various journals of international repute including *Nature Genetics, PNAS-USA* and *Molecular Plant*. He was also among the top 2% highly cited researchers in 2019 and 2020 compiled by Stanford University, USA. He was also listed among the top Biology and Biochemistry Scientists globally based on a meticulous examination of 166,880 scientists by Research.Com in 2022.

In recognition of his contribution, he has now been selected as Fellow of the Telangana Academy of Sciences (TAS), Telangana and Associate Fellow of the Andhra Pradesh Academy of Sciences (APAS), Andhra Pradesh; and the National Academy of Agricultural Sciences (NAAS), New Delhi, India. He is also Adjunct Associate Professor in University of Southern Queensland (USQ), Australia; Distinguished Professor in Institute of Crop Genetic Resources of Shandong Academy of Agricultural Sciences (SAAS), China; and Minjiang Scholar Lecturer Professor in Fujian Agriculture and Forestry University, Fujian, China. He is also Principal Coordinator of the International Climate Resilient Crop Genomics Consortium (ICRCGC, www.icrcgc.org).

List of contributors

Israr Ahmed
University of Delhi South Campus, Delhi, India

Abhishek Bhandawat
National Agri-Food Biotechnology Institute, Mohali, Punjab, India

Charupriya Chauhan
G. B. Pant University of Agriculture and Technology, Pantnagar, Uttarakhand, India

Krishna Kumar Choudhary
Mahila Mahavidyalaya, Banaras Hindu University, India

Ashish Gautam
G. B. Pant University of Agriculture and Technology, Pantnagar, Uttarakhand, India

Jinu Jacob
Indian Institute of Millets Research (ICAR-IIMR), Hyderabad, India

Kalant Jambaladinni
Central University of Karnataka, Karnataka, India

Priscilla Kagolla
Central University of Karnataka, Karnataka, India

Bini Karunakaran
Kerala Agricultural University, Kerala, India

Anirudh Kumar
Indira Gandhi National Tribal University, Amarkantak, Madhya Pradesh, India

Gaurav Kumar
University of Delhi South Campus, Delhi, India

Manoj Kumar
Agricultural Research Organization, Israel

Rakesh Kumar
Central University of Karnataka, Karnataka, India

Nikhil Malhotra
National Bureau of Plant Genetic Resources Regional Station, Phagli, India

Ankita Mishra
National Agri-Food Biotechnology Institute, Mohali, Punjab, India

Neelam Mishra
Department of Microbiology, Gulbarga University, Gulbarga, India

Ravindra K Panwar
G. B. Pant University of Agriculture and Technology, Pantnagar, Uttarakhand, India

Devade Pandurang Ramrao
Central University of Karnataka, Karnataka, India

Mallikarjuna Rao Puli
University of Hyderabad, Hyderabad, India

Gulab Khan Rohela
Central Sericultural Research and Training Institute, Pampore, J&K (UT), India

Joy Roy
National Agri-Food Biotechnology Institute, Mohali, Punjab, India

Pawan Saini
Central Sericultural Research and Training Institute, Pampore, J&K (UT), India

Abin Sebastian
St Georges College, Aruvithura, India

Himanshu Sharma
National Agri-Food Biotechnology Institute, Mohali, Punjab, India

Shivani Sharma
Indian Agricultural Research Institute Regional Station, Shimla, India

Vikas Sharma
Sant Baba Bhag Singh University, Jalandhar, Punjab, India

Vinay Sharma
International Crops Research Institute for the Semi-Arid Tropics (ICRISAT), Hyderabad, India

Vinita Sharma
National Agri-Food Biotechnology Institute, Mohali, Punjab, India

Pawan Shukla
Seri-Biotech Research Laboratory (SBRL), Bangalore, India

Shweta
Banaras Hindu University, Banaras, India

Kadambot H M Siddique
University of Western Australia, Perth, Australia

Naveen Kumar Singh
Central University of South Bihar, Gaya, India

Nitesh Singh
Indira Gandhi National Tribal University, Amarkantak, Madhya Pradesh, India

Pooja Singh
Mahila Mahavidyalaya, Banaras Hindu University, Banaras, India

K V Sumiya
Kerala Agricultural University, Kerala, India

Srinivas Suresh
Central University of Karnataka, Karnataka, India

Salil K Tewari
G. B. Pant University of Agriculture and Technology, Pantnagar, Uttarakhand, India

Kamal Tyagi
Agricultural Research Organization, Israel

Chibuikem Unamba
Imo State University, Owerri, Nigeria

Himabindu Vasuki
School of Biological Sciences, NISER, Jatni, Bhubaneswar, India

S K Verma
G. B. Pant University of Agriculture and Technology, Pantnagar, Uttarakhand, India

Abhaypratap Vishwakarma
Deshbandhu College, University of Delhi, Delhi, India

Deepanker Yadav
Guru Ghasidas Vishwavidyalaya, Bilaspur, India

IOP Publishing

Molecular Response and Genetic Engineering for Stress in Plants, Volume 1
Abiotic stress
Pawan Shukla, Anirudh Kumar, Rakesh Kumar and Manish K Pandey

Chapter 1

Understanding environmental associated abiotic stress response in plants under changing climate

Vinay Sharma[1], Kalant Jambaladinni[2], Nitesh Singh[3], Neelam Mishra[4], Anirudh Kumar[3], Rakesh Kumar[2*]

[1]*International Crops Research Institute for the Semi-Arid Tropics (ICRISAT), Patancheru, Hyderabad-502324, Telangana, India*
[2]*Department of Life Science, Central University of Karnataka, Kadaganchi, Karnataka-585367, India*
[3]*Department of Botany, Indira Gandhi National Tribal University, Amarkantak-484887, Madhya Pradesh, India*
[4]*Department of Microbiology, Gulbarga University, Gulbarga, India*

The world population has been increasing exponentially, and it is expected that by 2050 food demand will be increased by 70%. However, the impact of abiotic stress has also increased significantly, which poses a major threat to food security. According to environmentalists and agronomists, the detrimental effect of this abiotic stress will be much higher than expectations because of global climate change. Therefore, to meet the future food supply demand and maintain guaranteed food security, the development of superior varieties of crop plants is essential. High and low temperatures, drought, salinity, and heavy metals are among the key abiotic stresses expected to increase as a result of global climate change, according to 'adapting climate-smart crops on the face of the climate science community.' In this chapter we summarise various aspects of abiotic stresses with respect to climate change.

1.1 Introduction

The world population is growing exponentially, which is a threat to human society as it will lead to excessive demand of various resources including food. The growing

* Author to whom any correspondence should be addressed Email: rakeshgupta.hcu@gmail.com.

doi:10.1088/978-0-7503-4921-5ch1

population has severely impacted on the ecosystem by increasing arable land through deforestation. These anthropogenic activities have led to multidimensional environmental degradation and triggered global climate change, which includes unpredictable rain and expansion of semiarid regions [1]. Additionally, it has increased the incidences of terminal drought and heat waves, and also changed the composition of soil microbial consortia. Finally, it could lead to the emergence of abiotic-stress-derived biotic stress, causing enhanced damage to agricultural plants and the associated yields [2]. Presently, drought, heat, cold, salinity and metal stress are drawing significant attention from crop biologists because they pose major challenges for sustainable food security and growing food demand [3].

Although drought is considered to be most common abiotic stress, presently heat stress has also been considered an important stressor in the face of global climate change. Heat stress significantly affects seed germination, fertility, fertilisation, seed-setting, and overall grain yield [4]. Additionally, it reduces the photosynthetic ability of the leaf and overall biomass of the crop plants. Presently, the detrimental effects of heat stress are expanding in several countries, with developing countries expected to experience more exposure to heat stress. Together, drought and heat stress cause substantial reductions in development and yield of important crop plants; however, the extent of loss depends upon on the crop. Unlike heat stress, cold stress is caused due to low temperatures, which leads to changes in membrane fluidity [5]. Under low temperatures the cytoplasm freezes and leads to cell death, therefore plant cells often activate a metabolic process to trigger the accumulation of osmoprotectants. Cold stress instigates change in the cellular membrane fluidity and physiolog, which leads to pollen abortion and seed loss, delayed blooming, and reduced growth rate. The cold stress induces Ca^{2+} signalling, and invokes cascades of the expression of cold-inducible genes triggered by the *CAMTA* transcription factors (TFs), which in turn instigates the expression of another class of TF encoded by the gene *DREB1* [6]. Kidokoro *et al* suggested the existence of a two-part mechanism for cold sensing during day and night, coupled to *CAMTA* and *DREB1* TF-based induction of cold-responsive gene expression [6]. Additionally, the assimilation of protective proteins including anti-freezing proteins (AFPs), late embryogenesis abundant (LEA) proteins and cold shock proteins (CSPs) is essential for cold-stress tolerance in plants [7].

Salt stress is considered as one of the foremost abiotic stressors that can limit the productivity of dominant staple crops globally such as rice, maize, soybean, wheat, etc. In future, the impact of salinity is expected to increase significantly. The salinity of arable lands is increasing consistently due to the excessive use of contaminated and deep underground water for irrigation [8]. The major culprit for salinity stress is the increase of Na^+ and Cl^- ions in the cells of various plant organs, leading to severe ion imbalance due to excessive uptake and causing major physiological disorders [8]. Increased concentrations of Na^+ ions influence the uptake of essential K^+ ions, which are indispensable for the usual growth and development of a plant [9]. Genetic variation for salt tolerance has been examined in model and crop plants to better identify genetic variation related to salt tolerance [10, 11]. In monocots, sorghum is shown to have better tolerance compared to rice and wheat; however, among dicots the model plant *Arabidopsis* was found to be very sensitive when compared with halophytes such as *Atriplex spp., Mesembryanthemum crystallinum, Thellungiella salsuginea,* etc [10, 11].

In the past decade heavy metal stress has been gained significant attention, especially in developing countries. The demand for minerals has substantially increased, which has led in turn to uncontrollable destruction of the Earth's crust. Furthermore, the incidences of excessive use of chemicals, improper disposal of toxic metal waste, including batteries, electronic chips, etc, has also increased. All this activity has significantly deteriorated soil heath through increased abundances of heavy metals like aluminium, arsenic, cadmium, lead, chromium, mercury, etc [12]. These heavy metals slowly accumulate in plant tissue, an assimilation process that can be increased multifold if the soil acidity increases, finally leading to reduced biomass and crop productivity [13]. All the abovementioned abiotic stressors are capable of generating reactive oxygen species (ROS), which can lead to the death of the plant cell [14]. In this chapter, we will describe the major abiotic stresses, genes associated with tolerance, and recent important breakthroughs relating to enhancing crop plant tolerance.

1.2 Aspects of abiotic stress

1.2.1 Types

Drought, salinity, oxidative factors, and high and low temperatures are abiotic stressors that can have a substantial impact on plant growth, development, and

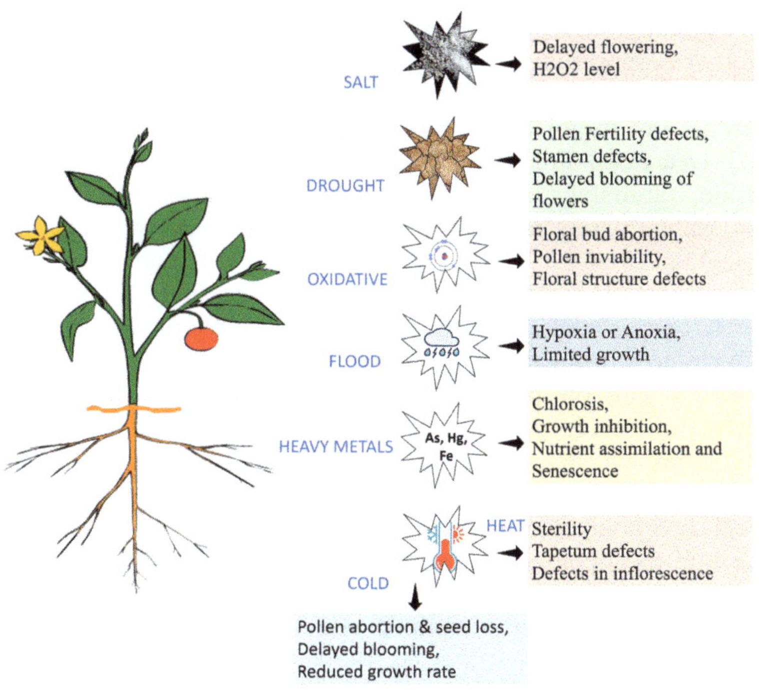

Figure 1.1. A systematic diagram representing various types of abiotic stress and associated physiological and phenotypic changes.

productivity (figure 1.1). Researchers have spent years trying to figure out what causes abiotic stress in order to improve a plants' tolerance to it, particularly for improving crop production in agriculture. Frequent environmental changes have a major influence on the vegetative development of plants from a range of perspectives, leading to molecular, cellular, physiological, and morphological changes. However, recent breakthroughs in this discipline have begun to elucidate on these key processes, which may possibly contribute to crop yield enhancement in unfavourable environments.

1.2.1.1 Drought stress

Drought is amongst the foremost constraints on agricultural production, adversely affecting crop yields. Additionally, as a major constraint on plant development, dehydration can affect plant respiration, photosynthesis, and stomatal movement, which limits plant growth and physiological functioning [15]. Drought stress occurs when the accessibility of moisture for plants is decreased due to a low abundance of water in the soil. On the flip side, water stress occurs in plants when the rate of transpiration from leaf surfaces exceeds the rate of water uptake by roots. Such disparity in water uptake and loss from plants materialises mostly when the water potential of a plant's roots is higher than the soil's water potential. Several plants, including water spenders, water collectors, and water saver xerophytes, thrive in deserts' arid conditions without succumbing to water stress [16]. As a result, researchers have realised that drought is not synonymous with plant water deficiency. Typically, atmospheric conditions lead to a continual water deficit via transpiration or evaporation. Stomatal closure is an early plant response that occurs during mild to moderate water deficit conditions, occurring along with the lowered water potential and turgor accompanying even a minor change in relative water content [17]. Decreased stomatal conductance significantly reduces water loss and CO_2 diffusion, ultimately limiting photosynthetic assimilation. To cope with a shortage of water, plants normally exhibit four different response mechanisms, namely drought avoidance (DA), drought escape (DE), drought tolerance (DT), and drought recovery (DR) [15]. DT refers to a plant's capacity to maintain a specific level of physiological function by differential gene expression and altered metabolic processes in order to mitigate or repair the stress damage caused by the drought. Plants primarily employ protoplasmic tolerance by elevating the amount of osmoregulatory molecules in plant cells in order to uphold cell turgor pressure and by regulating the activity of cell defence enzymes in order to prevent the aggregation of toxic matter. However, DA refers to a plant's ability to maintain key physiological functions during mild or moderate drought stress situations by adapting certain morphological structures or rate of growth in order to avoid the detrimental consequences of drought stress. It is primarily defined by the capability of plants to uphold high water potential in the context of water scarcity. To achieve DA, plants often employ three strategies: (i) reducing moisture loss through stomatal regulation, leaf rolling, and cutical deposition on the leaf's surface; (ii) increasing the ability of a well-developed root system to absorb water (particularly expanding root depth, root density, or root/shoot ratio) and developing specific organs to store water; (iii) promoting or delaying the transition from vegetative to reproductive development in order to reduce total abortion during severe drought stress [18]. DE is the capability of plants to restructure their life span (short life cycle to avoid drought stress). DR refers to a plant's ability to regain growth and increase production (for crops) during an extended period of drought stress that results in the total loss of turgor pressure and dehydration of leaves. Furthermore, natural drought stress is a complicated and unexpected event. As a result, it is quite difficult to completely and properly assess a plant species' overall drought resistance. Depending on the goal of the study, researchers commonly use one trait or a combination of several attributed traits to assess a plant's drought resistance. Such attributed traits are categorised into three categories: related to DA, DT, and integrative mechanisms [19]. These traits are the most important and crucial in drought resistance breeding, and they provide an insight into the physiological and molecular processes of drought resistance.

1.2.1.2 Heat stress

Temperature is an important environmental variable that manipulates the growth, development, and yield of crops. Persistently above-optimal temperatures for plants cause heat stress and lower yields. It becomes lethal at some threshold. Heat stress has shown a detrimental effect on plant physiology (by reducing photosynthesis and increasing respiration), and plant productivity [20]. Another unfavourable consequence of heat stress is its negative impact on the plant root system, which is responsible for providing support, nutrition, water intake, and transport to other plant organs, resulting in interruptions of pollination, blooming, and root development [21]. Heat stress has affected a number of physiological processes associated with seed size and quality. Heat stress during the filling of grain significantly reduces the accumulation of starch in wheat, rice, and maize [22]. Heat stress also lowers the amounts of fructose, sugar nucleotides, and hexose phosphate [23]. In general, declines in total yield are primarily the result of a changing balance of source and sink processes that occur during conditions of heat stress. Plants' ability to withstand temperature stress has been viewed in terms of avoidance, escape, or tolerance [24]. Avoidance is any process that enables the plant to develop or remain tolerant to the perceived stress impact, therefore mitigating the majority of the stressor's negative effects. In escape, plants adjust their development cycle in advance of experiencing severe stress, for example by anticipating reproduction or reducing vegetative elements. From an agronomic perspective, tolerance is unquestionably more intriguing. Because it is based on the plant's tolerance to stress, it does not require significant alteration of the growth habit, which might have a detrimental impact on productivity. Heat-tolerant plants are more likely to continue their growth cycle regardless of the stress conditions.

1.2.1.3 Cold stress

Another abiotic stressor that affects plant growth and development is cold stress. Chilling (0 °C–15 °C) and freezing (0 °C) are two types of cold stress that have a negative impact on a plant's development and yield [25]. Generally, chilling stress inhibits plant growth and development and has a series of foremost consequences for plant cells. In the beginning, chilling stress has an effect on plant cell membrane rigidification, which is thought to be the leading event that initiates downstream cold-stress responses. Likewise, chilling stress destabilises proteins or protein complexes and disturbs the activity of enzymes such as ROS-scavenging enzymes. Such events cause photoinhibition and disrupt photosynthesis, along with incurring severe membrane damage. Ultimately, freezing stress has an effect on gene expression and protein synthesis as it promotes RNA secondary structure formation [26]. Freezing stress is more harmful to plants than chilling stress, and it can even lead to plant death. In natural conditions, freezing injury starts with the formation of extracellular ice nucleation [27]. After the formation of ice nuclei, they start to expand and crystallise, and their spreading into the apoplast, in which they trigger water efflux, leads to dehydration of the cells. Once ice crystals enter cells, irreversible damage ensues [27]. Plants have adapted many mechanisms to mitigate the damage caused by cold. For example, in the cold-acclimatisation mechanism, plants subjected to nonlethal low temperatures for certain days establish an improved capability to bear further freezing stress [28]. Plants enhance their tolerance for cold stress throughout this process by producing a number of protective components (like soluble sugars and proline) and proteins (like LEA, AFPs, and CSPs) [29]. These components are involved in the regulation of osmotic potential, integrity of cell membranes, and scavenging of ROS in plants exposed to cold stress [30]. Various elements of cold-stress signalling pathways have been revealed over the last two decades, counting many messenger molecules, TFs, protein kinases, and phosphatases. Nevertheless, the complex network of molecular mechanisms regulating signal perception and subsequently transduction in plants remains unclear. Low temperatures are amongst the utmost severity and most variable environmental stresses. Thus, the plant's ability to adapt to an unfavourable environment in its life cycle requires it being capable to achieve equilibrium between endogenous developmental cues and exogenous signals and to combine stress reactions with developmental patterning. Investigation is needed to understand the mechanism through which plant cells decide their fate. At the molecular level, a better knowledge of the poise between organogenesis and cold tolerance can aid in the improvement of crop cold-stress responses.

1.2.1.4 Salinity stress

Salinity is a key constraint on plant growth and crop yield, particularly in arid and semiarid areas. Saline soils have an electrical conductivity of more than 4 dS m^{-1} at 25 °C, in the saturation soil-paste extract, which equates to around 40 mM NaCl and produces an osmotic pressure of nearly 0.2 MPa [31]. If crops are grown on soils with an electrical conductivity value greater than 4, the overall yield is considerably reduced. Salts include chlorides, sulphates, carbonates, and bicarbonates of sodium, potassium, magnesium, and calcium. Because salt-affected soils have a complex ionic constitution, which exhibits a wide range of physiochemical characteristics, it is also important to draw a distinction between salinisation and the sodicity of the soil. The term 'soil salinisation' refers to the deposition of soluble salts in soils [32]. It is feasible in arid and semiarid conditions, where evapotranspiration volumes exceed precipitation volumes throughout the year. The term 'sodicity' refers to the amount of Na^+ retained in the soil. Clay with a high sodicity (more than 5% Na+ in the total cation content) expands excessively when moist, obstructing air and water flow and resulting in poor drainage. Two distinct phases have been found in plants' responses to salinity. An ion-independent development arrest, which occurs between minutes and days, results in stomatal closure and cell expansion inhibition, mostly in the shoot [33]. The second phase occurs over a period of days or even weeks and involves the accumulation of cytotoxic ions, which inhibit metabolic activities, result in early senescence, and finally in cell death [34]. Several physiological and molecular processes regulate this stress tolerance, including osmotic tolerance, ionic tolerance, and tissue tolerance [34]. The osmotic tolerance response occurs rapidly and involves a quick change in stomatal conductance in order to maintain water. It relies on quick long-distance (root to shoot) signalling processes which generally bypass the osmotic effects of NaCl, KCl, mannitol, or polyethylene glycol [35]. Salt entry into the root system activates multiple signalling pathways which facilitate ionic tolerance by limiting (net) Na^+ influx into the root and lowering (net) Na^+ translocation. Third, tissue tolerance is strengthened by the compartmentalisation of harmful ions into vacuoles, which prevents them from interfering with cytoplasmic functions. The processes described above have been discovered in a wide range of plant species, and the reported differences in tolerance between glycophytes and halophytes are mostly related to the reliability of the mechanisms used [36]. Although significant progress has been made in recent decades, several of the key mechanisms that contribute to tolerance remain to be elucidated. Future research needs to uncover the exact nature of Na^+ uptake mechanisms. Assessing toxicity at the cellular and tissue level will enable the establishment of aims for tolerance improvements. While perception, sensing, and signalling cascades are lacking key aspects, especially at the commencement of pathways, a better knowledge of the function of further minerals such as K^+ might allow researchers to minimise salt stress by manipulation of their absorption and distribution.

1.2.1.5 Heavy metals stress

Plants have evolved unique mechanisms for controlling physiological metal ion concentrations and restricting nonessential heavy metal access. Heavy metals are all nonbiodegradable and affect the biological structure and functions in such a way that they cause harm or death. Metal toxicity symptoms are observable in plants as a consequence of prior metal-induced structural and ultrastructural alterations. Such alterations at the cellular level are the effect of toxic metals' direct interaction with structural components, which affects the signal transduction and/or metabolic activity. Various studies have shown heavy metals' cytogenetic consequences on plants. Heavy metals impair plant development through interfering with several biochemical, morphological, and molecular events. They induce variations at the transcriptional level of several genes that encode proteins which act as a defensive system for stress-induced damage [37, 38]. Heavy metal toxicity in plants has been studied through three different molecular processes based on their physical and chemical functions: (i) generation of ROS via auto-oxidation and the Fenton reaction; (ii) obstructing of key functional groups in biomolecules; (iii) removal of key metal ions from biomolecules [39]. Earlier study of *Ocimum tenuiflorum* reported accumulation of chromium, which results in the formation of malondialdehyde via an overabundance of free radicals [40]. Likewise, previous studies reported copper interaction with a number of biochemical pathways in plants via raising the generation of ROS [41, 42]. Such compounds, in turn, cause damage to cell walls by adhering to the sulfhydryl groups of protein complexes or through enhancing the rate of lipid peroxidation. Several strategies are employed by plants to mitigate damage due to high heavy metal concentrations via detoxification, which impart tolerance in response to heavy metal stress. Some mechanisms target specific metal ions (evidently, a few plants possess multiple strategies to prevent metal accumulation), and comprise the exclusion of specific metals from the intracellular environment. A plant's first line of defence is to minimise the uptake of metals. It is proficient with the aid of cellular and root exudates. Such exudates act as a barrier to metal entry into the cell by enhancing efflux. Plants facing stress also engage in symbiotic associations with other organisms, like mycorrhizal fungi, that accumulate metals inside the rhizosphere, thus restricting the plant from reaching them. Plants have evolved signalling cascades that involve multiple genes simultaneously for defensive and regulatory functions in response to heavy metal stress. The mitogen-activated protein kinase (MAPK) cascade is one of the most critical pathways implicated in plant abiotic stress [43]. Under such scenarios, efforts need to be made to find ways to cope with such stressors.

1.2.1.6 Waterlogging, submergence and hypoxia

The excessive precipitation of rainwater often causes waterlogging, which results in submergence of the crops causing hypoxia. The stagnant water causes hypoxia leading to the production of significant amounts of ROS, which can affect plant survival. Waterlogging may initiate numerous morphological traits and cellular acclimation responses [44]. On agricultural land it causes the release of excessive minerals that can further cause severe stress. In *Arabidopsis*, flooding and low-oxygen stress is regulated by a TF, encoding gene ethylene response factors (ERF-VIIs), which includes five homologous ERF-VII genes, e.g. *RAP2.2, RAP2.3, RAP2.12*, *HRE1,* and *HRE2,* [45, 46]. In rice, one of the most important homologues of the ERF-VII TF gene, *SUB1A*, the master regulator of submergence tolerance, allowing plants to endure complete submergence for 14–16 days [47]. Other important submergence tolerance ERF-VII family genes, *SNORKEL1 (SK1)* and *SNORKEL2 (SK2)*, were identified in deep-water rice [48].

1.2.2 Abiotic stress and oxidative stress

Crop growth, development, and production are all negatively affected by environmental stresses. Oxidative stress is caused by an imbalance in the manufacture and elimination of ROS and finally results in cell death. ROS are continuously formed during regular cell metabolism, mostly by respiratory and photosynthetic processes.

Oxidative stress is commonly a factor in responses to a number of abiotic stresses, like drought, salinity, high heat, low temperature, and heavy metal stress. Oxidative stress is a controlled process, and the balance between ROS and their quenching determines a plant's survival. When the equilibrium between the formation of ROS and their quenching by antioxidants becomes unbalanced, oxidative damage is caused to the cells. The degree to which oxidative stress occurs is determined by the kind of ROS generated, its amount and location of release, its contact with other cellular components, as well as the developmental stage and effectiveness of the cell [49]. Excessive and continuous ROS generation can harm plant cell structures, causing lipid peroxidation, protein oxidation, nucleic acid impairment, enzyme inhibition, and programmed cell death [50]. Nevertheless, in response to these ROS, plants have developed defensive scavenging processes. ROS levels are kept under control by dynamic and synergistic processes involving ROS-scavenging antioxidants which regulate the intracellular ROS concentration [51]. An antioxidant is defined as a substance that is proficient at scavenging ROS without being converted into a harmful radical [52]. These molecules play critical and interrelated roles in nonenzymatic detoxification reactions as well as substrates in enzyme-catalysed detoxification reactions [53]. Plant stress tolerance has been significantly improved due to recent advancements in molecular and genetic techniques, which have resulted in the formation of transgenic plants with increased antioxidant enzyme activity. Overexpressions of antioxidant enzyme genes in transgenic plants show a positive response to stress tolerance. Additionally, chemical priming is a suitable alternative to genetic engineering for accomplishing comparable aims, which simultaneously regulates the antioxidant defence system in response to stress [54]. Moreover, identification and modification of pathways linked with ROS-detoxifying regulators can be enhanced to develop stress tolerance genotypes which will be helpful for crop yield improvement under stressful conditions.

1.3 Major affected crops and resistant varieties

Legumes are one of the most vital cash crops providing a good quantity of nourishment because they are rich in free amino acids, protein, fats, minerals, and vitamins [55]. Chickpeas, groundnuts and soybeans are leading leguminous crops. The groundnut and soybean are grown as oilseed crops, used for both protein and oil; in particular, demand for such crops has increased significantly due to the availability of varieties with high oleic acids [56]. The groundnut, also known as a 'poor man's almond', contains good amounts of vitamin E, free amino acids, iron and zinc [57]; they are now being utilised as a ready-to-eat nutrition supplement in many developing countries to support malnourished children [58]. Chickpeas are mostly used as a source of free amino acids, protein, and minerals [55]. Similarly, sunflowers are used as an important oilseed crop because of the good oil composition [59]. Amongst oilseed crops, mustard is one of the most important because its oil is used for cooking and its oil cakes are used for feeding livestock [59]. Likewise, several important monocot crop plant species are economically significant as well as being crucial for sustainable food requirements, as they form part of staple foods such as maize, wheat, rice, sorghum, etc. These crop plants experience a wide range of abiotic stresses, which incur huge losses in farmers' fields. Presently several tolerant varieties for these crop plants are available; we here provide a list of several important of the most important (table 1.1).

1.4 Key genes identified for abiotic tolerance

Several candidate genes identified through mutant screening or expression profiling have been investigated to determine the potential roles in abiotic stress responsiveness. Regulatory proteins are seen to be important in plants' responses to drought stress. In plants, phosphorylation and dephosphorylation of proteins are frequent processes triggered by drought stress. Various kinases have been involved in drought response, including calcineurin B-like (CBL) interacting protein kinase (CIPK), calcium-dependent protein kinases (CDPKs), mitogen-activated protein kinases (MAPKs), and sucrose nonfermenting protein (SNF1)-related kinase 2 (SnRK2). In response to drought stress, a *CDPK* gene, *CPK10,* is known to regulate stomatal closing and opening through the phytohormones ABA and a Ca^{2+} signalling pathways in *Arabidopsis* [60]. In rice, the *OsCDPK7* gene has been proven to be involved in drought and salt stress tolerance [61]. Similarly, expression of the *CIPK23* gene has been identified under various abiotic stress responses, and RNAi suppression of the *CIPK23* gene reports drought susceptibility in plants [62]. Likewise, the regulatory function of the *OsMPK5* and *DSM1* genes in triggering the drought resistance response has been reviewed. In *Arabidopsis*, Umezawa *et al* described the participation of the *SnRK2C* gene in regulating the stress-responsive gene expression which facilitates DT under stress conditions [63]. TFs are another important type of regulatory protein that function at the transcriptional level to control gene expression in reaction to drought stress. Various members of different TF classes have been identified in response to drought, including APETALA2/ethylene-responsive element binding protein (AP2/EREBP), basic leucine zipper (bZIP), MYB, NAM-ATAF1/2-CUC2 (NAC), and zinc finger. Considering the contribution of TFs in the account, a SHN TF with AP2/EREBP domain has been identified in *Arabidopsis*, which triggers wax synthesis to modulate cuticle properties, thus imparting DT [64]. In rice, the regulatory functions of the *OsDREBs* and *ARAG1* genes (the *AP2* domain) have been reported in response to drought stress [65]. In another study, the rice *SNAC1* gene was engineered into wheat and cotton, which resulted in drought resistance [66, 67]. Similarly, overexpression of the *OsNAC10* gene using a *RCc3* root-specific promoter resulted in a substantial increase in root diameter, which imparts DT in transgenic rice [68].

A recent study in *Arabidopsis* reported the overexpression of the *AtZEP* gene, which showed vigorous growth in transgenic plants [69]. In rice, the Drought-induced *Wax Accumulation 1* (*DWA1*) and *Glossy 1* (*GL1–2*) genes have been reported, which are linked with leaf wax deposition and exhibited DT [70]. Additionally, in rice the reported overexpression of the *Os-OAT* gene offered a resistance to drought and osmotic stress [71]. Although it is well established that in

Table 1.1. List of selected crop plants and the associated major abiotic stressors that affect yield.

Crops plants	Stress	Tolerant genotype/varieties/cultivars
Dicot plants		
Soybean (*Glycine max* L.)	Drought	EC 538828, JS 97–52, EC 456548, EC602288, LD24, JD36, TF31
Black mustard (*Brassica juncea* L.)	Drought	IC 0583386, IC 0583448
	Heat	IC 0583386, IC 0583448
	Salinity	IC 401570, IC 0583386
	Cadmium (Cd) toxicity	Pusa Jai Kisan
Groundnut (*Arachis hypogaea*)	Drought	
	Salinity	
	Heat	
Chickpea (*Cicer arietinum* L.)	Drought	IPC L4–14, BGM 4005, IPCMB 19–3, Pusa Chickpea 10216
	Cadmium (Cd) toxicity	IC8, NC2, 4-PDG
	Copper (Cu) toxicity	RSG 888
Sunflower (*Helianthus annuus*)	Drought	Naturally adopted for drought
	Heat	Naturally adopted for heat
	Salinity	BRS323, Catisol, EXP11–26, EXP44–49, EXP60050, EXP887, HLA860HO and Olisun 5
Monocot plants		
Maize (*Zea mays* L.)	Drought	2004 TZE-WDT STR C4, EVDT-W 2000 STR C0, 2013 DTE STR-Y Syn, DT-Y STR Synthetic, 2012 TZE-WDT C4 STR C5, 2014 TZE-WDT STR, Obatanpa
	Heat	Zhengdan 958 (T)
	Arsenic (As) heavy metal stress	
Wheat (*Triticum aestivum* L.)	Drought	NI-5439, WH-1021, HD-2733, HB4
	Waterlogging	Ducula, Savannah, CSW 540–1, Coker9877
Rice (*Oryza sativa* L.)	Drought	Sahbhagi dhan, and DRR dhan 42, 43 and 44; Sukhadhan-1, 2, 3, 4, 5 and 6
	Heat	IR36, IR24, and IR64 (Indica cultivar); Akitakomachi and Koshihikari (Japonica cultivar); Nagina22 in India
Sorghum (*Sorghum bicolour*)	Drought	Naturally adopted for drought
	Heat	Naturally adopted for heat
Sorghum (*Sorghum bicolour*)	Aluminium (Al as heavy metal stress)	MCSR T33, Wahi, MCSR T69, MCSR T11

Arabidopsis salinity stress is responsible for delay in blooming [72], a mutant of *AtBFT,* which belongs to the FT/TERMINAL FLOWER 1 (TFL1) family, is unable to prolong blooming in response to salinity stress [73]. While salt-stressed plants exhibit decreased expression of *AP1*, an essential regulator of floral development, Further, the repression of *APETALA1* (*AP1*) expression is less severe in the *atbft* mutant than in the wild type [73]. As a result, *AtBFT* acts as a floral suppressor during salinity stress. Additionally, during salinity stress, *AtBFT* inhibits the association of FT and FD, both of which are decisive regulators of blooming [74]. *AtBFT* causes interruption by attaching to the C-terminal region of FD, which is similar to the region associated in the FD–FT interaction. Thus, when FD is attached to *AtBFT* during salinity stress, the FD–FT complex is depleted, which leads to delayed blooming. This work helps to bridge the gap between certain TFs triggered in reaction to drought/salt stress and recognised floral development regulatory pathways. In most cases, delayed blooming is induced by inhibition of flowering time genes. Such an approach enables the plant to maintain resources and energy in order to thrive in stressful environments (table 1.2).

Additionally, it is known that cold-activated signalling cascades interact with drought and salinity stress signal transduction pathways [75]. Heat stress triggers physiological adaptation, metabolism, and alterations in the carbon balance in *Arabidopsis* [76]. Low- and high-temperature stresses affect the reproductive development of plants. Plants detect low temperature through alterations in the fluidity of cellular membranes and metabolic functions. After that, cell Ca^{2+} level, calmodulin, phytohormone, and ROS pathways are altered, and the C-repeat binding factor (CBF) family of TFs is triggered to control downstream components during reproductive development [77]. Likewise, in rice, low-temperature-induced gibberellin depletion affects pollen formation [78]. In rice, low temperatures (19 °C) result in decreased endogenous gibberellic acid (GA) levels and expression of GA biosynthesis genes but increased expression of stress-responsive genes such as *DREB* and *DELLA* [78]. Another study, in blueberries, reported the function of a *SOC1*-like gene (*SOC1*) in reproductive development [79]. In comparison to the wild type, overexpression of *VcSOC1K* led to a rise in canes number, floral buds, and clusters of fruits upon low-temperature treatment. In comparison to plants that are exposed to low-temperature stress, those which are exposed to high-temperature stress need to eliminate excess heat energy. High temperatures, 28 °C in *Arabidopsis* and 32 °C in rice, have an effect on plant developmental processes, most notably male fertility [80]. The heat-stress response requires transcriptional control; specifically, heat shock transcription factors (HSFs) are essential for triggering the fast transcriptional activation of genes downstream [81]. Heat shock proteins (HSPs) are molecular chaperones that are required for all organisms to cope with the unfolding of proteins caused by heat. Small HSPs (sHSPs) are one subclass of HSPs; in *Arabidopsis* one sHSP, namely the BOBBER1 (BOB1) protein, which has a Nuclear Distribution C (NudC), is a chaperone which is required for folding of proteins and interaction [82]. Another study reported that *Atbob1* mutants are heat intolerant and involved in alterations of flowering and inflorescence meristems, which indicate its role in developmental and thermotolerance [83]. High-temperature stressors alter the

Table 1.2. List of genes associated with abiotic-stress-induced phenotype and tolerance.

Abiotic stress	Gene name	Biological function	Molecular function	Mutant phenotype function	Reference
Drought	*DREB1A*	Female reproductive development	Transcription factor	Arrested flower opening	[92]
	MYB21	Male reproductive development	Transcription factor	Stamen development defects	[92]
	ANAC019	Development of floral organ	Transcription factor	Drought sensitivity has increased, and floral organ recovery has been delayed	[93]
	*MYB*37	Development of seed	Transcription factor	Drought tolerance is improved, blooming is delayed, and seed productivity is enhanced.	[94]
	DIV2	Development of anthers	Transcription factor	Drought tolerance improved, anther defects reduced, and pollen fertility increased	[95]
	RAP2.6L	Development of pollen	Transcription factor	Improve drought tolerance and pollen fertility	[96]
	*EMF*1	Development of floral organ	Epigenetic regulator (H3K27me3)	Enhance drought tolerance and early blooming	[97]
	MPK6	Opening of a flower	Protein kinase	Reduced production of ethylene	[98]
	Fer1	Development of petal	Ion transport	Reduced level of free ferrous iron	[99]
Heat	*IRE1*	Development of pollen	Protein kinase/ribonuclease	Defects in the tapetum and sterility	[85]
	BOB1	Time of flowering and meristem identity	Protein chaperone	Reduced heat tolerance, slowed development rate, and defects in the leaf and inflorescence meristems	[83]
	*HSFA*2	Male reproductive development	Transcription factor	Increased tolerance to heat; lower pollen viability and germination rate	[100]
	HSFB2a	Female reproductive development	Transcription factor	Female gametophyte impaired	[101]
Cold	*SOC1*	Development of floral organs	Transcription factor	More floral buds, flowers, and fruit clusters, as well as increased cold tolerance	[79]
	HOS9	Time of flowering	Transcription factor	Reduced growth rate, delayed blooming, and enhanced sensitivity to chilling	[102]

(*Continued*)

Table 1.2. (*Continued*)

Abiotic stress	Gene name	Biological function	Molecular function	Mutant phenotype function	Reference
	GA3ox3	Development of pollen	Enzyme	Extensive pollen abortion and seed loss	[78]
	GA20ox1	Development of pollen	Enzyme	Extensive pollen abortion and seed loss	[78]
Salt	*NUC1*	High expression	Nucleolin	Increased growth rate and root length, as well as a reduction in H_2O_2 level	[103]
	CDKG2	Time of flowering	Protein kinase	Increased tolerance to salt and early blooming	[104]
	PAT10	Development of floral organs	Protein modification	Salt sensitivity increased, leaf size, plant height, and sterility decreased.	[105]
	SKB1	Development of anthers	Epigenetic regulator (H3K27me3)	Salt sensitivity is increased, growth rate is lowered, and flowering is delayed.	[106, 107]
	BFT	Time of flowering	Transcription factor	In response to stress no change in flowering time	[73, 74]
	*bHLH1*12	Time of flowering	Transcription factor	Raise in salt-induced H2O2 level, shortened roots, and a delay in flowering.	[108]
Oxidative	*SPL7*/*KIN*17	Development of flowers and seeds	Transcription factor	Reduce growth rate, increased sensitivity to oxidative stress, floral bud abortion, and pollen inviability (double mutant)	[109]
	*PIC*1	Development of flowers and seeds	Ion transport	Chlorosis of the leaves, impaired flowering, and seed development	[91]
	ROXY1	Development of petal	Oxidoreductase (metabolic enzyme)	Defects in the floral structure	[88]
	ROXY1/2	Development of petal	Oxidoreductase (metabolic enzyme)	Defects in the floral structure, higher H_2O_2 levels, and increased sensitivity to pathogens	[89]
	RD20	Time of flowering	Metabolic enzyme	Enhanced H_2O level and reduced flowering time	[110]
	ADR	Development of anthers		Indehiscence of the anther and male sterility	[87]

folding of proteins and activate a process called the unfolded protein response (UPR) [84]. Additionally, it has been revealed that genes associated with the UPR have a function in reproduction. Deng *et al* reported that, at ambient temperature, the *ire1a-2 ire1b-4* double mutant showed fertility, but showed male sterility at 27.5 ° C, due to tapetum morphological defects [85]. Another study, in cotton, reported that *GhCKI* (casein kinase I) has an important role in male reproduction during heat stress [86]. There are very few findings on the influence of oxidative stress in the reproductive development of plants. Oxidoreductases play a key role in floral growth during oxidative stress [87]. In the cellular response to oxidative stress, GRXs have been demonstrated to play antioxidant roles. In *Arabidopsis*, two genes, namely *AtROXY1* and *AtROXY2* (CC-type GRX), have been reported for functioning in petal and anther progression and differentiation [88]. In rice, a similar function of two homologues, *OsROXY1* and *OsROXY2*, was reported, which reveals a similar pattern of expression in floral organs, and their overexpression lines showing similar developmental alterations in flowers [89]. In *Arabidopsis*, the *membrane-spanning protein permease in chloroplasts 1* (*AtPIC1*) encodes for a transporter protein which facilitates transport of iron through the inner membrane of the chloroplast, aids ferritins in accumulating free iron, and decreases the levels of ROS in the cell [90]. Another study reported *AtPIC1* overexpression, which facilitates high accumulation of iron in the chloroplasts [91]. These findings extend the concept of ion homeostasis in plant cells in response to oxidative stress.

1.5 Recent examples of molecular approaches and outcomes: transgenics and CRISPR/Cas9

The aquaporins (AQPs) are intrinsic transmembrane proteins which are significantly diversified in higher plants, believed to originate and be related with bacterial membrane protein [111]. AQPs have been extensively explored for their role in water uptake and mineral uptakes; they are seldom involved with the uptake of heavy metals. For instance, overexpression of AQPs in crop plants has improved DT in model plants and crop plants, such as in overexpression of *TaAQP7* in *Nicotiana tabacum* and *OsRWC3* in *Oryza sativa* [111]. Further, the accumulation of auxin, GA, and abscisic acid (ABA) led to higher accumulation of AQPs in *Arabidopsis* and rice, respectively [111–113].

The clustered regularly interspaced short palindromic repeats (CRISPR) gene editing technique is a gene/genome engineering technique in molecular biology by which a living organism's genomes can be modified. Recently, the ABA receptors *pyrabactin resistance* (*PYL1-PYL13*) were edited through CRISPR/Cas9 technology to enhance the DT in rice [114]. In another study, CRISPR/Cas9-targeted mutagenesis of *Oryza sativa* enhanced response to *ABA1* (*OsERA1*), which encodes for the β-subunit of farnesyltransferase and controls ABA signalling, resulted in improved sensitivity to ABA, and increased response to drought stress through stomatal regulation and root development [115]. Recently, a CRISPR/Cas9 system was used to produce a mutant of the gene *OPEN STOMATA 2* (*OST2*), encoding a H^+ ATPase which controls the stomatal response in *Arabidopsis* [116]. Likewise, the

Semi-rolled leaf 1 (*SRL1*) and *SRL2* genes, which encode for glycosyl phosphatidyl inositol-anchored proteins, was targeted through CRISPR/Cas9 to develop mutant lines of rice for drought-tolerant varieties [117]. Recently, to address submergence tolerance, a key gene, *SUBMERGENCE 1A-1* (*SUB1A*), of the rice cultivar Ciherang-Sub1 was disrupted through CRISPR/Cas9 [118].

Transporter proteins play a very important role in metal ion transportation. Among them, the ZIP family transporters are solely involved in the uptake and augmentation of iron and zinc in plants [119]. Interestingly, the proteins of this family appeared to get delimited in response to cadmium, chromium, and nickel stress response [120]. Similarly, in *Arabidopsis*, overexpression of metal ATPase encoding gene *AtHMA4* can progress the growth and development by enhancing root growth under cadmium, cobalt, and zinc [121]. Likewise, the two *Multidrug and Toxic Compound Extrusion Transporter* (*MATE*) and *Aluminium-Activated Malate Transporters 4* (*ALMT4*) encoding genes, *MATE* and *ALMT4,* were overexpressed in the *Lotus japonicas* to induce heavy metal tolerance [122, 123]. Recently, *OsNramp5*, a natural resistance associated macrophage proteins (NRAMP) family gene which is located in the vacuole tonoplast and is essential for iron transportation, was knocked out using the CRISPR/Cas9 method to induce cadmium stress tolerance in rice without compromising yield [124]. At present, several salinity-tolerant genes have been discovered and cloned such as *DST, OsHAL3, OsRR22, SKC1, SNAC2, P5CS*, and *OsNAP*. To enhance salinity tolerance in rice a TF encoding gene, *OsRR22,* which controls cytokinin signal transduction and metabolism, was edited through CRISPR/Cas9-targeted mutagenesis [125]. A schematic representation of the genes associated with the various abiotic stresses and resulting phenotypes through genome-editing technology is depicted in figure 1.2. The present research involves combined studies of genomics, transcriptomics, proteomics, metabolomics, and phenomics [126–129]. Therefore, it is expected that forthcoming research will have a huge impact on abiotic stress management in crop plants.

1.6 Conclusions

The incidences of biotic and abiotic stresses are increasing due to anthropogenic activities and global climate change. The impact of abiotic stressors is expected to increase drastically in the next 30 years, especially drought, heat, cold, salinity, and heavy metal stress. Although significant work has been done in these areas from various institutions, a clear understanding of the underlying mechanisms controlling abiotic tolerance is obscured because these traits are controlled by several genomic loci and multiple regulatory TFs. These TFs broadly interfere at DNA, epigenetic and transcript, protein and metabolite levels that cause changes at the phenome level. The main reason for this lag in understanding is the lesser availability of genetic resources, genome sequencing, and gene annotation data availability: as a result very few crops have been explored. Additionally, relatively little focus has been paid to abiotic stressors other than drought. Presently, with the advancement and availability of high-throughput next-generation sequencing, liquid chromatography–mass spectrometry (LC–MS), gas chromatography–mass spectrometry (GC–MS), and phonemics, much

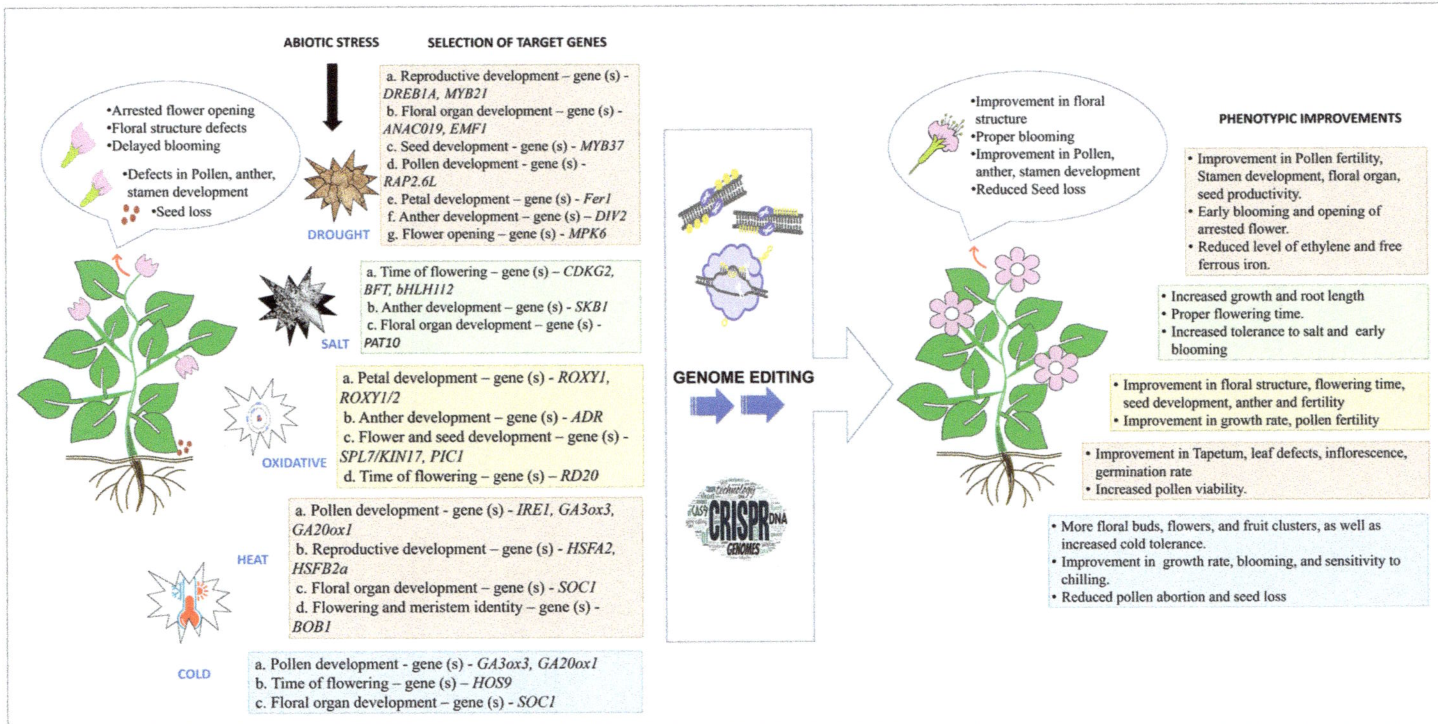

Figure 1.2. A detailed representation of the genes associated with the various abiotic stresses and resulting phenotypes. These genes can be targeted through gene editing and transgenic methods to develop climate-smart crops.

faster dissection of these complex trait is expected. Further, we anticipate in coming years that the genome-editing tool CRISPR/Cas9 will play a vital role in trait discovery and dissection.

Acknowledgement

R.K. thanks University Grant Commission, New Delhi, India for providing UGC Start-up research grant and support.

Multiple choice questions

1. Which of the following does NOT belongs to plant abiotic stresses?
    (a) Accumulation of cadmium in soil;
    (b) Abortion of gynoecium due to heat;
    (c) Freezing of cellular cytoplasm and change in membrane permeability;
    (d) Feeding of aphids on young leaves.
2. Which of the following can scale up the impact abiotic stress?
    (a) Proper rain precipitation;
    (b) Global climate change;
    (c) Growing legumes with cereals;
    (d) Excessive population growth of herbivores.
3. Which of the following metals is associated with heavy metal stress?
    (a) Cadmium;
    (b) Arsenic;
    (c) Lead;
    (d) All of the above.
4. Which of the following genes is overexpressed in plants for drought tolerance?
    (a) Aquaporins encoding genes;
    (b) RUSBISCO encoding gene;
    (c) Lycopene synthase encoding gene;
    (d) All of the above.
5. Which of the following genes is associated with submergence tolerance?
    (a) Phytoene synthase;
    (b) *SUB1A*, *SNORKEL1*, and *SNORKEL2*;
    (c) Chalcone synthase;
    (d) All of the above.

Answers: 1 (d); 2 (b); 3 (d); 4 (a); 5 (b).

Descriptive-type questions

1. Climate change will positively drive abiotic stress associated with plants. Discuss this by providing evidence to support this statement.
2. How can submergence tolerance affect the plant? Describe the genes identified which can be used to induce submergence tolerance?

3. What are the different types of abiotic stress? Considering the next 20 years, in your opinion suggest the types of abiotic stress which have the most significant potential to cause extensive damage?
4. Provide a detailed summary of the molecular regulation of abiotic stress tolerance by citing the genes and transcription factors associated with different types of abiotic stress.
5. You have been assigned to develop a climate-smart crop tolerant to both salinity and drought stress. What are the genes which would be selected to generate a tolerant rice plant through a genome-editing technique?

References

[1] Peleg Z and Blumwald E 2011 Hormone balance and abiotic stress tolerance in crop plants *Curr. Opin. Plant Biol.* **14** 290–5

[2] Fujita M, Fujita Y, Noutoshi Y, Takahashi F, Narusaka Y and Yamaguchi-Shinozaki K *et al* 2006 Crosstalk between abiotic and biotic stress responses: a current view from the points of convergence in the stress signaling networks *Curr. Opin. Plant Biol.* **9** 436–42

[3] Singh N, Mansoori A, Dey D, Kumar R and Kumar A 2021 Potential of metabolomics in plant abiotic stress management ed A Kumar, R Kumar, P Shukla and H K Patel *Omics Technologies for Sustainable Agriculture and Global Food Security* II (Singapore: Springer) pp 193–214

[4] Hasanuzzaman M, Bhuyan M H, Zulfiqar F, Raza A, Mohsin S M and Mahmud J A *et al* 2020 Reactive oxygen species and antioxidant defense in plants under abiotic stress: revisiting the crucial role of a universal defense regulator *Antioxidants* **9** 681

[5] Sanghera G S, Wani S H, Hussain W and Singh N B 2011 Engineering cold stress tolerance in crop plants *Curr. Genom.* **12** 30

[6] Kidokoro S, Yoneda K, Takasaki H, Takahashi F, Shinozaki K and Yamaguchi-Shinozaki K 2017 Different cold-signaling pathways function in the responses to rapid and gradual decreases in temperature *Plant Cell* **29** 760–74

[7] Ding Y, Shi Y and Yang S 2019 Advances and challenges in uncovering cold tolerance regulatory mechanisms in plants *New Phytol.* **222** 1690–704

[8] Gupta B and Huang B 2014 Mechanism of salinity tolerance in plants: physiological, biochemical, and molecular characterization *Int. J. Genomics*

[9] James R A, Blake C, Byrt C S and Munns R 2011 Major genes for Na^+ exclusion, Nax1 and Nax2 (wheat HKT1; 4 and HKT1; 5), decrease Na^+ accumulation in bread wheat leaves under saline and waterlogged conditions *J. Exp. Bot.* **62** 2939–47

[10] Pang Q, Chen S, Dai S, Chen Y, Wang Y and Yan X 2010 Comparative proteomics of salt tolerance in *Arabidopsis thaliana* and *Thellungiella halophila J. Proteome Res.* **9** 2584–99

[11] Ábrahám E 2011 Identification of *Arabidopsis* and *Thellungiella* genes involved in salt tolerance by novel genetic system *Acta Biol. Szeged.* **55** 53–7

[12] Pierart A, Shahid M, Séjalon-Delmas N and Dumat C 2015 Antimony bioavailability: knowledge and research perspectives for sustainable agricultures *J. Hazard. Mat.* **289** 219–34

[13] Tiwari S and Lata C 2018 Heavy metal stress, signaling, and tolerance due to plant-associated microbes: an overview *Front. Plant Sci.* **9** 452

[14] Iqbal M, Ahmad A, Ansari M K, Qureshi M I, Aref I M and Khan P R *et al* 2015 Improving the phytoextraction capacity of plants to scavenge metal(loid)-contaminated sites *Environ. Rev.* **23** 44–65

[15] Fang Y and Xiong L 2015 General mechanisms of drought response and their application in drought resistance improvement in plants *Cell. Mol. Life Sci.* **72** 673–89
[16] Salehi-Lisar S Y and Bakhshayeshan-Agdam H 2016 Drought stress in plants: causes, consequences, and tolerance *Drought Stress Tolerance in Plants* vol 1 (Cham: Springer) pp 1–16
[17] Lawlor D W and Tezara W 2009 Causes of decreased photosynthetic rate and metabolic capacity in water-deficient leaf cells: a critical evaluation of mechanisms and integration of processes *Annals Bot.* **103** 561–79
[18] Mitra J 2001 Genetics and genetic improvement of drought resistance in crop plants *Curr. Sci.* **25** 758–63
[19] Luo L J 2010 Breeding for water-saving and drought-resistance rice (WDR) in China *J. Exp. Bot.* **61** 3509–17
[20] Prasad P V, Bheemanahalli R and Jagadish S K 2017 Field crops and the fear of heat stress opportunities, challenges and future directions *Field Crops Res.* **200** 114–21
[21] Cho R 2018 *How Climate Change Will Alter Our Food. State of the Planet* (New York: Earth Institute, Columbia University)
[22] Janni M, Gullì M, Maestri E, Marmiroli M, Valliyodan B and Nguyen H T *et al* 2020 Molecular and genetic bases of heat stress responses in crop plants and breeding for increased resilience and productivity *J. Exp. Bot.* **71** 3780–802
[23] Yang H, Gu X, Ding M, Lu W and Lu D 2018 Heat stress during grain filling affects activities of enzymes involved in grain protein and starch synthesis in waxy maize *Sci. Rep.* **8** 1–9
[24] Osmond C B, Austin M P, Berry J A, Billings W D, Boyer J S and Dacey J W *et al* 1987 Stress physiology and the distribution of plants *BioScience* **37** 38–48
[25] Liu J, Shi Y and Yang S 2018 Insights into the regulation of C-repeat binding factors in plant cold signaling *J. Integr. Plant Biol.* **60** 780–95
[26] Ruelland E, Vaultier M N, Zachowski A and Hurry V 2009 Cold signalling and cold acclimation in plants *Adv. Bot. Res.* **49** 35–150
[27] Pearce R S 2001 Plant freezing and damage *Ann. Bot.* **87** 417–24
[28] Thomashow M F 1999 Plant cold acclimation: freezing tolerance genes and regulatory mechanisms *Ann. Rev. Plant Biol.* **50** 571–99
[29] Kaplan F, Kopka J, Sung D Y, Zhao W, Popp M and Porat R *et al* 2007 Transcript and metabolite profiling during cold acclimation of *Arabidopsis* reveals an intricate relationship of cold-regulated gene expression with modifications in metabolite content *Plant J.* **50** 967–81
[30] Dong C H, Zolman B K, Bartel B, Lee B H, Stevenson B and Agarwal M 2009 Disruption of *Arabidopsis* CHY1 reveals an important role of metabolic status in plant cold stress signaling *Mol. Plant* **2** 59–72
[31] Munns R and Tester M 2008 Mechanisms of salinity tolerance *Annu. Rev. Plant Biol.* **59** 651–81
[32] Bockheim J G and Gennadiyev A N 2000 The role of soil-forming processes in the definition of taxa in soil taxonomy and the world soil reference baseGeoderma **95** 53–72
[33] Rajendran K, Tester M and Roy S J 2009 Quantifying the three main components of salinity tolerance in cereals *Plant Cell Environ.* **32** 237–49
[34] Roy S J, Negrao S and Tester M 2014 Salt resistant crop plants *Curr. Opin. Biotechnol.* **26** 115–24

[35] Chazen O, Hartung W and Neumann P M 1995 The different effects of PEG 6000 and NaCI on leaf development are associated with differential inhibition of root water transport *Plant Cell Environ.* **18** 727–35

[36] Maathuis F J 2014 Sodium in plants: perception, signalling, and regulation of sodium fluxes *J. Exp. Bot.* **65** 849–58

[37] Didierjean L, Frendo P, Nasser W, Genot G, Marivet J and Burkard G 1996 Heavy-metal-responsive genes in maize: identification and comparison of their expression upon various forms of abiotic stress *Planta* **199** 1–8

[38] Murphy A, Zhou J, Goldsbrough P B and Taiz L 1997 Purification and immunological identification of metallothioneins 1 and 2 from *Arabidopsis thaliana Plant Physiol.* **113** 1293–301

[39] Schutzendubel A and Polle A 2002 Plant responses to abiotic stresses: heavy metal-induced oxidative stress and protection by mycorrhization *J. Exp. Bot.* **53** 1351–65

[40] Rai V, Vajpayee P, Singh S N and Mehrotra S 2004 Effect of chromium accumulation on photosynthetic pigments, oxidative stress defense system, nitrate reduction, proline level and eugenol content of *Ocimum tenuiflorum* L. *Plant Sci.* **167** 1159–69

[41] Jouili H and El Ferjani E 2003 Changes in antioxidant and lignifying enzyme activities in sunflower roots (*Helianthus annuus* L.) stressed with copper excess *C.R. Biol.* **326** 639–44

[42] Drążkiewicz M, Skórzyńska-Polit E and Krupa Z 2004 Copper-induced oxidative stress and antioxidant defence in *Arabidopsis thaliana Biometals* **17** 379–87

[43] Tena G, Asai T, Chiu W L and Sheen J 2001 Plant mitogen activated protein kinase signaling cascades *Curr. Opin. Plant Biol.* **4** 392–400

[44] Fukao T, Barrera-Figueroa B E, Juntawong P and Peña-Castro J M 2019 Submergence and waterlogging stress in plants: a review highlighting research opportunities and understudied aspects *Front. Plant Sci.* **10** 340

[45] Bui L T, Giuntoli B, Kosmacz M, Parlanti S and Licausi F 2015 Constitutively expressed ERF-VII transcription factors redundantly activate the core anaerobic response in *Arabidopsis thaliana Plant Sci.* **236** 37–43

[46] Gasch P, Fundinger M, Müller J T, Lee T, Bailey-Serres J and Mustroph A 2016 Redundant ERF-VII transcription factors bind to an evolutionarily conserved cis-motif to regulate hypoxia-responsive gene expression in *Arabidopsis Plant Cell* **28** 160–80

[47] Fukao T, Xu K, Ronald P C and Bailey-Serres J 2006 A variable cluster of ethylene response factor–like genes regulates metabolic and developmental acclimation responses to submergence in rice *Plant Cell* **18** 2021–34

[48] Hattori Y, Nagai K, Furukawa S, Song X J, Kawano R and Sakakibara H *et al* 2009 The ethylene response factors SNORKEL1 and SNORKEL2 allow rice to adapt to deep water *Nature* **460** 1026–30

[49] Møller I M, Jensen P E and Hansson A 2007 A Oxidative modifications to cellular components in plants *Annu. Rev. Plant Biol.* **58** 459–81

[50] Pérez-Pérez M E, Lemaire S D and Crespo J L 2012 Reactive oxygen species and autophagy in plants and algae *Plant Physiol.* **160** 156–64

[51] Singh N, Mansoori A, Jiwani G, Solanke A K U, Thakur T K, Kumar R, Chaurasiya M and Kumar A 2021 Antioxidant and antimicrobial study of *Schefflera vinosa* leaves crude extracts against rice pathogens *Arabian J. Chem.* **14** 103243

[52] Noctor G and Foyer C H 1998 Ascorbate and glutathione: keeping active oxygen under control *Annu. Rev. Plant Biol.* **49** 249–79

[53] Chaudière J and Ferrari-Iliou R 1999 Intracellular antioxidants: from chemical to biochemical mechanisms *Food Chem. Toxico.* **37** 949–62

[54] Hasanuzzaman M, Nahar K, Alam M, Roychowdhury R and Fujita M 2013 Physiological, biochemical, and molecular mechanisms of heat stress tolerance in plants *Inter. J. Mol. Sci.* **14** 9643–84

[55] Maphosa Y and Jideani V A 2017 The role of legumes in human nutrition *Functional Food-Improve Health through Adequate Food* (London: IntechOpen) 1 p 13

[56] Deshmukh D B, Marathi B, Sudini H K, Variath M T, Chaudhari S and Manohar S S *et al* 2020 Combining high oleic acid trait and resistance to late leaf spot and rust diseases in groundnut (*Arachis hypogaea* L.) *Front. Genet.* **11** 514

[57] Gangurde S S, Kumar R, Pandey A K, Burow M, Laza H E and Nayak S N *et al* 2019 Climate-smart groundnuts for achieving high productivity and improved quality: current status, challenges, and opportunities *Genomic Designing of Climate-Smart Oilseed Crops* (Cham: Springer) pp 133–72

[58] Bahwere P, Balaluka B, Wells J C, Mbiribindi C N, Sadler K and Akomo P *et al* 2016 Cereals and pulse-based ready-to-use therapeutic food as an alternative to the standard milk- and peanut paste–based formulation for treating severe acute malnutrition: a noninferiority, individually randomized controlled efficacy clinical trial *Amer. J. Clin. Nutri.* **103** 1145–61

[59] Konuskan D B, Arslan M and Oksuz A 2019 Physicochemical properties of cold pressed sunflower, peanut, rapeseed, mustard and olive oils grown in the Eastern Mediterranean region *Saudi J. Biol. Sci.* **26** 340–4

[60] Zou J J, Wei F J, Wang C, Wu J J, Ratnasekera D and Liu W X *et al* 2010 *Arabidopsis* calcium-dependent protein kinase CPK10 functions in abscisic acid- and Ca^{2+} mediated stomatal regulation in response to drought stress *Plant Physiol.* **154** 1541232–43

[61] Saijo Y, Hata S, Kyozuka J, Shimamoto K and Izui K 2000 Over-expression of a single Ca^{2+} dependent protein kinase confers both cold and salt/drought tolerance on rice plants *Plant J.* **23** 319–27

[62] Yang W, Kong Z, Omo-Ikerodah E, Xu W, Li Q and Xue Y 2008 Calcineurin B-like interacting protein kinase OsCIPK23 functions in pollination and drought stress responses in rice (*Oryza sativa* L.) *J. Genet. Genom.* **35** 531–2

[63] Umezawa T, Yoshida R, Maruyama K, Yamaguchi-Shinozaki K and Shinozaki K 2004 SRK2C, a SNF1-related protein kinase 2, improves drought tolerance by controlling stress-responsive gene expression in *Arabidopsis thaliana Proc. Natl. Acad. Sci. U.S.A.* **101** 17306–11

[64] Aharoni A, Dixit S, Jetter R, Thoenes E, Van Arkel G and Pereira A 2004 The SHINE clade of AP2 domain transcription factors activates wax biosynthesis, alters cuticle properties, and confers drought tolerance when overexpressed in *Arabidopsis Plant Cell* **16** 2463–80

[65] Chen J Q, Meng X P, Zhang Y, Xia M and Wang X P 2008 Over-expression of OsDREB genes lead to enhanced drought tolerance in rice *Biotechnol. Lett* **30** 2191–8

[66] Hu H, Dai M, Yao J, Xiao B, Li X and Zhang Q *et al* 2006 Overexpressing a NAM, ATAF, and CUC (NAC) transcription factor enhances drought resistance and salt tolerance in rice *Proc. Natl. Acad. Sci. U.S.A.* **103** 12987–92

[67] Liu G, Li X, Jin S, Liu X, Zhu L and Nie Y *et al* 2014 Overexpression of rice NAC gene SNAC1 improves drought and salt tolerance by enhancing root development and reducing transpiration rate in transgenic cotton *PLoS One* **9** e86895

[68] Jeong J S, Kim Y S, Baek K H, Jung H, Ha S H and Do Choi Y *et al* 2010 Root-specific expression of OsNAC10 improves drought tolerance and grain yield in rice under field drought conditions *Plant Physiol.* **153** 185–97

[69] Park H Y, Seok H Y, Park B K, Kim S H, Goh C H and Lee B H *et al* Overexpression of *Arabidopsis* ZEP enhances tolerance to osmotic stress *Biochem. Biophys. Res. Commun.* **375** 80–5

[70] Zhu X and Xiong L 2013 Putative megaenzyme DWA1 plays essential roles in drought resistance by regulating stress-induced wax deposition in rice *Proc. Natl. Acad. Sci. U.S.A.* **110** 17790–5

[71] You J, Hu H and Xiong L 2012 An ornithine δ-aminotransferase gene OsOAT confers drought and oxidative stress tolerance in rice *Plant Sci.* **197** 59–69

[72] Riboni M, Robustelli Test A, Galbiati M, Tonelli C and Conti L 2014 Environmental stress and flowering time: the photoperiodic connection *Plant Signal. Behav.* **9** e29036

[73] Ryu J Y, Park C M and Seo P J 2011 The floral repressor BROTHER of FT and TFL1 (BFT) modulates flowering initiation under high salinity in *Arabidopsis Mol. Cell.* **32** 295–303

[74] Ryu J Y, Lee H J, Seo P J, Jung J H, Ahn J H and Park C M 2014 The *Arabidopsis* floral repressor BFT delays flowering by competing with FT for FD binding under high salinity *Mol. Plant* **7** 377–87

[75] Zhu J K 2016 Abiotic stress signaling and responses in plants *Cell* **167** 313–24

[76] Quint M, Delker C, Franklin K A, Wigge P A, Halliday K J and Van Zanten M 2016 Molecular and genetic control of plant thermos morphogenesis *Nat. Plants* **2** 1–9

[77] Zinn K E, Tunc-Ozdemir M and Harper J F 2010 Temperature stress and plant sexual reproduction: uncovering the weakest links *J. Exp. Bot.* **61** 1959–68

[78] Sakata T, Oda S, Tsunaga Y, Shomura H, Kawagishi-Kobayashi M and Aya K *et al* 2014 Reduction of gibberellin by low temperature disrupts pollen development in rice *Plant Physiol.* **164** 2011–9

[79] Song G Q and Chen Q 2018 Overexpression of the MADS-box gene K-domain increases the yield potential of blueberry *Plant Sci.* **276** 22–31

[80] Hedhly A, Hormaza J I and Herrero M 2009 Global warming and sexual plant reproduction *Trends Plant Sci.* **14** 30–6

[81] Guo M, Liu J H, Ma X, Luo D X, Gong Z H and Lu M H 2016 The plant heat stress transcription factors (HSFs): structure, regulation, and function in response to abiotic stresses *Front. Plant Sci.* **9** 114

[82] Jurkuta R J, Kaplinsky N J, Spindel J E and Barton M K 2009 Partitioning the apical domain of the *Arabidopsis* embryo requires the *BOBBER1* NudC domain protein *Plant Cell* **21** 1957–71

[83] Perez D E, Hoyer J S, Johnson A I, Moody Z R, Lopez J and Kaplinsky N J 2009 BOBBER1 is a noncanonical *Arabidopsis* small heat shock protein required for both development and thermotolerance *Plant Physiol.* **151** 241–52

[84] Bao Y and Howell S H 2017 The unfolded protein response supports plant development and defense as well as responses to abiotic stress *Front. Plant Sci.* **8** 344

[85] Deng Y, Srivastava R, Quilichini T D, Dong H, Bao Y and Horner H T 2016 *IRE 1*, a component of the unfolded protein response signaling pathway, protects pollen development in *Arabidopsis* from heat stress *Plant J.* **88** 193–204

[86] Min L, Zhu L, Tu L, Deng F, Yuan D and Zhang X 2013 Cotton *Gh CKI* disrupts normal male reproduction by delaying tapetum programmed cell death via inactivating starch synthase *Plant J.* **75** 823–35

[87] Dai S Y, Hsu W H and Yang C H 2019 The gene *ANTHER DEHISCENCE REPRESSOR* (ADR) controls male fertility by suppressing the ROS accumulation and anther cell wall thickening in *Arabidopsis Sci. Rep.* **9** 1–4

[88] Xing S, Rosso M G and Zachgo S 2005 ROXY1, a member of the plant glutaredoxin family, is required for petal development in *Arabidopsis thaliana Development* **132** 1555–65

[89] Wang Z, Xing S, Birkenbihl R P and Zachgo S 2009 Conserved functions of *Arabidopsis* and rice CC-type glutaredoxins in flower development and pathogen response *Mol. Plant.* **2** 323–35

[90] Duy D, Wanner G, Meda A R, von Wirén N, Soll J and Philippar K 2007 PIC1, an ancient permease in *Arabidopsis* chloroplasts, mediates iron transport *Plant Cell* **19** 986–1006

[91] Duy D, Stübe R, Wanner G and Philippar K 2011 The chloroplast permease PIC1 regulates plant growth and development by directing homeostasis and transport of iron *Plant Physiol.* **155** 1709–22

[92] Su Z, Ma X, Guo H, Sukiran N L, Guo B and Assmann S M 2013 Flower development under drought stress: morphological and transcriptomic analyses reveal acute responses and long-term acclimation in *Arabidopsis Plant Cell* **25** 3785–807

[93] Sukiran N L, Ma J C, Ma H and Su Z 2019 ANAC019 is required for recovery of reproductive development under drought stress in *Arabidopsis Plant Mol. Biol.* **99** 161–74

[94] Yu Y T, Wu Z, Lu K, Bi C, Liang S and Wang X F *et al* 2016 Overexpression of the MYB37 transcription factor enhances abscisic acid sensitivity, and improves both drought tolerance and seed productivity in *Arabidopsis thaliana Plant Mol. Biol.* **90** 267–79

[95] Guo C, Yao L, You C, Wang S, Cui J and Ge X *et al* 2016 MID1 plays an important role in response to drought stress during reproductive development *Plant J.* **88** 280–93

[96] Jin Y, Pan W, Zheng X, Cheng X, Liu M and Ma H *et al* 2018 *OsERF101*, an ERF family transcription factor, regulates drought stress response in reproductive tissues *Plant Mol. Biol.* **98** 51–65

[97] Pu L, Liu M S, Kim S Y, Chen L F, Fletcher J C and Sung Z R 2013 EMBRYONIC FLOWER1 and ULTRAPETALA1 act antagonistically on *Arabidopsis* development and stress response *Plant Physiol.* **162** 812–30

[98] Meng Y, Ma N, Zhang Q, You Q, Li N and Ali Khan M *et al* 2014 Precise spatio-temporal modulation of ACC synthase by MPK 6 cascade mediates the response of rose flowers to rehydration *Plant J.* **79** 941–50

[99] Liu J, Fan Y, Zou J, Fang Y, Wang L and Wang M *et al* 2017 RhABF2/Ferritin module affects rose (*Rosa hybrida*) petal dehydration tolerance and senescence by modulating iron levels *Plant J.* **92** 1157–69

[100] Fragkostefanakis S, Mesihovic A, Simm S, Paupière M J, Hu Y and Paul P *et al* 2016 HsfA2 controls the activity of developmentally and stress-regulated heat stress protection mechanisms in tomato male reproductive tissues *Plant Physiol.* **170** 2461–77

[101] Wunderlich M, Groß-Hardt R and Schöffl F 2014 Heat shock factor HSFB2a involved in gametophyte development of *Arabidopsis thaliana* and its expression is controlled by a heat-inducible long non-coding antisense RNA *Plant Mol. Bio.* **85** 541–50

[102] Zhu J, Shi H, Lee B H, Damsz B, Cheng S and Stirm V *et al* 2004 An *Arabidopsis* homeodomain transcription factor gene, HOS9, mediates cold tolerance through a CBF-independent pathway *Proc. Natl. Acad. Sci. U.S.A.* **101** 9873–8
[103] Sripinyowanich S, Chamnanmanoontham N, Udomchalothorn T, Maneeprasopsuk S, Santawee P and Buaboocha T *et al* 2013 Overexpression of a partial fragment of the salt-responsive gene OsNUC1 enhances salt adaptation in transgenic *Arabidopsis thaliana* and rice (*Oryza sativa* L.) during salt stress *Plant Sci.* **213** 67–78
[104] Ma X, Qiao Z, Chen D, Yang W, Zhou R and Zhang W *et al* 2015 CYCLIN-DEPENDENT KINASE G2 regulates salinity stress response and salt mediated flowering in *Arabidopsis thaliana Plant Mol. Biol.* **88** 287–99
[105] Zhou L Z, Li S, Feng Q N, Zhang Y L, Zhao X and Zeng Y L *et al* 2013 PROTEIN S-ACYL TRANSFERASE10 is critical for development and salt tolerance in *Arabidopsis Plant Cell* **25** 1093–107
[106] Schmitz R J, Sung S and Amasino R M 2008 Histone arginine methylation is required for vernalization-induced epigenetic silencing of FLC in winter-annual *Arabidopsis thaliana Proc. Natl. Acad. Sci. U.S.A.* **105** 411–6
[107] Zhang Z, Zhang S, Zhang Y, Wang X, Li D and Li Q *et al* 2011 *Arabidopsis* floral initiator SKB1 confers high salt tolerance by regulating transcription and pre-mRNA splicing through altering histone H4R3 and small nuclear ribonucleoprotein LSM4 methylation *Plant Cell* **23** 396–411
[108] Chen H C, Hsieh-Feng V, Liao P C, Cheng W H, Liu L Y and Yang Y W 2017 The function of *OsbHLH068* is partially redundant with its homolog, *AtbHLH112*, in the regulation of the salt stress response but has opposite functions to control flowering in *Arabidopsis Plant Mol. Biol.* **94** 531–48
[109] Garcia-Molina A, Xing S and Huijser P 2014 A conserved KIN17 curved DNA-binding domain protein assembles with SQUAMOSA PROMOTER-BINDING PROTEIN-LIKE7 to adapt *Arabidopsis* growth and development to limiting copper availability *Plant Physiol.* **164** 828–40
[110] Blée E, Boachon B, Burcklen M, Le Guédard M, Hanano A and Heintz D *et al* 2014 The reductase activity of the *Arabidopsis* caleosin RESPONSIVE TO DESSICATION20 mediates gibberellin-dependent flowering time, abscisic acid sensitivity, and tolerance to oxidative stress *Plant Physiol.* **166** 109–24
[111] Ahmad M, Ali Q, Hafeez M M and Malik A 2021 Improvement for biotic and abiotic stress tolerance in crop plants *Biol. Clin. Sci. Res. J.* **2021**
[112] Shi H, Chen L, Ye T, Liu X, Ding K and Chan Z 2014 Modulation of auxin content in *Arabidopsis* confers improved drought stress resistance *Plant Physiol. Biochem.* **82** 209–17
[113] Pang Y, Li J, Qi B, Tian M, Sun L and Wang X *et al* 2017 Aquaporin AtTIP5; 1 as an essential target of gibberellins promotes hypocotyl cell elongation in *Arabidopsis thaliana* under excess boron stress *Funct. Plant Biol.* **45** 305–14
[114] Miao C, Xiao L, Hua K, Zou C, Zhao Y and Bressan R A 2018 Mutations in a subfamily of abscisic acid receptor genes promote rice growth and productivity *Proc. Natl. Acad. Sci. U.S.A.* **115** 6058–63
[115] Ogata T, Ishizaki T, Fujita M and Fujita Y 2020 CRISPR/Cas9-targeted mutagenesis of *OsERA1* confers enhanced responses to abscisic acid and drought stress and increased primary root growth under nonstressed conditions in rice *PLoS One* **15** e0243376

[116] Osakabe Y, Watanabe T, Sugano S S, Ueta R, Ishihara R and Shinozaki K *et al* 2016 Optimization of CRISPR/Cas9 genome editing to modify abiotic stress responses in plants *Sci. Rep.* **6** 1–10

[117] Liao S, Qin X, Luo L, Han Y, Wang X and Usman B *et al* 2019 CRISPR/Cas9-induced mutagenesis of *semi-rolled Leaf1, 2* confers curled leaf phenotype and drought tolerance by influencing protein expression patterns and ROS scavenging in rice (*Oryza sativa* L.) *Agronomy* **9** 728

[118] Liang Y, Biswas S, Kim B, Bailey-Serres J and Septiningsih E M 2021 Improved transformation and regeneration of indica rice: disruption of SUB1A as a test case via CRISPR-Cas9 *Int. J. Mol. Sci.* **22** 6989

[119] Manara A 2012 *Plant Responses to Heavy Metal Toxicity Plants and Heavy Metals* (Dordrecht: Springer) pp 27–53

[120] Thapa G, Sadhukhan A, Panda S K and Sahoo L 2012 Molecular mechanistic model of plant heavy metal tolerance *Biometals* **25** 489–505

[121] Barabasz A, Klimecka M, Kendziorek M, Weremczuk A, Ruszczyńska A, Bulska E and Antosiewicz D M 2016 The ratio of Zn to Cd supply as a determinant of metal-homeostasis gene expression in tobacco and its modulation by overexpressing the metal exporter AtHMA4 *J. Exp. Bot.* **17** erw389

[122] Siemianowski O, Mills R F, Williams L E and Antosiewicz D M 2011 Expression of the P1B-type ATPase *AtHMA4* in tobacco modifies Zn and Cd root to shoot partitioning and metal tolerance *Plant Biotechnol. J.* **9** 64–74

[123] Wang J, Hou Q, Li P, Yang L, Sun X and Benedito V A *et al* 2017 Diverse functions of multidrug and toxin extrusion (MATE) transporters in citric acid efflux and metal homeostasis in *Medicago truncatula Plant J.* **90** 79–95

[124] Tang L, Mao B, Li Y, Lv Q, Zhang L and Chen C *et al* 2017 Knockout of OsNramp5 using the CRISPR/Cas9 system produces low Cd-accumulating indica rice without compromising yield *Sci. Rep.* **7** 1–2

[125] Zhang A, Liu Y, Wang F, Li T, Chen Z and Kong D *et al* 2019 Enhanced rice salinity tolerance via CRISPR/Cas9-targeted mutagenesis of the *OsRR22* gene *Mol. Breed.* **39** 1–10

[126] Kumar R, Bohra A, Pandey A K, Pandey M K and Kumar A Metabolomics for plant improvement: status and prospects *Front. Plant Sci.* **8** 1302

[127] Kumar R, Sharma V, Suresh S, Ramrao D P, Veershetty A and Kumar S *et al* 2021 Understanding omics driven plant improvement and de novo crop domestication: some examples *Front. Genet.* **12** 415

[128] Sharma V, Gupta P, Priscilla K, Hangargi B, Veershetty A and Ramrao D P *et al* 2021 Metabolomics intervention towards better understanding of plant traits *Cells* **10** 346

[129] Kumar R, Nizampatnam N R, Alam M, Thakur T K and Kumar A 2021 Genome editing technologies for plant improvement: advances, applications and challenges *Omics Technologies for Sustainable Agriculture and Global Food Security* ed A Kumar, R Kumar, P Shukla and M K Pandey vol 1 1st edn (Singapore: Springer) pp 213–40

Chapter 2

Metabolic engineering for understanding abiotic stress tolerance in plants

Manoj Kumar[1], Kadambot H M Siddique[2*]

[1]Institute of Plant Sciences, Agricultural Research Organization, Volcani Center, Rishon LeZion 7505101, Israel
[2]The UWA Institute of Agriculture, and UWA School of Agriculture and Environment, The University of Western Australia, Perth, WA 6001, Australia

Abiotic stresses such as drought, salinity, heat, cold, and heavy metals severely affect the growth and production of plants. Metabolomics is an efficient approach for extracting complete information on the metabolites and metabolic pathways implicated in abiotic stress tolerance. Furthermore, metabolic engineering can improve secondary metabolite production, thus enhancing plant tolerance to different environmental stresses. Several metabolic genes have been identified and engineered to generate stress-tolerant transgenic crops; however, there has been limited achievement in the development of abiotic stress-tolerant cultivars. Hence, there is significant demand to identify novel metabolic genes and pathways for efficiently enhancing plant tolerance to abiotic stresses. Here, we review the application of primary and secondary metabolic genes using metabolic engineering technology for developing crop plants that tolerate abiotic stresses.

2.1 Introduction

Adverse climatic factors such as drought, salinity, heat, cold, and heavy metals affect agro-ecological conditions, food safety, and food security [1]. These abiotic stress factors are the key growth stressors that restrict crop productivity and quality globally [2] through various changes in molecular, biochemical, physiological, and morphological parameters [3]. Metabolites play a very significant role in plant

* Author to whom any correspondence should be addressed kadambot.siddique@uwa.edu.au.

doi:10.1088/978-0-7503-4921-5ch2

growth, development, and productivity. They have been implicated in membrane configuration and scaffolding, cell signaling, whole-plant resource distribution, and energy storage under stress conditions [4]. Numerous abiotic stresses perturb the metabolism of the plant by inhibiting metabolic enzymes, substrate deficiency, the excess requirement of specific compounds, or their combination. Therefore, metabolic engineering should sustain indispensable metabolism, and adjust by adopting a novel steady-state in light of the existing environmental stresses [5]. Metabolic engineering advances cellular characteristics by adjusting particular biochemical reactions or by encountering new ones through recombinant DNA technology [6]. It is an outstanding tool that provides an extensive understanding of abiotic stress responses in plants. In plants, abiotic stress tolerance is a multifaceted mechanism that needs to be fully understood and interpreted [7, 8].

Metabolic engineering alters endogenous pathways, redirecting one or more enzymatic reactions in plants to generate novel compounds, enhance the manufacture of existing compounds, or impede the manufacture of known metabolic compounds [9]. Several connections among DNA, proteins, and metabolites, known as the interactome, are the main factors involved in any cellular process. Proteins are the main functional substances facilitating the metabolic process by interchanging metabolites. Metabolic engineering requires an understanding of such associations to precisely manage biological systems [10]. Reprogramming of primary metabolites (PMs) or secondary metabolites (SMs) and genetically engineered metabolic genes play a significant role in plant adaptation and responses to environmental stresses, including drought [11]. Identifying novel genetic engineering approaches and potential metabolic engineering targets for abiotic stress tolerance will influence feasible crop yields and productivity.

Metabolomics is used to identify and/or quantify PMs and SMs in plants under abiotic stress [12, 13]. Environmental stresses can trigger hyperaggregation of many plant metabolites [14]. Under drought, the stimulation of PMs or SMs can adjust the rigidity and turgidity of plant cells and tissues, ion transport, redox homeostasis, and enzyme activity [15, 16]. Plant SMs are derivatives of PMs to endure various adverse physiological changes due to stressors [17, 18].

In plants, two key procedures (targeted and nontargeted) are used to decipher metabolic reprogramming during abiotic stress conditions [19]. Targeted metabolomics identify and evaluate known metabolites under different environmental stresses. Nontargeted metabolomics imparts an overview of the ample metabolites [20, 21]. Thus, metabolomic techniques can disclose the significant functions for metabolic readjustment, including accumulation and regulation of PM and SM levels in plants for stress management of agriculturally important crops [22, 23]. In this chapter, we discuss current findings on the application of primary and secondary metabolic genes using metabolic engineering technology to develop abiotic-stress-tolerant crop plants.

2.2 Metabolomics-mediated interpretation of abiotic stress tolerance in plants

In plants, physiological responses to environmental stresses involve transcriptional-level alterations, post-translational modifications, and metabolite

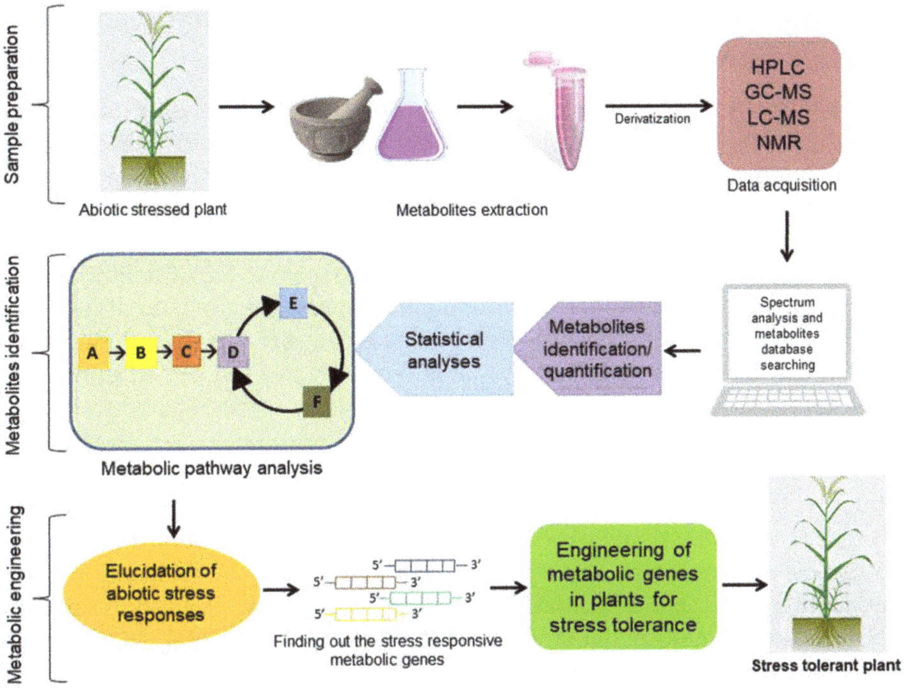

Figure 2.1. Workflow for metabolomics analysis and metabolic engineering for abiotic stress tolerance in plants. These images have been obtained by the author from the Pixabay website where they were made available under the Pixabay License. They are included within this article on that basis. Test tube stock vector courtesy Gungner/Shutterstock.com.

alterations and/or accumulation that lead to a specific physiological response or phenotype [24]. Under abiotic stress conditions, plants maintain homeostasis through an adaptable multifarious metabolic network comprising genes coded for proteins and metabolites that work toward phenotypes. Therefore, metabolic profiling is generally considered a bridge connecting phenotypes and genotypes that can reproduce the physiological position of growing plants [25, 26]. The plant metabolic network is rearranged under stress conditions to protect metabolic homeostasis and synthesize compounds that ameliorate the stress [5]. Most abiotic stresses disturb plant metabolism due to substrate shortage, metabolic enzyme inhibition, excessive requirements for a particular compound, or a combination of the previous. Metabolomics offers an insight into new metabolites that improve our knowledge of the mechanisms underlying tolerance to abiotic stress. Figure 2.1 illustrates the metabolomic workflow for interpreting abiotic stress tolerance in plants. The metabolomic analysis comprises sample extraction, analysis, data reduction, and metabolite identification [27] and generates large amounts of data to achieve maximum functional information. In the metabolomics workflow, the selected signals from LC-MS-based metabolomics are interpreted to identify metabolites through computational spectral analysis [28].

2.3 Abiotic-stress-induced adjustment of primary and secondary metabolites

Primary metabolites are essential for the appropriate development of plant cells and are directly involved in growth, reproduction, respiration, and photosynthesis in plants [29]. They comprise amino acids, sugars, lipids, and polyols that allocate plants to acclimate and survive abiotic stresses [30]. Abiotic stresses influence the levels of PMs, including sugars and polyols (e.g. trehalose, sorbitol), amino acids (e.g. proline), and lipids and fatty acids (e.g. diacylglycerols, sterols) (figure 2.2). Carbohydrates and sugars are the predominant factors for enhancing plant tolerance to abiotic stresses [31, 32]. Sugars contribute to different biochemical, structural, and physiological properties [33]. Sugars are chemically active biomolecules implicated in respiration, photosynthesis, flowering, seed germination, senescence, etc. Thus, modulation of sugar content in plants would enhance their adaptation and responses to abiotic stress [34]. Plant responses to abiotic stress involve various changes in amino acid metabolism. Plants have developed diverse approaches for decreasing the negative effects of abiotic stress, many of which involve amino acid metabolism [35].

Plants sense abiotic stress and convey the signal downstream, usually via the plasma membrane; therefore, lipid signaling or intermediary lipid biosynthesis and metabolism are critical for the environmental stress response [36]. Membrane lipid peroxidation under abiotic stress restricts plant growth and development [37]. In plants, lipid signaling molecules include sphingolipids, phosphatidic acid (PA), oxylipins, lysophospholipids, and free fatty acids [36]. These lipid molecules are

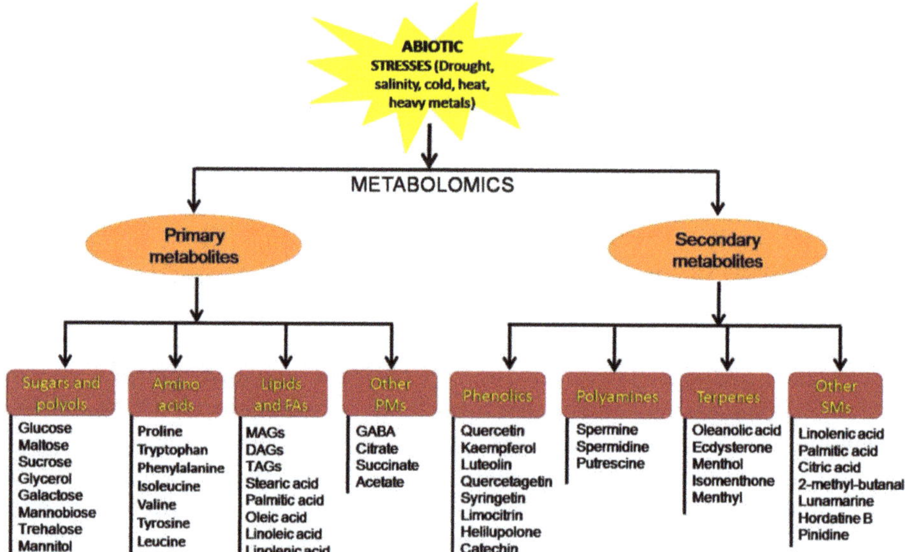

Figure 2.2. Various classes of key primary and secondary metabolites identified in plants under abiotic stress conditions.

frequently activated under abiotic stress conditions, including drought stress [38], salinity stress [39], and temperature stress [40].

Primary metabolites such as sugars and amino acids increase under eCO_2 stress and combined eCO_2 and drought stress in plants. The amino acids include proline, γ-aminobutyric acid (GABA), valine, lysine, isoleucine, arginine, phenylalanine, tryptophan, and tyrosine, and the sugars include trehalose and fructose [41]. Amino acids are important metabolites for cellular functioning and protein synthesis [42]. The aromatic amino acids tyrosine, tryptophan, and phenylalanine are necessary constituents of protein synthesis in plants and function as precursors for several SMs essential for plant growth [43]. Osmolytes and amino acids retain cells' turgor pressure and guard cellular processes by detoxifying reactive oxygen species (ROS) components in plant cells. Metabolites that provide osmoprotection to plants include amino acids (proline, asparagine, serine, and glutamine), GABA, soluble carbohydrates, glycine betaine (GB), and polyamines. These metabolites also reduce deterioration levels and retain proteins in their native tertiary structures [30]. Various key PMs identified in several plants under several abiotic stresses are listed in table 2.1.

Secondary metabolites play a significant role in adjusting plants to the environment and coping with stress conditions. A wide variety of SMs is synthesized from PMs but is not usually essential for plant survival. However, SMs protect plants from diverse environmental stresses. Plants comprise a broad range of protection mechanisms that enable them to deal with adverse conditions, accumulate SMs, and alleviate environmental stresses at the metabolomic level [11, 14]. In plants, several key SMs identified under abiotic stress conditions are listed in table 2.2. These SMs are mainly classified into three groups: phenolic compounds (e.g. flavonoids, tannins, lignins), nitrogen-containing metabolites including alkaloids (e.g. glycosides, cyanogenic glucosinolates), and terpenes (e.g. isoprenoids or terpenoids) [14, 44]. Secondary metabolite pathways comprise shikimic acid, mevalonic acid, methylerythritol phosphate, and phenylpropanoic acid [45]. Secondary metabolites mainly scavenge ROS molecules for defense of the plant cell from lipid peroxidation and achieve a crucial role in a variety of defense-related activities [46]. The major enzymes of the phenylpropanoid pathway implicated under stress conditions are phenylalanine ammonia-lyase (PAL), coumarate-4-hydroxylase (C4H), and 4-coumaroyl CoA ligase (4CL) [47]. Increased accumulation of SMs improves tolerance against stressful environments by regulating biochemical and physiological attributes in plants [48]. Phenolic compounds such as coumarins, flavonoids, lignins, and tannins are an important class of plant SMs [49]. Modulations in flavonoid content in plants under abiotic stress were associated with ROS production due to the defensive function of flavonols [50]. Lignins are the main component of plant cells with high-molecular-weight phenolic polymers and are involved in the synthesis of plants' cell walls [51]. Lignin accumulation plays a significant function in plant abiotic stress tolerance [52].

2.4 Engineering metabolic genes and pathways to improve abiotic stress tolerance

Metabolic engineering alters the endogeneous metabolic pathway by enhancing or degrading the level of endogenous compounds [53]. Enhancing targeted metabolites

Table 2.1. Key metabolites implicated in several plant species under abiotic stresses.

Plant species	Method of analysis	Tissue	Key metabolites	References
			Drought stress	
Avena sativa	GC	Leaves	FA: linolenic acid, palmitic acidLipids: free fatty acids, monoacylglycerols (MAGs), diacylglycerols (DAGs), triacylglycerols (TAGs)	[146]
Brachypodium distachyon	GC-MS	Leaves	AA: norvaline CH: glucose, maltose, sucrose, glycerol, galactose mannobiose	[110]
Hordeum vulgare	HPLC-DAD-MSn	Leaves	SM: chlorogenic acids, flavone glycosides, caffeoyl-hexose, hydroxycinnamic acids, sinapoyl-hexoses, feruloyl-hexose	[147]
Hordeum vulgare	GC-MS-EI	Fifth leaf, palea	AA: aspartate, glutamine, serine, proline, glycine, threonine, aromatic amino acids	[148]
Oryza sativa	GC/EI-TOF-MS	Leaves	AA: arginine, glutamate, proline PA: spermine, spermidine, putrescine OM: GABA	[149]
Triticum aestivum	GC-MS	Flag leaves	AA: asparagines, glutamine, lysine, methionine, serine	[150]
Zea mays	^{1}H-NMR	Leaves	AA: alanine Lipids: triacylglyceride OM: glutamate, malate, formate	[151]
Zea mays	GC-MS	Leaf blades	AA: glycine, myo-inositol	[152]
African eggplant	GC-MS	Leaves	AA: proline, glutamate CH: fructose, sucrose OM: tricarboxylic cycle metabolite	[153]
A. hypogaea	GC-MS	Leaves, roots	CH: D-mannitol, xylonic acid, myo-inositol, glucose, L-mano-pyranose, D-ribose, succinic acid, D-xylopyranose, α-D-glucopyranose, D-fructose, 2-deoxyribose, galactosoxime, D-turanose, 2-butenedoic acids, malic acid, 2-deoxyribose FA: palmitic acid, stearic acid, pentadecanoic acid, 8,11-octadecadienoic acid, pentadecanoic acid	[154]

Cicer arietinum	UPLC-HRMS	Leaves	AA: l-arginine, l-proline, l-isoleucine, l-histidine, tryptophan OM: allantoin	[155]
Craterostigma plantagineum	HPLC	Leaves	PA: spermidine, putrescine, spermine	[156]
Glycine max	NMR	Leaves, roots	AA: alanineCH: sucrose OM: citrate, succinate, acetate	[157]
Lentils	GC/EI-TOF-MS	Cotyledons, radicles, shoots	CH: erythronic acid PA: cadaverine, putrescine OM:isocitric acid, nicotinic acid	[158]
Nicotiana tabacum	GC-MS, LC-MS	Leaves, roots	CH: trehalose, mannitol, myo-inositol, galactinol OM: GABA	[158]
Nigella sativa	GC	Seeds	FA: palmitic acid, stearic acid, myristic acid, linoleic acid, oleic acid, linolenic acid, arachidic acid	[159]
Portulaca oleracea	GC	Leaves	FA: arachidic acid, oleic acid, linolenic acid, palmitic acid, behenic acid, linoleic acid, stearic acid	[160]
Vigna unguiculata	GC-TOF	Leaves	CH: raffinose, rhamnose,	[161]
Vitis vinifera	SPME-GC-MS	Leaves	SM: kaempferol-3-O-glucoside, quercetin-3-O-glucoside OM: 2-methyl-butanal phenylacetaldehyde, citric acid	[162]
Salinity stress				
Aeluropus lagopoides	CE-MS	Shoots	AA: asparagine, alanine, glutamine, lysine, proline, arginine, phenylalanine, histidine, valine, serine, tyrosine OM: adenine, 2-oxoglutarate, aconitate, adenosine, glucose-6-phosphate, succinate, adenosine 5′-monophosphate, ribulose-5-phosphate, fumarate, ribose-5-phosphate, citrate	[163]
Hordeum vulgare	HPLC	Leaves, roots	CH: pinitol, mannitol, sorbitol AA: proline	[164]
Zea mays	^{1}H-NMR	Shoots, roots	CH: glucose, sucroseAA: asparagines, alanine, glutamine OM: γ-amino-N-butyric acid, glycine betaine, succinate, malic acid, *trans*-aconitic acid, malic acid, citrate, acetoacetate, 2-oxoglutarate	[165]

(*Continued*)

Table 2.1. (*Continued*)

Plant species	Method of analysis	Tissue	Key metabolites	References
Arabidopsis thaliana	GC-MS	Roots, shoots	CH: trehalose, sucrose, galactose, melibiose, fructose, gentiobiose, erythritol, threonate galactinol, xylitol, galactonate AA: β-alanine betaine, proline, threonine, asparagines, glutamine, isoleucine OM: myo-inositol-1-phosphate, urea, citrate, glycerol-3-phosphate, sinapinate, succinate, propanoate, fumarate, aconitate, glycerate, butyrate, malate	[166]
Limonium latifolium	HPLC, ^{1}H-NMR	Shoots, roots	CH: fructose, sucrose, meso-inositol, glucose, chiro-inositol AA: glutamine, β-Alanine, glutamic acid, proline OM: β-alanine betaine, malate, choline-O-sulfate, citrate, choline,	[167]
Thellungiella halophila	GC-HP5890	Whole plant	CH: glucose, fructose, sorbitol, raffinose, galactinol, glycerol, maltose, sucrose, inositol, trehalose AA: aspartate, glutamine, proline, glycine, serine,threonine FA: stearic acid, palmitic acid OM: malic acid, citric acid, phosphoric acid, fumaric acid	[168]
Cicer arietinum	GC-QqQ-MS,LC-MS	Flowers, pods	CH: gentiobiose, frucose, sucrose, ery-ol, arabinose, inositol AA: glycine, trptophan, hydroxyproline, lysine, treonine, glutamic acid, histidine, cysteine, homoserine, methionine, isoleucine, valine, leucine, arginine, proline OM: putrescine, fumarate, pipecolate, GABA cis-aconitate, isocitrate, citrate, malate, 2-oxoglutarate, citrate	[169]
Glycine max, G. soja	GC-MS, LC-FT/MS	Leaves	AA: alanine CH: sucrose, lactitol, maltitolFA: linolenic acid OM: caffeic acid, abscisic acid, glutathione	[170]
Nicotiana tabacum	^{1}H-NMR	Aerial parts	CH: sucrose, glucose, myo-inositol, fructose AA: proline, tyrosine, glutamine, trptophan, asparagine, alanine, isoleucine, valine, phenylalanine	[171]

			OM: formate, uridine, nicotine, succinate, malate, choline, allantoin, uracil, ethanolamine, N-methylnicotinamide, hypoxanthine, γ-amino-n-butyrate, dimethylamine	
Solanum lycopersicon	UHPLC-QTOF-MS	Leaves	SM:1,3-dicaffeoylquinic acid, 1-feruoyl-5-caffeoylquinic acid, 3-caffeoyl-1–5 quinolactone, quercetin-3-hexodide	[92]
Fagopyrum esculentum	UFLC	Sprouts	SM: rutin, isoorientin, orientin, vitexin	[172]
Nitraria tangutorum	GC-MS	Cell suspensions	AA: asparagines, alainine, serine, valine, proline FA: 9,12-octadecadienoic acid, octadecanoic acid, hexadecanoic acid, OM: malic acid, succinate, acetamide, galactofuranose SM: sitosterol	[173]
Olea europaea	HPLC	Leaves, roots	SM: oleuropein SM: hydroxytyrosol	[174]
Rosmarinus officinalis	HPLC	Leaves	SM: camphene, cineole, α-terpineol, nopol, camphor,borneol	[175]
Salvia mirzayanii	HGC-MS	Leaves	OM: linalyl acetate, bicyclogermacrene, 1,8-cineole, α-terpinyl acetate	[176]
Atriplex halimus	GC-FID, UPLC	Seedlings	AA: alanine, arginine threonine, glycine, valine, proline, leucine, phenylalanine CH: malate, saccharose, myo-inositol	[177]
			Heat stress	
Populus tomentosa	GC/TOF-MS	Leaves	AA: proline CH: raffinose, melibiose	[178]
Solanum tuberosum	LC-MS	Leaves	AA: histidine, L-proline, tyrosine SM: quercetin, apigenin, isoprenoids, flavonoids, phenylpropanoids, simple phenols CH: mannose, xylose, arabinose, fucose, raffinose OM: jasmonic acid	[179]
Arabidopsis thaliana	GC-MS	Aerial tissues	AA: alanine, glycine, isoleucine, valine, asparagine, ornithine CH: erythritol, fructose, galactinol, glycerol, trehalose, xylose, myo-inositol SM: putrescine	[180]

(Continued)

Table 2.1. (*Continued*)

Plant species	Method of analysis	Tissue	Key metabolites	References
			OM: GABA, citramalic acid, beta-citric acid, glyceric acid, uracil, D-(2)-quinic acid, fumaric acid, dehydroascorbic acid, succinic acid, 2-ketoglutaric acid	
Arabidopsis thaliana	RP/UPLC-MS/ MSESIHILIC/UPLC-MS/MS	Leaves	CH: sucrose, RFOs Lipids: 2-aminoheptanoate, glycerol-3-phosphate, glycerophosphorylcholine (GPC)	[181]
Triticum aestivum	LC-HRMS	Leaves	AA: L-tryptophan, L-histidine, 4-aminobutanoate (GABA), L-arginine, leucine, L-phenylalanine, L-threonine, L-aspartate, OM: pipecolate, drummondol, anthranilate	[182]
Hordeum vulgare	HPLC, MS (IC-MS/MS)	Leaves	OM: glutathione	[183]
Pinus radiata	GC-MS, LC-Orbitrap-MS	Seedlings	AA: L-phenylalanine OM: hexadecanoic acid, dihydromyricetin, zeatin riboside, isopentenyl adenosine (iPA)	[184]
Solanum lycopersicum	QTOFMS	Pollen LC	PA: spermine, spermidine SM: quercetin, kaempferol dihexoside OM: tomatin, hydroxycinnamic, hydroxytomatine	[185]
Cold stress				
Arabidopsis thaliana	GC-TOF-MS	Seedlings	AA: glutamine, proline, asparagines, glycine, homoserine, tryptophan, ornithine CH: fructose, inositol, galactinol, glucose, raffinose, melibiose, threitol, trehalose PA: putrescine OM: aconitate, alpha ketoglutarate, ascorbate, gluconate, gluconate lactone, isosuccinate, citrulline, cycloserine	[186]
Oryza sativa	GC-MS	Leaves	AA: alanine, valine, glutamic acid, aspartic acid, tyrosine, cysteine, isoleucine, glycine, tryptophan, leucine, lysine, serine, proline, threonine	[187]

			CH: glucose, arabinose cellobiose, fructofuranose, glucopyranose, fructose, mannitose, melibiose, rhamnose, ribose, sorbinose, fucose, benzyl glucopyranoside, galactinol, 2-o-glycerol-beta-d-galalctopyranoside, methyl-beta-d-mannopyranoside, ethylglucopyranoside PA: putrescine, ethanolamine, triethanolamine, tryptamine-5-hydroxy OM: palmic acid, heptadecoic acid, oxalic acid, octadecanoic acid, succinic acid, quinic acid, GABA, eicosanoic acid, aminobutyric acid, citric acid, benzoic acid, cinnamic acid, 4-hydroxy, phosphoric acid, pyruvic acid, shikimic acid	
Oryza sativa	LC-MS/MS	Seedlings	CH: inositolOM: N-acetyl-L-glutamic acid, L-ascorbate, lysine butyrate, N-caffeoyl putrescine, anthranilic acid	[188]
Wheat	GC-QqQ-MS	Leaves	AA: alanine, arabinose, arabitol, arginine, asparagines, phenylalanine, proline, glycine, histidine, homoserine, isoleucine, leucine, lysine CH: fructose, galactitol, glucose, maltose mannose, inositol PA: putrescine OM: aconitate, 2-oxoglutarate, glucose-6-P, fructose 6-P, citrate, citrulline, GABA, isocitrate	[189]
Ipomoea batatas	Tuber	LC-MS/MS	OM: glutaric acid, D-glucoronic acid, 3-hydroxy-3-methylpentane-1,5-dioic acid, 5-methylcytosine, apigenin 7-O-glucoside (cosmosiin), apigenin O-malonylhexoside, sinapoylcholine, N-acetyl-5-hydroxytryptamine	[190]
Vitis amurensis, V. vinifera	GC-TOF-MS	Leaves	AA: aspartate, ornithine CH: xylose, raffinose, sucrose OM: malate, ascorbate, glycerate, malonate, tyramine, citrate, erythronate, threonate	[191]

(*Continued*)

Table 2.1. (*Continued*)

Plant species	Method of analysis	Tissue	Key metabolites	References
Heavy metal stress				
Arabidopsis thaliana	GC-MS	Seedlings	AA: proline, ascorbate CH: glycerol, trehalose OM: glycine betaine, glutathione	[192]
Arabidopsis thaliana	31P NMR	Cell suspension	AA: alanine, glutamine, glutamate CH: fructose, sucrose, glucose OM: citrate, malate	[193]
Raphanus sativus	GC-MS	Roots	CH: glucose, α-D-glucopyranoside, turanose, β-D-glucopyranose, galactose, inositol, maltose, fructose AA: proline, threonine, alanine OM: gluconate, octadecanoic acid, hexadecanoic acid, oxalic acid	[194]
Cynodon dactylon	ICP-MS, GC-MS	Leaves	AA: glycine, norvaline, proline, serine, threonine, glutamic acid, gulonic acid CH: xylulose, galactose, talose OM: glyceric acid, oxoglutaric acid, citric acid, malic acid	[195]
Solanum lycopersicum	^{1}H-NMR, HPLC-PDA	Leaves	AA: alanine, aspartate, asparagine, tyrosine, valine proline, phenylalanine, glutamate, glutamine, isoleucine CH: glucose, fructose, sucrose OM: citrate, malate, GABA, choline, glucopyranosyl, trigonelline	[196]
Silene cucubalus	^{1}H-NMR	Plant cell extracts	AA: glutamine, branched chain amino acids OM: malic acid, acetate	[197]

AA, amino acid; EI, electrospray ionization; CH: carbohydrate; GC-MS, gas chromatography-mass spectrometry; FA, fatty acid; GABA, γ-aminobutyric acid; HPLC-DAD-MS, high-performance liquid chromatography coupled with diode-array detection and multiple-stage mass spectrometry; HPLC, high-performance liquid chromatography; ICP-MS, inductively coupled plasma-mass-spectrometry; LC-HRMS, liquid chromatography high-resolution mass spectrometry; LC-MS, liquid chromatography-mass spectrometry; PAs, polyamines; SM, secondary metabolites; ^{1}H-NMR, nuclear magnetic resonance; OM, other metabolites; TOCSY, total correlation spectroscopy; TOF, time-of-flight; SPME-GC-MS, solid phase micro extraction-gas chromatography-mass spectrometry; UPLC-HRMS, ultra-performance liquid chromatography high-resolution mass spectrometry.

can be attained by overexpressing and/or repressing one or more genes [54]. Overexpression of these metabolic genes results in the metabolite biosynthesis that is responsible for osmolyte synthesis. Gene products may be suitable metabolite compounds/molecules or enzymes occupied in producing the targeted metabolites. Key metabolic compounds, such as polyamines, proline, flavonoids, GB, mannitol, trehalose, and lipids, accumulate in transgenic plants under abiotic stress (table 2.2), leading to cellular activities such as modulation of antioxidant activities, ROS detoxification, improved relative water contents, reduced malondialdehyde (MDA) content, ion leakage, and membrane stability that enhance growth and abiotic stress tolerance (figure 2.3). The functions of metabolic gene expression in abiotic stress tolerance are given in table 2.2. Overall, we conclude that increased biosynthesis of metabolites imparts a level of defense from stress and increases growth and abiotic stress tolerance in plants.

2.4.1 Drought tolerance

Drought affects the growth of the plant, water accessibilities, photosynthesis, pigment content, membrane integrity, yield, and biomass [55, 56]. Several drought-related genes have been characterized from diverse sources and characterized in transgenic plants to improve drought stress tolerance in the plant [57–59]. We proposed a schematic model showing the application of metabolic genes involved in drought tolerance through the accumulation of several metabolites in figure 2.3 and table 2.2. Putrescine, spermine, and spermidine are the most abundant polyamines found in plants, animals, and bacteria [60]. Arginine decarboxylase (ADC) catalyzes the biosynthesis of polyamines to generate putrescine from arginine. Transgenic *Arabidopsis thaliana* overexpressing the *ADC* gene from *A. thaliana* [61] and *Poncirus trifoliate* [62] provide drought stress tolerance through an increased accumulation of putrescine (table 2.2). Similarly, transgenic rice plants expressing the *Datura stramonium ADC* gene exhibited more tolerance to drought stress than wild-type plants [63]. The enzyme choline monooxygenase (CMO) catalyzes the biosynthesis of GB in plants which accumulated during salt and drought stress. Transgenic *Nicotiana tabacum* expressing *Beta vulgaris BvCMO* gene increased salt and drought stress tolerance through GB accumulation [64]. The overexpression of the betaine aldehyde dehydrogenase gene *BADH* from *Ammopiptanthus nanus* in transgenic *Arabidopsis* showed drought and salinity stress tolerance through increased GB accumulation [65].

In plant cells, flavonoids are essential SMs with major functions in balancing the cellular redox stability. Chalcone synthase (CHS) is the principal enzyme involved in the flavonoid biosynthetic pathway, which is regulated under drought stress. Overexpression of the tobacco *NtCHS* gene increased drought and oxidative stress tolerance in transgenic *N. tabacum* compared to the wild-type plant [66]. Overexpression of dehydrin genes from diverse plant species conferred improved drought tolerance through elevated proline accumulation [57, 67–69] (table 2.2). Proline functions as a compatible osmolyte normally found in plants especially under stress conditions; its accumulation provides drought stress tolerance in numerous plant species [70]. Proline biosynthesis normally occurs in the glutamate

Table 2.2. Metabolic genes' application to generate transgenic crops with enhanced abiotic stress tolerance.

Gene	Locus ID	Source	Transgenic plants	Metabolite accumulation	References
		Drought stress			
Arginine decarboxylase (*PtADC*)	HQ008237	*Poncirus trifoliata*	*Arabidopsis thaliana*	Enhanced putrescine	[62]
Arginine decarboxylase (*AtADC*)	BT000682	*Arabidopsis thaliana*	*Arabidopsis thaliana*	Improved putrescine	[61]
Betaine aldehyde dehydrogenase (*AnBADH*)	KJ841914	*Ammopiptanthus nanus*	*Arabidopsis thaliana*	Enhanced glycine betaine	[65]
Choline monooxygenase (*BvCMO*)	AB221007	*Beta vulgaris*	*Nicotiana tabacum*	Enhanced glycine betaine	[64]
Chalcone synthase (*NtCHS*)	LOC107801774	*Nicotiana tabacum*	*Nicotiana tabacum*	Enhanced flavanoids content (quercetin, rutin, naringenin)	[66]
Cysteine protease (*TaCP*)	AY841792	*Triticum aestivum*	*Arabidopsis thaliana*	Enhanced activity of *Cysteine protease*	[74]
Choline oxidase (*AgcodA*)	AY589052	*Arthrobacter globiformis*	*Solanum tuberosum*	Enhanced glycine betaine	[198]
Dehydrin (*ShDHN*)	AK319970	*Solanum habrochaites*	*Solanum lycopersicum*	Enhanced proline	[67]
Dehydrin (*OesDHN*)	KR349290	*Olea europaea*	*Arabidopsis thaliana*	Enhanced proline	[69]
Dehydrin (*PmLEAs*)	XM_016796383	*Prunus mume*	*Nicotiana tabacum*	Enhanced proline	[68]
Flavanone 3-hydroxylase (*RsF3H*)	JQ043380	*Reaumuria soongorica*	*Reaumuria soongorica*	Enhanced flavonoids, anthocyanin	[199]
Mannitol-1-phosphate dehydrogenase (*EcmtlD*)	EFF7369098	*Escherichia coli*	*Triticum aestivum*	Enhanced mannitol	[200]
Ornithine δ-aminotransferase (*Atδ-OAT*)	NM_123987	*Arabidopsis thaliana*	*Oryza sativa*	Enhanced proline	[201]

Ornithine δ-aminotransferase (*OsOAT*)	LOC Os03g44150	*Oryza sativa*	*Oryza sativa*	Enhanced proline	[202]
Spermidine synthase (*CfSPDS*)	BD142348	*Cucurbita ficifolia*	*Arapidopsis thaliana*	Enhanced spermidine synthase activity, spermidine content	[93]
Trehalose-6-phosphate synthase (*EcTPS*; *otsA*), trehalose-6-phosphate phosphatase (*EcTPP*; *otsB*)	NC_000913	*Escherichia coli*	*Oryza sativa*	Enhanced trehalose	[78]
Trehalose-6-phosphate synthase1 (*OsTPS1*)	HM050424	*Oryza sativa*	*Oryza sativa*	Enhanced trehalose, proline	[90]
Trehalose-6-phosphate synthase1 (*ScTPS1*), trehalose-6-phosphate synthase2 (*ScTPS2*)	NC_001134	*Saccharomyces cerevisiae*	*Nicotiana tabacum*	Enhanced trehalose	[77]
Wax synthase/acyl-CoA:diacylglycerol acyltransferase (*AtWSD1*)	AT5G37300	*Arabidopsis thaliana*	*Arapidopsis thaliana*, *Camelina sativa*	Enhanced deposition of epicuticular wax crystals, leaf, and stem wax loading	[75]
WRI4-like gene (*CeWRI4*)	MW039149	*Cyperus esculentus*	*Arabidopsis thaliana*	Enhanced cuticular wax biosynthesis and deposition	[203]
Δ1-pyrroline-5-carboxylate synthetase genes (*OsP5CS*)	D49714	*Oryza sativa*	*Petunia hybrida*	Enhanced proline	[72]
Δ1-pyrroline-5-carboxylate synthetase (*VaP5CS*)	VIRPYRR	*Vigna aconitifolia*	*Nicotiana tabacum*	Enhanced proline	[73]
		Salinity stress			
Arginine decarboxylase (*AsADC*)	X56802	*Avena sativa*	*Oryza sativa*	Enhanced putrescine, polyamine, ADC levels	[204]
Δ 1-pyrroline-5-carboxylate synthetase (*P5CSF129A*)	P5CS_VIGAC	*Vigna aconitifolia*	*Oryza sativa*	Enhanced proline	[205]
Δ 1-pyrroline-5-carboxylate synthetase (*P5CS*)	VIRPYRR	*Vigna aconitifolia*	*Oryza sativa*	Enhanced proline	[206]

(*Continued*)

Table 2.2. (*Continued*)

Gene	Locus ID	Source	Transgenic plants	Metabolite accumulation	References
S-adenosyl methionine decarboxylase (*SAMDC*)	CAA58762	*Tritordeum*	*Oryza sativa*	Biosynthesis of polyamine for spermidine and spermine production	[86]
Trehalose-6-phosphate synthase (*EcTPS; otsA*), trehalose-6-phosphate phosphatase (*EcTPP; otsB*)	NC_000913	*Escherichia coli*	*Oryza sativa*	Enhanced trehalose	[78]
Trehalose-6-phosphate synthase (*EcTPS*), trehalose-6-phosphate phosphatase (*EcTPP*)	EU070413, NC_002695	*Escherichia coli*	*Oryza sativa*	Enhanced trehalose	[89]
Trehalose-6-phosphate synthase1 (*OsTPS1*)	HM050424	*Oryza sativa*	*Oryza sativa*	Enhanced proline and trehalose	[90]
Mannitol-1-phosphate dehydrogenase (*mtlD*)	EFF7369098	*Escherichia coli*	*Triticum aestivum*	Enhanced mannitol accumulation	[200]
Flavanone 3-hydroxylase (*PnF3H*)	MK036761	*Pohlia nutans*	*Arabidopsis thaliana*	Flavanoids accumulation	[85]
Betaine aldehyde dehydrogenase (*AnBADH*)	KJ841914	*Ammopiptanthus nanus*	*Arabidopsis thaliana*	Enhanced glycine betaine	[65]
Choline monooxygenase (*BvCMO*)	AB221007	*Beta vulgaris*	*Nicotiana tabacum*	Enhanced glycine betaine	[64]
Dehydrin (*TdDhn-5*)	AY619566	*Triticum durum*	*Arabidopsis thaliana*	Enhanced proline	[57]
Dehydrin (*ShDHN*)	AK319970	*Solanum habrochaites*	*Solanum lycopersicum*	Enhanced proline	[67]
Spermidine synthase (*CfSPDS*)	BD142348	*Cucurbita ficifolia*	*Arabidopsis thaliana*	Enhanced spermidine synthase activity, spermidine content	[93]

		Heat stress			
Galactinol synthase1 *(AtGolS1)*	NM_130286	*Arabidopsis thaliana*	*Arabidopsis thaliana*	Raffinose accumulation	[100]
Δ 1-pyrroline-5-carboxylate synthetase *(AtP5CS1)*	NM_001202786	*Arabidopsis thaliana*	*Arabidopsis thaliana*	Proline accumulation	[98]
Heat Shock Factor Gene *(AtHsfA3)*	NM_120453	*Arabidopsis thaliana*	*Arabidopsis thaliana*	Galactinol accumulation	[102]
Heat stress transcription factors *(FaHsfA2c)*	KU674348	*Festuca arundinacea*	*Arabidopsis thaliana*	Galactinol accumulation	[52]
Heat stress transcription factors *(LlHsfA3A, LlHsfA3B)*	KF870404	*Lilium longiflorum*	*Arabidopsis thaliana*	Proline accumulation	[103]
Glutathione transferase *(Pvgstu3–3)*	JQ316125	*Phaseolus vulgaris*	*Nicotiana tabacum*	Increased accumulation of metabolites (proline, oxoproline, mannitol)	[104]
Sterol glycosyl trnasferases *(WsSGTL1)*	DQ356887	*Withania somnifera*	*Arabidopsis thaliana*	*Sterol accumulation*	[109]
		Cold stress			
Glutathione S-transferases *(OsGSTL2)*	Os03g17470	*Oryza sativa*	*Arabidopsis thaliana*	Glutathione	[145]
Arginine decarboxylase *(PtADC)*	HQ008237	*Poncirus trifoliata*	*Arabidopsis thaliana*	Enhanced putrescine	[62]
Dehydrin *(PmLEAs)*	XM_016796383	*Prunus mume*	*Nicotiana tabacum*	Enhanced proline	[68]
Dehydrin *(SiDHN)*	AHN92202	*Saussurea involucrata*	*Solanum lycopersicum*	Enhanced proline	[126]
Spermidine synthase *(CfSPDS)*	BD142348	*Cucurbita ficifolia*	*Arabidopsis thaliana*	Enhanced spermidine content, spermidine synthase activity	[93]

(Continued)

Table 2.2. (*Continued*)

Gene	Locus ID	Source	Transgenic plants	Metabolite accumulation	References
Trehalose-6-phosphate synthase1 (*OsTPS1*)	HM050424	*Oryza sativa*	*Oryza sativa*	Increased trehalose, proline	[90]
EcTPS; EcTPP	NC_000913	*Escherichia coli*	*Oryza sativa*	Increased trehalose	[78]
		Heavy metal stress			
γ-glutamylcysteine synthetase-glutathione synthetase (*StGCS-GS*)	GQ848551	*Streptococcus thermophilus*	*Beta vulgaris*	Glutathione	[143]
Phytochelatin synthase (*AsPCS1*)	AF384110	*Allium sativum*	*Allium sativum*	Phytochelatins	[138]
Metallothioneins (MTs) (*AsMT2a*)	AY821676	*Allium sativum*	*Allium sativum*	Metallothioneins	[138]
Ferrochelatase-1 (*AtFC1*)	AT5G26030	*Arabidopsis thaliana*	*Arabidopsis thaliana*	Glutathione	[141]
Phytochelatin synthase (*AtPCS1*)	NP_199220	*Arabidopsis thaliana*	*Nicotiana tabacum*	Phytochelatins	[207]
Phytochelatin synthase (*CePCS*)	AAK62992	*Caenorhabditis elegans*	*Nicotiana tabacum*	g-glutamylcysteine	[207]
Phytochelatin synthase (*AtPCS1*)	AF085230	*Arabidopsis thaliana*	*Arabidopsis thaliana*	γ-glutamylcysteine	[142]
Glutathione S-transferases (*OsGSTL*2)	Os03g17470	*Oryza sativa*	*Arabidopsis thaliana*	Glutathione	[145]

pathway, catalyzed by Δ1-pyrroline-5-carboxylate reductase (P5CR) and Δ1-pyrroline-5-carboxylate synthase (P5CS) in the chloroplasts or cytoplasm [71]. Expression of *P5CS* genes in plants showed better drought tolerance through higher proline accumulation (table 2.2), including expression of the rice *OsP5CS* gene in a transgenic *Petunia hybrid* [72] and *VaP5CS* gene from *Vigna aconitifolia* in transgenic *N. tabacum* [73]. Overexpression of the *TaCP* gene from *T. Aestivum* in transgenic *Arabidopsis* increased tolerance to drought stress [74]. *A. thaliana* and *Camelina sativa* plants overexpressing the wax synthase/acyl-CoA: diacylglycerol acyltransferase (*WSD1*) gene from *Arabidopsis* showed increased drought tolerance, mannitol, ABA, and salt stress by accumulating epicuticular wax crystals and stem and leaf wax loading [75]. Trehalose has an important role in osmotic regulation [76]. Overexpression of trehalose-6-phosphate synthase1 *(ScTPS1)* and trehalose-6-phosphate synthase2 *(ScTPS2)* genes from *Saccharomyces cerevisiae* showed increased drought tolerance in transgenic *N. tabacum* [77]. In addition, the *TPS* gene from *Oryza sativa* showed increased tolerance to drought, salt, and cold stress in transgenic rice [78].

2.4.2 Salinity stress tolerance

Soil salinity negatively affects crop productivity worldwide. Genetic engineering technologies have been applied to develop salt-stress-tolerant plants through changes in enzymatic activities involved in metabolite biosynthesis [79], antioxidant enzymatic activities, and enzymes involved in the biosynthesis of membrane lipid transport [80]. In plants, proline is a key endogenous osmolyte that accumulates under several abiotic stresses, including salt stress [81]. Δ1-pyrroline-5-carboxylate synthetase genes play key roles in salinity stress tolerance by directing proline biosynthesis [82]. Overexpression of the *ADC* gene from *Avena sativa* in transgenic *Lotus tenuis* enhanced salt tolerance by accumulating proline, which helps in cell membrane stability [83] (table 2.2). Expression of the wheat dehydrin *(TdDHN-5)* gene in transgenic *A. thaliana* provided tolerance against salt stress by increasing proline metabolism and decreasing H_2O_2 levels [84]. Similarly, overexpression of the dehydrin gene *ShDHN* from *Solanum habrochaites* in transgenic tomato showed salt-stress tolerance through higher proline accumulation [67].

Flavanone 3-hydroxylase (*F3H*) is one of the main facilitator genes with a significant role in flavonoid biosynthesis in the phenylalanine ammonia-lyase (PAL) pathway. Transgenic *A. thaliana* expressing the *F3H* gene from *Pohlia nutans* conferred enhanced salt and oxidative stress tolerance [85]. Overexpression of the S-adenosyl methionine decarboxylase *(SAMDC)* gene from *Tritordeum* enhanced polyamine (spermidine and spermine) accumulation in transgenic rice to increase salinity stress tolerance [86]. Overexpression of the *BADH* gene isolated from *Atriplex canescens* in transgenic *Solanum tuberosum* enhanced salinity stress tolerance by enhancing chlorophyll and proline accumulation, whereas there was a reduction of MDA in comparison to the wild-type plant [87].

In plants, biosynthesis of trehalose normally occurs via the trehalose-6-phosphate synthase (TPS)/trehalose-6-phosphate phosphatase (TPP) pathway, where TPS

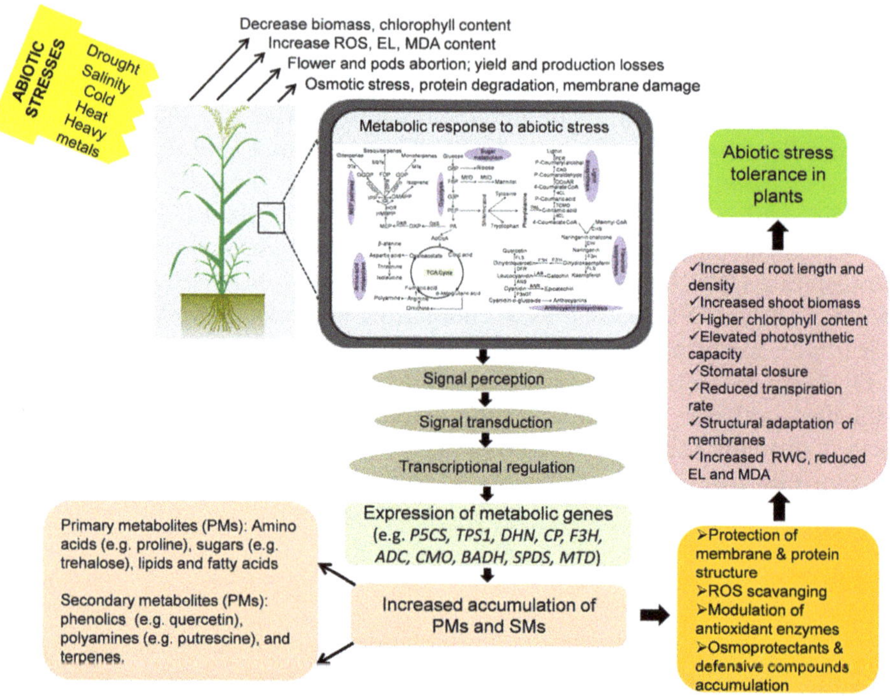

Figure 2.3. Schematic model displaying the metabolic engineering for abiotic stress tolerance in plants. The proposed model depicts an abiotic-stress-mediated decrease in chlorophyll content and shoots and roots biomass enhanced reactive oxygen species (ROS), electrolytic leakage (EL), malondialdehyde (MDA), and flower and pod abortion, osmotic stress generation, protein degradation, and membrane damage, and reduced yield. Primary metabolites (PMs) and secondary metabolites (SMs) readjust under abiotic stress in plant cells to retain osmotic balance and trigger a range of PM and SM pathways to survive. Plants respond and adjust to abiotic stresses such as drought, salinity, cold, heat, and heavy metals using different pathways for signal perception, transduction, transcriptional regulation, and expression of various metabolic genes. Abiotic-stress-induced expression of metabolic genes, such as arginine decarboxylase gene (*ADC*), Δ1-pyrroline-5-carboxylate synthetase (*P5CS*), cysteine protease (*CP*), flavanone 3-hydroxylase (*F3H*), dehydrin (*DHN*), trehalose-6-phosphate synthase1 (*TPS1*), choline monooxygenase (*CMO*), betaine aldehyde dehydrogenase (*BADH*), mannitol dehydrogenase (*MTD*), spermidine synthase (*SPDS*), and wax synthase/acyl-CoA: diacylglycerol acyltransferase (*WSD*), result in the accumulation of PMs and SMs and also various osmoprotectants, defensive compounds, and ROS-scavenging molecules inside plant cells. Changes in antioxidant activities restrict cell injury and sustain homeostasis. Morphological changes take place that improve plant growth and tolerance to abiotic stresses. AcCoA, acetyl-CoA; 4CL, 4-coumarate-CoA ligase; CCoAR, cinnamyl alcohol dehydrogenase; CAD, cinnamoyl-CoA reductase; CHS, chalcone synthase; CHI, chalcone isomerase; F3H, flavanone 3-hydroxylase; F3'H, flavonoid 3'-hydroxylase; DTs, diterpene synthase; DXP, 1-deoxy-D-xylulose-5-phosphate; F6P, fructose 6-phosphate; FDP, farnesyl diphosphate; G3P, glyceraldehyde 3-phosphate; FLS, flavonol synthase; G6P, glucose-6-phosphate; GGDP, geranyl geranyl diphosphate; IDI, isopentenyl diphosphate isomerase; MtlD, mannitol-1-phosphate dehydrogenase; PER, peroxidase; PAL, phenylalanine ammonia-lyase; IPP, isopentenyl diphosphate; MTs, monoterpene synthase; PA, pyruvic acid; SQTs, sesquiterpene synthase; TCA;tricarboxylic acid PEP, phosphoenolpyruvate. Figure created with BioRender.com (https://app.biorender.com/biorender-templates).

catalyzes the conversion of glucose-6-phosphate (Glc6P) and UDP-glucose to synthesize, and then TPP dephosphorylates T6P to produce trehalose. Trehalose-6-phosphate synthase is the main enzyme involved in the biosynthesis of trehalose since it modulates trehalose-6-phosphate (T6P) contents, thus influencing plant metabolic and developmental activities [88]. Overexpression of *TPS* and *TPP* genes of *Escherichia coli* in transgenic rice plants accumulated enhanced trehalose content that improved salt-stress tolerance [78, 89]. Similarly, Li *et al* disclosed that the overexpression of the rice *OsTPS* gene enhanced salt-stress tolerance in transgenic rice plants by accumulating trehalose and proline [90] (table 2.2).

Sorbitol belongs to the sugar alcohols group; it is synthesized from glucose-6-phosphate by sorbitol-6-phosphate dehydrogenase (S6PDH) and sorbitol-6-pyrophosphatase (S6PP). The enzyme sorbitol-6-phosphate dehydrogenase, encoded by the *S6PDH* gene, catalyzes the biosynthesis of sorbitol and has been used to develop abiotic-stress-tolerant transgenic plants. Transgenic Japanese persimmon expressing the apple (*Malus domestica*) gene (*S6PDH*) encoding sorbitol-6-phosphate dehydrogenase had improved tolerance against salt stress because of increased sorbitol accumulation [91]. The gene encoding chalcone synthase (*CHS*) overexpression in transgenic tomato (*Solanum lycopersicon*) provides salinity stress tolerance due to flavanoid accumulation [92]. Transgenic *A. thaliana* expressing the metabolic gene spermidine synthase (*CfSPDS*) from *Cucurbita ficifolia* accumulated spermidine, which enhanced tolerance to salinity stress [93] (table 2.2).

2.4.3 Heat stress tolerance

Heat stress adversely affects plants by altering several physiological processes, including photosynthesis, water relations, lipid metabolism, and primary and secondary metabolism [94]. It also damages cell membranes, increases ROS production, and causes senescence and cell death. Anthocyanin usually accumulates in vegetative tissues during heat stress [95] to maintain leaf transpiration due to the lower osmotic potential [96]. Engineering metabolic genes for metabolite accumulation in plants' overexpression of these genes in transgenic plants resulted in tolerance against heat stress (figure 2.3, table 2.2). P5CS is the principal enzyme involved in proline biosynthesis. It controls the proline content in plant cells and increases osmotic stress resistance. Glutamic acid is the main precursor of the biosynthesis of proline in unfavorable environments. It is catalyzed by Δ1-pyrroline-5-carboxylate synthetase genes and Δ1-pyrroline-5-carboxylate reductase into proline [97]. Overexpression of Δ1-pyrroline-5-carboxylate synthetase gene *AtP5CS1* increased proline accumulation in *A. thaliana*, increasing heat stress tolerance [98]. Galactinol functions as a precursor to the raffinose family of oligosaccharides (RFOs, e.g. stachyose, verbacose, and raffinose), and in this process sucrose acquires the galactosyl residue and yields larger myo-inositol and oligosaccharides [99]. Expression of the galactinol synthase1 (*AtGolS1*) gene in transgenic *A. thaliana* accumulated higher levels of raffinose, providing plant heat tolerance [100]. In plants, heat shock factors (Hsfs) play a fundamental role and are controllers of responses towards abiotic stresses, particularly high-temperature stress. The heat stress response is directed at the stage of transcription by heat shock factors which are regulated by high-temperature stress. Activated/triggered heat shock factors exclusively attach to palindromic heat shock elements that are important for the transcription of Hsfs genes; this procedure works under acquired and basal heat stress responses [101]. Engineering heat shock factor gene *HsfA3* from *A. thaliana, FaHsfA2c* gene from *Festuca arundinacea*, and *LlHsfA3A* and *LlHsfA3B* from *Lilium longiflorum* for their overexpression increased galactinol accumulation in transgenic *A. thaliana*, resulting in heat tolerance [52, 102, 103] (table 2.2). Overexpression of *Pvgstu3–3* gene isolated from *Phaseolus vulgaris* enhanced heat tolerance in transgenic *N. tabacum* by increasing metabolite accumulation such as proline, oxoproline, and mannitol [104]. Sterol glycosylation is advanced by a key enzyme, sterol glycosyltransferase (SGT), belonging to a multigene family [105]. It is involved in the accumulation of sterol glycosides, which increase metabolite stability, transport, and solubility in plants [106–108]. Overexpression of sterol glycosyltransferases gene *WsSGTL1* from *Withania somnifera* enhanced heat stress tolerance in transgenic *A. thaliana* by accumulating sterols [109].

2.4.4 Cold stress tolerance

Cold stress has detrimental effects on the growth, development, and production of agriculturally important crop plants [110]. Cold stress alters physiological parameters, causing chilling injury, restricted growth, and molecular changes, including the expression of miRNA and stress-related genes [111]. Plants can produce and consume energy under cold stress conditions [112]. Sugars and adenosine triphosphate (ATP) are considered the most significant energy metabolites for cold tolerance [113]. Increased sucrose biosynthesis was associated with cold tolerance in *Arabidopsis* by averting a metabolism bottleneck during cold stress [114]. Trehalose may play a significant role as an osmolyte to alleviate the damage of membranes and is also involved in the cold stress response [88]. Xu *et al* reported that cold stress upregulated TPS and TPP in tobacco plants [115]. Cold-tolerant *Pinus halepensis* seedlings increased glucose and fructose accumulation, suggesting that they are involved in cold tolerance [116].

In plants, amino acid accumulation is common under cold stress [117, 118]. Enhanced amino acids contents such as proline, glutamic acid, glutathione, histidine, arginine, and tryptophan were observed in cold-tolerant *P. halepensis* seedlings compared to cold-sensitive seedlings [116]. Proline is a well-known osmoprotectant with a dynamic function in ROS scavenging, protein structure stabilization, and cellular homeostasis adjustment [119]. Increased accumulation of aromatic amino acids (phenylalanine and tyrosine), precursors of the shikimate-phenylpropanoid pathway, was observed under cold stress [120]. Exogenous application of phenylalanine to tomato plants enhanced chilling stress tolerance by activating ROS-scavenging systems [121]. Cold stress produces phenolic compounds, which are subsequently absorbed in cell walls as lignin and suberin [122]. Adaptation of apple trees to cold environments was associated with increased chlorogenic acid production [123]. For instance, high and low temperatures affected

SM production in somatic embryos of *Eleutherococcus senticosus* by eliciting oxidative stress [124].

Metabolic gene expression results in the accumulation of various metabolites that protect plants against cold stress (figure 2.3, table 2.2). Overexpression of the *ADC* gene isolated from *Poncirus trifoliata* showed tolerance to cold stress in transgenic *A. thaliana* by accumulating the polyamine compound putrescine [62]. Dehydrins (DHNs), or group 2 late embryogenesis abundant proteins, have an essential role in abiotic stress management in plants. Accumulation of DHN proteins characteristically occurs in mature seeds or vegetative tissues under dehydration, salinity, cold, or freezing stress [125]. The dehydrin genes *PmLEAs* from *Prunus mume, ShDHN* from *S. habrochaites*, and *SiDHN* from *Saussurea involucrate* provided tolerance against cold stress by accumulating proline in *N. tabacum* [68] and tomato [67, 126], respectively (table 2.2). Overexpression of the *EcTPS* and *EcTPP* genes from *E. coli* and *OsTPS1* from rice enhanced cold stress tolerance in *O. sativa* by accumulating trehalose and proline [78, 90]. Similarly, rice glutathione S-transferases gene *OsGSTL2* overexpression in transgenic *A. thaliana* conferred tolerance against cold stress by increasing glutathione (GSH) accumulation [90].

2.4.5 Heavy metals stress tolerance

In plants, the development of tolerance towards heavy metals stress by genetic engineering technology requires one to pinpoint the principal genetic factors involved in metal stress tolerance and introduce these metabolic genes. Stress tolerance is generally expressed by stimulating the mechanisms implicated in radical scavenging and DNA repair at the cellular level. The most common approach for understanding heavy metals' stress tolerance is investigating antioxidant activities [127]. Heavy metals can be extremely reactive, causing toxicity in plant cells. Heavy metal stress changes several physiological processes by deactivating and denaturing enzymes and proteins, stopping the functional group of molecules essential for metabolic processes, replacing/dislocating important metal ions through functional cellular units and biomolecules, disturbing rigidity of membrane and conformational modifications [128, 129] that affects plant metabolism, restricts respiration and photosynthesis, and activates various key enzymatic activities [130–132].

Molecular, biochemical, and physiological approaches are used to identify the mechanisms underlying heavy metal aggregation to fight the detrimental properties of heavy metals stress. Adaptive mechanisms developed by stress-tolerant plants include limiting heavy metal absorption and transport, synthesizing transporters involved in heavy metal stress, chelating metals through specific ligands such as metallothioneins (MTs) and phytochelatins (PCs) stimulating stress-responsive proteins, biosynthesizing metabolites, specifically proline and polyamines [133–135]. Exogenous proline treatment in *Vigna radiata* elevated heavy metal stress tolerance by maintaining an elevated GSH level and GSH-metabolizing enzymes [136]. Hence, metabolic genes' expression involved in proline accumulation would be useful for developing plants with heavy metal stress tolerance.

Plant MTs perform an important role in the stability of essential metals (e.g. Ni, Se, Cu, and Zn) and they also detox xenobiotic metals (e.g. Ag, Cd, and Hg) [137]. Expression of the metallothioneins (*AsMT2a*) gene in *Allium sativum* evolved in tolerance against heavy metals stress [138] (table 2.2). Chaturvedi *et al* reported that overexpression of the *SbMt2* gene from *Salicornia brachiata* in tobacco showed increased heavy metal tolerance [139]. Phytochelatins (PCs) are cysteine-rich thiol-reactive peptides that combine with a variety of toxic metalloids and metals, establishing their expression as useful for increasing the capacity of plants for phytoremediation [140]. Transgenic *A. sativum* expressing the phytochelatin synthase *(AsPCS1)* gene resulted in tolerance to heavy metals stress [138]. The *A. thaliana* expressing *AtFC1* gene provides tolerance against cadmium stress [141]. Similarly, expression of the *AtPCS1* gene showed tolerance against heavy metals stress in transgenic *A. thaliana* [142]. Transgenic *Beta vulgaris* expressing the γ-glutamylcysteine synthetase-glutathione synthetase *(StGCS-GS)* gene isolated from a probiotic bacteria *Streptococcus thermophilus* showed tolerance to cadmium, zinc, and copper stress by accumulating glutathione [143]. A multifunctional enzyme, glutathione S-transferases, belongs to the ubiquitous family that protects against a variety of biotic and abiotic stresses, including heavy metals stress [144]. The overexpression of the glutathione S-transferases gene *OsGSTL2* isolated from rice enhanced heavy metals' stress tolerance as well as tolerance against other abiotic stresses such as osmotic stress, salt, and cold by accumulating glutathione [145].

2.5 Conclusions

In plants, metabolomics is used to disclose responses to various environmental stresses. Metabolic engineering with the aim to develop abiotic stress tolerance in crops may vastly enhance plant productivity. This chapter has discussed metabolic engineering for developing abiotic stress tolerance in plants, including salinity, drought, heat, heavy metals, and cold. Researchers have made considerable efforts to insert metabolic genes in transgenic plants that accumulate various metabolites, resulting in abiotic stress tolerance. However, further investigations are needed to identify the genes, enzymes, and pathways involved in metabolite biosynthesis. Studies are also needed to identify the key metabolic regulators for genetic engineering of various biotic and abiotic stress tolerances in agriculturally important crops.

Multiple choice questions

1. Which is the method used for metabolite identification in plants?
   (a) SDS-PAGE;
   (b) LC-MS;
   (c) PCR;
   (d) Electrophoresis.
2. Expression of *P5CS* genes in plants results in …
   (a) Decrease in proline content;
   (b) Increase in proline content;

    (c) decrease in sorbitol;
    (d) Increase in trehalose.
3. The precursor for the raffinose family of oligosaccharides is …
    (a) Trehalose;
    (b) Sorbitol;
    (c) Galactinol;
    (d) Maltose.
4. Chalcone synthase (CHS) is the principal enzyme involved in …
    (a) The osmolyte biosynthetic pathway;
    (b) The flavonoid biosynthetic pathway;
    (c) The ABA biosynthetic pathway;
    (d) The amino acid biosynthetic pathway.
5. The enzyme choline monooxygenase (CMO) catalyzes the biosynthesis of …
    (a) Proline;
    (b) Mannitol;
    (c) Glycine betaine;
    (d) Glutamine.

Answers: 1 (b); 2 (b); 3 (c); 4 (b); 5 (c).

Descriptive-type questions

1. Define metabolomics and its role in understanding abiotic stress tolerance in plants?
2. Describe secondary metabolites identified in plants under abiotic stress conditions.
3. Describe the mechanism by which polyamines, betaine aldehyde dehydrogenase, and proline provide abiotic stress tolerance to the plants.
4. What are the important metabolic pathways engineered to improve abiotic stress tolerance?
5. Write short notes on the role of sorbitol, glycine betaine, and trehalose in plants.

References

[1] Schmidhuber J and Tubiello F N 2007 Global food security under climate change *PNAS* **104** 19703–8

[2] Boyer J S 1982 Plant productivity and environment *Science* **218** 443–8

[3] Wang W, Vinocur B, Shoseyov O and Altman A 2000 Biotechnology of plant osmotic stress tolerance physiological and molecular considerations *IV Int. Symp. on In Vitro Culture and Horticultural Breeding* **560** 285–92

[4] Wen W, Li K, Alseekh S, Omranian N, Zhao L, Zhou Y, Xiao Y, Jin M, Yang N and Liu H 2015 Genetic determinants of the network of primary metabolism and their relationships to plant performance in a maize recombinant inbred line population *The Plant Cell* **27** 1839–56

[5] Obata T and Fernie A R 2012 The use of metabolomics to dissect plant responses to abiotic stresses *Cell. Mol. Life Sci.* **69** 3225–43
[6] Stephanopoulos G 1999 Metabolic fluxes and metabolic engineering *Metab. Eng.* **1** 1–11
[7] Vickers C E, Possell M, Laothawornkitkul J, Ryan A C, Hewitt C N and Mullineaux P M 2011 Isoprene synthesis in plants: lessons from a transgenic tobacco model *Plant, Cell Environ.* **34** 1043–53
[8] Ahmed I, Shukla P and Gautam R 2021 *Omics Technologies for Sustainable Agriculture and Global Food Security* vol I (Berlin: Springer) pp 169–89
[9] Kumar A 2015 *Plant Biology and Biotechnology* (Berlin: Springer) pp 517–26
[10] Vemuri G N and Aristidou A A 2005 Metabolic engineering in the-omics era: elucidating and modulating regulatory networks *Microbiol. Mol. Biol. Rev.* **69** 197–216
[11] Kumar M, Patel M K, Kumar N, Bajpai A B and Siddique K H 2021 Metabolomics and molecular approaches reveal drought stress tolerance in plants *Int. J. Mol. Sci.* **22** 9108
[12] Isah T 2019 Stress and defense responses in plant secondary metabolites production *Biol. Res.* **52** 39
[13] Singh N, Mansoori A, Dey D, Kumar R and Kumar A 2021 *Omics Technologies for Sustainable Agriculture and Global Food Security* vol II (Berlin: Springer) pp 193–214
[14] Patel M K, Kumar M, Li W, Luo Y, Burritt D J, Alkan N and Tran L-S P 2020 Enhancing salt tolerance of plants: from metabolic reprogramming to exogenous chemical treatments and molecular approaches *Cells* **9** 2492
[15] Rodziewicz P, Swarcewicz B, Chmielewska K, Wojakowska A and Stobiecki M 2014 Influence of abiotic stresses on plant proteome and metabolome changes *Acta Physiol. Plant.* **36** 1–19
[16] Jaya P, Ranjan A, Afsana A N, Srivastava A K and Mishra L N 2021 *Omics Technologies for Sustainable Agriculture and Global Food Security* II (Berlin: Springer) pp 69–93
[17] Kinghorn A D 1994 The discovery of drugs from higher plants *Discovery of Novel Natural Products with Therapeutic Potential* (New York: Elsevier) pp 81–108
[18] Akula R and Ravishankar G A 2011 Influence of abiotic stress signals on secondary metabolites in plants *Plant Signal. Behav.* **6** 1720–31
[19] Mishra A, Patel M K and Jha B 2015 Non-targeted metabolomics and scavenging activity of reactive oxygen species reveal the potential of Salicornia brachiata as a functional food *J. Funct. Foods* **13** 21–31
[20] Roberts L, Souza A, Gerszten R and Clish C 2012 Targeted metabolomics *Curr. Protoc. Mol. Biol.* **98** 30.2.1–30.2.24
[21] Van Meulebroek L, Hanssens J, Steppe K and Vanhaecke L 2016 Metabolic fingerprinting to assess the impact of salinity on carotenoid content in developing tomato fruits *Int. J. Mol. Sci.* **17** 821
[22] Zagorchev L, Seal C E, Kranner I and Odjakova M 2013 A central role for thiols in plant tolerance to abiotic stress *Int. J. Mol. Sci.* **14** 7405–32
[23] Cusido R M, Onrubia M, Sabater-Jara A B, Moyano E, Bonfill M, Goossens A, Pedreño M A and Palazon J 2014 A rational approach to improving the biotechnological production of taxanes in plant cell cultures of *Taxus spp. Biotechnol. Adv.* **32** 1157–67
[24] Verslues P E, Agarwal M, Katiyar-Agarwal S, Zhu J and Zhu J K 2006 Methods and concepts in quantifying resistance to drought, salt and freezing, abiotic stresses that affect plant water status *Plant J.* **45** 523–39

[25] Sanchez D H, Redestig H, Krämer U, Udvardi M K and Kopka J 2008 Metabolome-ionome-biomass interactions: what can we learn about salt stress by multiparallel phenotyping? *Plant Signal. Behav.* **3** 598–600

[26] Fukushima A and Kusano M 2014 A network perspective on nitrogen metabolism from model to crop plants using integrated 'omics' approaches *J. Exp. Bot.* **65** 5619–30

[27] Choi Y H and Verpoorte R 2014 Metabolomics: what you see is what you extract *Phytochem. Anal.* **25** 289–90

[28] Ludwig C, Easton J M, Lodi A, Tiziani S, Manzoor S E, Southam A D, Byrne J J, Bishop L M, He S and Arvanitis T N 2012 Birmingham Metabolite Library: a publicly accessible database of 1-D 1 H and 2-D 1 HJ-resolved NMR spectra of authentic metabolite standards (BML-NMR) *Metabolomics* **8** 8–18

[29] Kumar R, Bohra A, Pandey A K, Pandey M K and Kumar A 2017 Metabolomics for plant improvement: status and prospects *Front. Plant Sci.* **8** 1302

[30] Fraire-Velázquez S and Balderas-Hernández V E 2013 Abiotic stress in plants and metabolic responses *Abiotic Stress—Plant Responses and Applications in Agriculture* (Rijeka: InTech) pp 25–48

[31] Sami F, Yusuf M, Faizan M, Faraz A and Hayat S 2016 Role of sugars under abiotic stress *Plant Physiol. Biochem.* **109** 54–61

[32] Gupta A K and Kaur N 2005 Sugar signalling and gene expression in relation to carbohydrate metabolism under abiotic stresses in plants *J. Biosci.* **30** 761–76

[33] Gangola M P, Jaiswal S, Kannan U, Gaur P M, Båga M and Chibbar R N 2016 Galactinol synthase enzyme activity influences raffinose family oligosaccharides (RFO) accumulation in developing chickpea (*Cicer arietinum* L.) seeds *Phytochemistry* **125** 88–98

[34] Gangola M P and Ramadoss B R 2018 *Biochemical, Physiological and Molecular Avenues for Combating Abiotic Stress Tolerance in Plants* (Amsterdam: Elsevier) pp 17–38

[35] Hildebrandt T M 2018 Synthesis versus degradation: directions of amino acid metabolism during *Arabidopsis* abiotic stress response *Plant Mol. Biol.* **98** 121–35

[36] Hou Q, Ufer G and Bartels D 2016 Lipid signalling in plant responses to abiotic stress *Plant, Cell Environ.* **39** 1029–48

[37] Zhang M, Deng X, Yin L, Qi L, Wang X, Wang S and Li H 2016 Regulation of galactolipid biosynthesis by overexpression of the rice MGD gene contributes to enhanced aluminum tolerance in tobacco *Front. Plant Sci.* **7** 337

[38] Gasulla F, Vom Dorp K, Dombrink I, Zähringer U, Gisch N, Dörmann P and Bartels D 2013 The role of lipid metabolism in the acquisition of desiccation tolerance in C raterostigma plantagineum: a comparative approach *Plant J.* **75** 726–41

[39] Munnik T, Meijer H J, Ter Riet B, Hirt H, Frank W, Bartels D and Musgrave A 2000 Hyperosmotic stress stimulates phospholipase D activity and elevates the levels of phosphatidic acid and diacylglycerol pyrophosphate *Plant J.* **22** 147–54

[40] Arisz S A, van Wijk R v, Roels W, Zhu J-K, Haring M A and Munnik T 2013 Rapid phosphatidic acid accumulation in response to low temperature stress in *Arabidopsis* is generated through diacylglycerol kinase *Front. Plant Sci.* **4** 1

[41] Rodrigues A M, Jorge T, Osorio S, Pott D M, Lidon F C, DaMatta F M, Marques I, Ribeiro-Barros A I, Ramalho J C and António C 2021 Primary metabolite profile changes in *Coffea* spp. promoted by single and combined exposure to drought and elevated CO_2 concentration *Metabolites* **11** 427

[42] Ahanger M A, Gul F, Ahmad P and Akram N A 2018 Environmental stresses and metabolomics—deciphering the role of stress responsive metabolites *Plant Metabolites and Regulation under Environmental Stress* (New York: Academic) pp 53–67
[43] Tzin V and Galili G 2010 The biosynthetic pathways for shikimate and aromatic amino acids in *Arabidopsis thaliana The Arabidopsis Book* (Rockville, MD: American Society of Plant Biologists) 8
[44] Wink M 2015 Modes of action of herbal medicines and plant secondary metabolites *Medicines* **2** 251–86
[45] Akhi M Z, Haque M M and Biswas M S 2021 *Antioxidants* (London: IntechOpen)
[46] Treml J and Šmejkal K 2016 Flavonoids as potent scavengers of hydroxyl radicals *Compr. Rev. Food Sci. Food Safety* **15** 720–38
[47] Bartwal A, Mall R, Lohani P, Guru S and Arora S 2013 Role of secondary metabolites and brassinosteroids in plant defense against environmental stresses *J. Plant Growth Regul.* **32** 216–32
[48] Kabera J N, Semana E, Mussa A R and He X 2014 Plant secondary metabolites: biosynthesis, classification, function and pharmacological properties *J. Pharm. Pharmacol.* **2** 377–92
[49] Cheynier V, Comte G, Davies K M, Lattanzio V and Martens S 2013 Plant phenolics: recent advances on their biosynthesis, genetics, and ecophysiology *Plant Physiol. Biochem.* **72** 1–20
[50] Havaux M and Kloppstech K 2001 The protective functions of carotenoid and flavonoid pigments against excess visible radiation at chilling temperature investigated in *Arabidopsis npq* and *tt* mutants *Planta* **213** 953–66
[51] Liu Q, Luo L and Zheng L 2018 Lignins: biosynthesis and biological functions in plants *Int. J. Mol. Sci.* **19** 335
[52] Wang A, Zhu M, Luo Y, Liu Y, Li R, Kou M, Wang X, Zhang Y, Meng X and Zheng Y 2017 A sweet potato cinnamate 4-hydroxylase gene, IbC4H, increases phenolics content and enhances drought tolerance in tobacco *Acta Physiol. Plant.* **39** 276
[53] Capell T and Christou P 2004 Progress in plant metabolic engineering *Curr. Opin. Biotechnol.* **15** 148–54
[54] Sinha R, Bala M, Kumar M, Sharma T R and Singh A K 2020 *Legume Genomics* (Berlin: Springer) pp 277–303
[55] Praba M L, Cairns J, Babu R and Lafitte H 2009 Identification of physiological traits underlying cultivar differences in drought tolerance in rice and wheat *J. Agron. Crop Sci.* **195** 30–46
[56] Kumar M, Mishra S, Dixit V, Agarwal L, Chauhan P S and Nautiyal C S 2016 Synergistic effect of Pseudomonas putida and Bacillus amyloliquefaciens ameliorates drought stress in chickpea (*Cicer arietinum* L.) *Plant Signal. Behav.* **11** e1071004
[57] Brini F, Hanin M, Lumbreras V, Amara I, Khoudi H, Hassairi A, Pages M and Masmoudi K 2007 Overexpression of wheat dehydrin DHN-5 enhances tolerance to salt and osmotic stress in *ARABIDOPSIS thaliana Plant Cell Rep* **26** 2017–26
[58] Kumar M, Yusuf M A, Yadav P and Narayan S 2019 Overexpression of chickpea defensin gene confers tolerance to water-deficit stress in *Arabidopsis thaliana Front. Plant Sci.* **10** 290
[59] Das A, Basu P S, Kumar M, Ansari J, Shukla A, Thakur S, Singh P, Datta S, Chaturvedi S K and Sheshshayee M 2021 Transgenic chickpea (*Cicer arietinum* L.) harbouring AtDREB1a are physiologically better adapted to water deficit *BMC Plant Biol.* **21** 1–17

[60] Kumar A, Taylor M, Altabella T and Tiburcio A F 1997 Recent advances in polyamine research *Trends Plant Sci.* **2** 124–30

[61] Alcázar R, Planas J, Saxena T, Zarza X, Bortolotti C, Cuevas J, Bitrián M, Tiburcio A F and Altabella T 2010 Putrescine accumulation confers drought tolerance in transgenic *Arabidopsis* plants over-expressing the homologous Arginine decarboxylase 2 gene *Plant Physiol. Biochem.* **48** 547–52

[62] Wang J, Sun P-P, Chen C-L, Wang Y, Fu X-Z and Liu J-H 2011 An arginine decarboxylase gene PtADC from *Poncirus trifoliata* confers abiotic stress tolerance and promotes primary root growth in *Arabidopsis J. Exp. Bot.* **62** 2899–914

[63] Capell T, Bassie L and Christou P 2004 Modulation of the polyamine biosynthetic pathway in transgenic rice confers tolerance to drought stress *Proc. Natl Acad. Sci.* **101** 9909–14

[64] Zhang J, Tan W, Yang X-H and Zhang H-X 2008 Plastid-expressed choline monooxygenase gene improves salt and drought tolerance through accumulation of glycine betaine in tobacco *Plant Cell Reports* **27** 1113–24

[65] Yu H-Q, Zhou X-Y, Wang Y-G, Zhou S-F, Fu F-L and Li W-C 2017 A betaine aldehyde dehydrogenase gene from *Ammopiptanthus nanus* enhances tolerance of *Arabidopsis* to high salt and drought stresses *Plant Growth Regul.* **83** 265–76

[66] Zhao M, Jin L, Hu B, Yao H, Gao Y, Wang R, Li F, Guo J and Li K 2019 Overexpression of chalcone synthase gene improves flavonoid accumulation and drought tolerance in tobacco *Research Square* https://doi.org/10.21203/rs.2.18297/v1

[67] Liu H, Yu C, Li H, Ouyang B, Wang T, Zhang J, Wang X and Ye Z 2015 Overexpression of ShDHN, a dehydrin gene from *Solanum habrochaites* enhances tolerance to multiple abiotic stresses in tomato *Plant Sci.* **231** 198–211

[68] Bao F, Du D, An Y, Yang W, Wang J, Cheng T and Zhang Q 2017 Overexpression of *Prunus mume* dehydrin genes in tobacco enhances tolerance to cold and drought *Front. Plant Sci.* **8** 151

[69] Chiappetta A, Muto A, Bruno L, Woloszynska M, Van Lijsebettens M and Bitonti M B 2015 A dehydrin gene isolated from feral olive enhances drought tolerance in *Arabidopsis* transgenic plants *Front. Plant Sci.* **6** 392

[70] Mafakheri A, Siosemardeh A, Bahramnejad B, Struik P and Sohrabi Y 2010 Effect of drought stress on yield, proline and chlorophyll contents in three chickpea cultivars *Aust. J. Crop Sci.* **4** 580–5

[71] Székely G *et al* 2008 Duplicated P5CS genes of *Arabidopsis* play distinct roles in stress regulation and developmental control of proline biosynthesis *Plant J.* **53** 11–28

[72] Yamada M, Morishita H, Urano K, Shiozaki N, Yamaguchi-Shinozaki K, Shinozaki K and Yoshiba Y 2005 Effects of free proline accumulation in petunias under drought stress *J. Exp. Bot.* **56** 1975–81

[73] Borgo L, Marur C J and Vieira L G E 2015 Effects of high proline accumulation on chloroplast and mitochondrial ultrastructure and on osmotic adjustment in tobacco plants *Acta Sci., Agron.* **37** 191–9

[74] Zang Q-W, Wang C-X, Li X-Y, Guo Z-A, Jing R-L, Zhao J and Chang X-P 2010 Isolation and characterization of a gene encoding a polyethylene glycol-induced cysteine protease in common wheat *J. Biosci.* **35** 379–88

[75] Abdullah H M, Rodriguez J, Salacup J M, Castañeda I S, Schnell D J, Pareek A and Dhankher O P 2021 Increased cuticle waxes by overexpression of WSD1 improves osmotic stress tolerance in *Arabidopsis thaliana* and *Camelina sativa Int. J. Mol. Sci.* **22** 5173

[76] Jain N K and Roy I 2009 Effect of trehalose on protein structure *Protein Sci.* **18** 24–36
[77] Karim S, Aronsson H, Ericson H, Pirhonen M, Leyman B, Welin B, Mäntylä E, Palva E T, Van Dijck P and Holmström K-O 2007 Improved drought tolerance without undesired side effects in transgenic plants producing trehalose *Plant Mol. Biol.* **64** 371–86
[78] Garg A K, Kim J-K, Owens T G, Ranwala A P, Do Choi Y, Kochian L V and Wu R J 2002 Trehalose accumulation in rice plants confers high tolerance levels to different abiotic stresses *Proc. Natl Acad. Sci.* **99** 15898–903
[79] Li Q, Yu H M, Meng X F, Lin J S, Li Y J and Hou B K 2018 Ectopic expression of glycosyltransferase UGT 76E11 increases flavonoid accumulation and enhances abiotic stress tolerance in *Arabidopsis Plant Biol.* **20** 10–9
[80] Zheng L, Chen S, Xie L, Lu Z, Liu M, Han X, Qiao G, Jiang J, Zhuo R and Qiu W 2018 Overexpression of cysteine protease gene from *Salix matsudana* enhances salt tolerance in transgenic *Arabidopsis Environ. Exp. Bot.* **147** 53–62
[81] Slama I, Abdelly C, Bouchereau A, Flowers T and Savouré A 2015 Diversity, distribution and roles of osmoprotective compounds accumulated in halophytes under abiotic stress *Annal. Bot.* **115** 433–47
[82] Nguyen H C, Lin K H, Ho S L, Chiang C M and Yang C M 2018 Enhancing the abiotic stress tolerance of plants: from chemical treatment to biotechnological approaches *Physiol. Plant.* **164** 452–66
[83] Espasandin F D, Calzadilla P I, Maiale S J, Ruiz O A and Sansberro P A 2018 Overexpression of the arginine decarboxylase gene improves tolerance to salt stress in *Lotus tenuis* plants *J. Plant Growth Regul.* **37** 156–65
[84] Saibi W, Feki K, Mahmoud R B and Brini F 2015 Durum wheat dehydrin (DHN-5) confers salinity tolerance to transgenic *Arabidopsis* plants through the regulation of proline metabolism and ROS scavenging system *Planta* **242** 1187–94
[85] Li C, Liu S, Yao X, Wang J, Wang T, Zhang Z, Zhang P and Chen K 2017 PnF3H, a flavanone 3-hydroxylase from the Antarctic moss *Pohlia nutans*, confers tolerance to salt stress and ABA treatment in transgenic *Arabidopsis Plant Growth Regul.* **83** 489–500
[86] Roy M and Wu R 2002 Overexpression of S-adenosylmethionine decarboxylase gene in rice increases polyamine level and enhances sodium chloride-stress tolerance *Plant Sci.* **163** 987–92
[87] Ali A, Ali Q, Iqbal M, Nasir I and Wang X 2020 Salt tolerance of potato genetically engineered with the *Atriplex canescens* BADH gene *Biol. Plant* **64** 271–9
[88] Lunn J E, Delorge I, Figueroa C M, Van Dijck P and Stitt M 2014 Trehalose metabolism in plants *Plant J.* **79** 544–67
[89] Jang I-C, Oh S-J, Seo J-S, Choi W-B, Song S I, Kim C H, Kim Y S, Seo H-S, Choi Y D and Nahm B H 2003 Expression of a bifunctional fusion of the *Escherichia coli* genes for trehalose-6-phosphate synthase and trehalose-6-phosphate phosphatase in transgenic rice plants increases trehalose accumulation and abiotic stress tolerance without stunting growth *Plant Physiol.* **131** 516–24
[90] Li H-W, Zang B-S, Deng X-W and Wang X-P 2011 Overexpression of the trehalose-6-phosphate synthase gene OsTPS1 enhances abiotic stress tolerance in rice *Planta* **234** 1007–18
[91] Gao M, Tao R, Miura K, Dandekar A M and Sugiura A 2001 Transformation of Japanese persimmon (*Diospyros kaki* Thunb.) with apple cDNA encoding NADP-dependent sorbitol-6-phosphate dehydrogenase *Plant Sci.* **160** 837–45

[92] Martinez V, Mestre T C, Rubio F, Girones-Vilaplana A, Moreno D A, Mittler R and Rivero R M 2016 Accumulation of flavonols over hydroxycinnamic acids favors oxidative damage protection under abiotic stress *Front. Plant Sci.* **7** 838

[93] Kasukabe Y, He L, Nada K, Misawa S, Ihara I and Tachibana S 2004 Overexpression of spermidine synthase enhances tolerance to multiple environmental stresses and up-regulates the expression of various stress-regulated genes in transgenic *Arabidopsis thaliana Plant Cell Physiol.* **45** 712–22

[94] Xu S, Li J, Zhang X, Wei H and Cui L 2006 Effects of heat acclimation pretreatment on changes of membrane lipid peroxidation, antioxidant metabolites, and ultrastructure of chloroplasts in two cool-season turfgrass species under heat stress *Environ. Exp. Bot.* **56** 274–85

[95] Wahid A and Ghazanfar A 2006 Possible involvement of some secondary metabolites in salt tolerance of sugarcane *J. Plant Physiol.* **163** 723–30

[96] Chalker-Scott L 2002 Do anthocyanins function as osmoregulators in leaf tissues? *Adv. Bot. Res.* **37** 103–6

[97] Hu C, Delauney A J and Verma D 1992 A bifunctional enzyme (delta 1-pyrroline-5-carboxylate synthetase) catalyzes the first two steps in proline biosynthesis in plants *Proc. Natl Acad. Sci.* **89** 9354–8

[98] Lv W-T, Lin B, Zhang M and Hua X-J 2011 Proline accumulation is inhibitory to *Arabidopsis* seedlings during heat stress *Plant Physiol.* **156** 1921–33

[99] Lehle L and Tanner W 1973 Biosynthesis of raffinose *Hoppe-Seylers Zeitschrift fur Physiologische Chemie* (Berlin: Walter De Gruyter & Co) p 1217

[100] Panikulangara T J, Eggers-Schumacher G, Wunderlich M, Stransky H and Schöffl F 2004 Galactinol synthase1. A novel heat shock factor target gene responsible for heat-induced synthesis of raffinose family oligosaccharides in *Arabidopsis Plant Physiol.* **136** 3148–58

[101] Xuan Y, Zhou S, Wang L, Cheng Y and Zhao L 2010 Nitric oxide functions as a signal and acts upstream of AtCaM3 in thermotolerance in *Arabidopsis* seedlings *Plant Physiol.* **153** 1895–906

[102] Song C, Chung W S and Lim C O 2016 Overexpression of heat shock factor gene HsfA3 increases galactinol levels and oxidative stress tolerance in *Arabidopsis Mol. Cells* **39** 477

[103] Wu Z, Liang J, Wang C, Zhao X, Zhong X, Cao X, Li G, He J and Yi M 2018 Overexpression of lily HsfA3 s in *Arabidopsis* confers increased thermotolerance and salt sensitivity via alterations in proline catabolism *J. Exp. Bot.* **69** 2005–21

[104] Stavridou E, Voulgari G, Michailidis M, Kostas S, Chronopoulou E G, Labrou N E, Madesis P and Nianiou-Obeidat I 2021 Overexpression of Abiotic stress-inducible *Pvgstu* gene activates early protective responses in tobacco under combined heat and drought *Int. J. Mol. Sci.* **22** 2352

[105] Tarkowská D and Strnad M 2016 Plant ecdysteroids: plant sterols with intriguing distributions, biological effects and relations to plant hormones *Planta* **244** 545–55

[106] Marler T E and Shaw C A 2009 Free and glycosylated sterol bioaccumulation in developing *Cycas micronesica* seeds *Food Chem.* **115** 615–9

[107] Pandey V, Dhar Y V, Gupta P, Bag S K, Atri N, Asif M H, Trivedi P K and Misra P 2015 Comparative interactions of withanolides and sterols with two members of sterol glycosyltransferases from *Withania somnifera BMC Bioinf.* **16** 1–10

[108] Sun W, Liang L, Meng X, Li Y, Gao F, Liu X, Wang S, Gao X and Wang L 2016 Biochemical and molecular characterization of a flavonoid 3-O-glycosyltransferase responsible for anthocyanins and flavonols biosynthesis in *Freesia hybrida Front. Plant Sci.* **7** 410

[109] Mishra M K, Chaturvedi P, Singh R, Singh G, Sharma L K, Pandey V, Kumari N and Misra P 2013 Overexpression of WsSGTL1 gene of *Withania somnifera* enhances salt tolerance, heat tolerance and cold acclimation ability in transgenic *Arabidopsis* plants *PLoS One* **8** e63064

[110] Shi Y, Ding Y and Yang S 2015 Cold signal transduction and its interplay with phytohormones during cold acclimation *Plant Cell Physiol.* **56** 7–15

[111] Koc I, Filiz E and Tombuloglu H 2015 Assessment of miRNA expression profile and differential expression pattern of target genes in cold-tolerant and cold-sensitive tomato cultivars *Biotechnol. Biotechnol. Equip.* **29** 851–60

[112] Huner N P, Öquist G and Sarhan F 1998 Energy balance and acclimation to light and cold *Trends Plant Sci.* **3** 224–30

[113] Fristedt R, Martins N F, Strenkert D, Clarke C A, Suchoszek M, Thiele W, Schöttler M A and Merchant S S 2015 The thylakoid membrane protein CGL160 supports CF1CF0 ATP synthase accumulation in *Arabidopsis thaliana PLoS One* **10** e0121658

[114] Strand Å, Foyer C, Gustafsson P, Gardeström P and Hurry V 2003 Altering flux through the sucrose biosynthesis pathway in transgenic *Arabidopsis thaliana* modifies photosynthetic acclimation at low temperatures and the development of freezing tolerance *Plant, Cell Environ.* **26** 523–35

[115] Xu J, Chen Z, Wang F, Jia W and Xu Z 2020 Combined transcriptomic and metabolomic analyses uncover rearranged gene expression and metabolite metabolism in tobacco during cold acclimation *Sci. Rep.* **10** 1–13

[116] Taïbi K, Del Campo A D, Vilagrosa A, Bellés J M, López-Gresa M P, López-Nicolás J M and Mulet J M 2018 Distinctive physiological and molecular responses to cold stress among cold-tolerant and cold-sensitive *Pinus halepensis* seed sources *BMC Plant Biol.* **18** 1–11

[117] Usadel B, Blaesing O E, Gibon Y, Poree F, Hoehne M, Guenter M, Trethewey R, Kamlage B, Poorter H and Stitt M 2008 Multilevel genomic analysis of the response of transcripts, enzyme activities and metabolites in *Arabidopsis* rosettes to a progressive decrease of temperature in the non-freezing range *Plant, Cell Environ.* **31** 518–47

[118] Zhang W-F, Gong Z-H, Wu M-B, Chan H, Yuan Y-J, Tang N, Zhang Q, Miao M-J, Chang W and Li Z 2019 Integrative comparative analyses of metabolite and transcript profiles uncovers complex regulatory network in tomato (*Solanum lycopersicum* L.) fruit undergoing chilling injury *Sci. Rep.* **9** 1–13

[119] Szabados L and Savouré A 2010 Proline: a multifunctional amino acid *Trends Plant Sci.* **15** 89–97

[120] Solecka D and Kacperska A 2003 Phenylpropanoid deficiency affects the course of plant acclimation to cold *Physiol. Plant.* **119** 253–62

[121] Aghdam M S, Moradi M, Razavi F and Rabiei V 2019 Exogenous phenylalanine application promotes chilling tolerance in tomato fruits during cold storage by ensuring supply of NADPH for activation of ROS scavenging systems *Sci. Horticult.* **246** 818–25

[122] Griffith M and Yaish M W 2004 Antifreeze proteins in overwintering plants: a tale of two activities *Trends Plant Sci.* **9** 399–405

[123] Pérez-Ilzarbe J, Hernández T, Estrella I and Vendrell M 1997 Cold storage of apples (cv. Granny Smith) and changes in phenolic compounds *Z. Lebensm. Forsch.* A **204** 52–5

[124] Shohael A M, Ali M B, Yu K-W, Hahn E-J and Paek K-Y 2006 Effect of temperature on secondary metabolites production and antioxidant enzyme activities in *Eleutherococcus senticosus* somatic embryos *Plant Cell, Tissue Organ Cult.* **85** 219–28

[125] Hanin M, Brini F, Ebel C, Toda Y, Takeda S and Masmoudi K 2011 Plant dehydrins and stress tolerance: versatile proteins for complex mechanisms *Plant Signal Behav* **6** 1503–9
[126] Guo X, Zhang L, Wang X, Zhang M, Xi Y, Wang A and Zhu J 2019 Overexpression of *Saussurea involucrata* dehydrin gene SiDHN promotes cold and drought tolerance in transgenic tomato plants *PLoS One* **14** e0225090
[127] Koźmińska A, Wiszniewska A, Hanus-Fajerska E and Muszyńska E 2018 Recent strategies of increasing metal tolerance and phytoremediation potential using genetic transformation of plants *Plant Biotechnol. Rep.* **12** 1–14
[128] Ramesh G 2008 Cloning and characterization of metallothionein genes of ectomycorrhizal fungus *Hebeloma cylindrosporum*
[129] Villiers F, Ducruix C, Hugouvieux V, Jarno N, Ezan E, Garin J, Junot C and Bourguignon J 2011 Investigating the plant response to cadmium exposure by proteomic and metabolomic approaches *Proteomics* **11** 1650–63
[130] Sharma P and Dubey R 2007 Involvement of oxidative stress and role of antioxidative defense system in growing rice seedlings exposed to toxic concentrations of aluminum *Plant Cell Reports* **26** 2027–38
[131] Hossain M A, Hossain M Z and Fujita M 2009 Stress-induced changes of methylglyoxal level and glyoxalase I activity in pumpkin seedlings and cDNA cloning of glyoxalase I gene *Aust. J. Crop Sci.* **3** 53
[132] Tan Y-F, O'Toole N, Taylor N L and Millar A H 2010 Divalent metal ions in plant mitochondria and their role in interactions with proteins and oxidative stress-induced damage to respiratory function *Plant Physiol.* **152** 747–61
[133] Yang X E, Jin X F, Feng Y and Islam E 2005 Molecular mechanisms and genetic basis of heavy metal tolerance/hyperaccumulation in plants *J. Integr. Plant Biol.* **47** 1025–35
[134] DalCorso G, Farinati S, Maistri S and Furini A 2008 How plants cope with cadmium: staking all on metabolism and gene expression *J. Integr. Plant Biol.* **50** 1268–80
[135] Sharma S S and Dietz K-J 2009 The relationship between metal toxicity and cellular redox imbalance *Trends Plant Sci.* **14** 43–50
[136] Hossain M A, Hasanuzzaman M and Fujita M 2010 Up-regulation of antioxidant and glyoxalase systems by exogenous glycinebetaine and proline in mung bean confer tolerance to cadmium stress *Physiol. Mol. Biol. Plants* **16** 259–72
[137] Tripathi P, Singh P K, Mishra S, Gautam N, Dwivedi S, Chakrabarty D and Tripathi R D 2015 Recent advances in the expression and regulation of plant metallothioneins for metal homeostasis and tolerance *Environmental Waste Management* (Boca Raton, FL: CRC Press) pp 551–64
[138] Zhang H, Xu W, Guo J, He Z and Ma M 2005 Coordinated responses of phytochelatins and metallothioneins to heavy metals in garlic seedlings *Plant Sci.* **169** 1059–65
[139] Chaturvedi A K, Patel M K, Mishra A, Tiwari V and Jha B 2014 The SbMT-2 gene from a halophyte confers abiotic stress tolerance and modulates ROS scavenging in transgenic tobacco *PLoS One* **9** e111379
[140] Cobbett C and Goldsbrough P 2002 Phytochelatins and metallothioneins: roles in heavy metal detoxification and homeostasis *Annu. Rev. Plant Biol.* **53** 159–82
[141] Song J, Feng S J, Chen J, Zhao W T and Yang Z M 2017 A cadmium stress-responsive gene AtFC1 confers plant tolerance to cadmium toxicity *BMC Plant Biol.* **17** 1–15

[142] Li Y, Dhankher O P, Carreira L, Lee D, Chen A, Schroeder J I, Balish R S and Meagher R B 2004 Overexpression of phytochelatin synthase in *Arabidopsis* leads to enhanced arsenic tolerance and cadmium hypersensitivity *Plant Cell Physiol.* **45** 1787–97

[143] Liu D, An Z, Mao Z, Ma L and Lu Z 2015 Enhanced heavy metal tolerance and accumulation by transgenic sugar beets expressing *Streptococcus thermophilus* StGCS-GS in the presence of Cd, Zn and Cu alone or in combination *PLoS One* **10** e0128824

[144] Bartling D, Radzio R, Steiner U and Weiler E W 1993 A glutathione S-transferase with glutathione-peroxidase activity from *Arabidopsis thaliana*: molecular cloning and functional characterization *Eur. J. Biochem.* **216** 579–86

[145] Kumar S, Asif M H, Chakrabarty D, Tripathi R D, Dubey R S and Trivedi P K 2013 Expression of a rice Lambda class of glutathione S-transferase, OsGSTL2, in *Arabidopsis* provides tolerance to heavy metal and other abiotic stresses *J. Hazard. Mater.* **248** 228–37

[146] Sánchez-Martín J, Canales F J, Tweed J K, Lee M R, Rubiales D, Gómez-Cadenas A, Arbona V, Mur L A and Prats E 2018 Fatty acid profile changes during gradual soil water depletion in oats suggests a role for jasmonates in coping with drought *Front. Plant Sci.* **9** 1077

[147] Piasecka A, Sawikowska A, Kuczyńska A, Ogrodowicz P, Mikołajczak K, Krystkowiak K, Gudyś K, Guzy-Wróbelska J, Krajewski P and Kachlicki P 2017 Drought-related secondary metabolites of barley (*Hordeum vulgare* L.) leaves and their metabolomic quantitative trait loci *Plant J.* **89** 898–913

[148] Hein J A, Sherrard M E, Manfredi K P and Abebe T 2016 The fifth leaf and spike organs of barley (*Hordeum vulgare* L.) display different physiological and metabolic responses to drought stress *BMC Plant Biol.* **16** 1–12

[149] Do P T, Degenkolbe T, Erban A, Heyer A G, Kopka J, Köhl K I, Hincha D K and Zuther E 2013 Dissecting rice polyamine metabolism under controlled long-term drought stress *PLoS One* **8** e60325

[150] Yadav A K, Carroll A J, Estavillo G M, Rebetzke G J and Pogson B J 2019 Wheat drought tolerance in the field is predicted by amino acid responses to glasshouse-imposed drought *J. Exp. Bot.* **70** 4931–48

[151] Sun C, Gao X, Fu J, Zhou J and Wu X 2015 Metabolic response of maize (*Zea mays* L.) plants to combined drought and salt stress *Plant Soil* **388** 99–117

[152] Obata T, Witt S, Lisec J, Palacios-Rojas N, Florez-Sarasa I, Yousfi S, Araus J L, Cairns J E and Fernie A R 2015 Metabolite profiles of maize leaves in drought, heat, and combined stress field trials reveal the relationship between metabolism and grain yield *Plant Physiol.* **169** 2665–83

[153] Mibei E K, Owino W O, Ambuko J, Giovannoni J J and Onyango A N 2018 Metabolomic analyses to evaluate the effect of drought stress on selected African Eggplant accessions *J. Sci. Food Agric.* **98** 205–16

[154] Gundaraniya S A, Ambalam P S and Tomar R S 2020 Metabolomic profiling of drought-tolerant and susceptible peanut (*Arachis hypogaea* L.) genotypes in response to drought stress *ACS Omega* **5** 31209–19

[155] Khan N, Bano A, Rahman M A, Rathinasabapathi B and Babar M A 2019 UPLC-HRMS-based untargeted metabolic profiling reveals changes in chickpea (*Cicer arietinum*) metabolome following long-term drought stress *Plant, Cell Environ.* **42** 115–32

[156] Alcázar R, Bitrián M, Bartels D, Koncz C, Altabella T and Tiburcio A F 2011 Polyamine metabolic canalization in response to drought stress in *Arabidopsis* and the resurrection plant *Craterostigma plantagineum Plant Signal. Behav.* **6** 243–50

[157] Coutinho I D, Henning L M M, Döpp S A, Nepomuceno A, Moraes L A C, Marcolino-Gomes J, Richter C, Schwalbe H and Colnago L A 2018 Identification of primary and secondary metabolites and transcriptome profile of soybean tissues during different stages of hypoxia *Data Brief* **21** 1089–100

[158] Muscolo A, Junker A, Klukas C, Weigelt-Fischer K, Riewe D and Altmann T 2015 Phenotypic and metabolic responses to drought and salinity of four contrasting lentil accessions *J. Exp. Bot.* **66** 5467–80

[159] Bayati P, Karimmojeni H and Razmjoo J 2020 Changes in essential oil yield and fatty acid contents in black cumin (*Nigella sativa* L.) genotypes in response to drought stress *Ind. Crops Prod.* **155** 112764

[160] Saheri F, Barzin G, Pishkar L, Boojar M M A and Babaeekhou L 2020 Foliar spray of salicylic acid induces physiological and biochemical changes in purslane (*Portulaca oleracea* L.) under drought stress *Biologia* **75** 2189–200

[161] Gomes A M, Rodrigues A P, António C, Rodrigues A M, Leitão A E, Batista-Santos P, Nhantumbo N, Massinga R, Ribeiro-Barros A I and Ramalho J C 2020 Drought response of cowpea (*Vigna unguiculata* (L.) Walp.) landraces at leaf physiological and metabolite profile levels *Environ. Exp. Bot.* **175** 104060

[162] Griesser M, Weingart G, Schoedl-Hummel K, Neumann N, Becker M, Varmuza K, Liebner F, Schuhmacher R and Forneck A 2015 Severe drought stress is affecting selected primary metabolites, polyphenols, and volatile metabolites in grapevine leaves (*Vitis vinifera* cv. Pinot noir) *Plant Physiol. Biochem.* **88** 17–26

[163] Sobhanian H, Motamed N, Jazii F R, Nakamura T and Komatsu S 2010 Salt stress induced differential proteome and metabolome response in the shoots of *Aeluropus lagopoides* (Poaceae), a halophyte C4 plant *J. Proteome Res.* **9** 2882–97

[164] Chen Z, Pottosin I I, Cuin T A, Fuglsang A T, Tester M, Jha D, Zepeda-Jazo I, Zhou M, Palmgren M G and Newman I A 2007 Root plasma membrane transporters controlling K^+/Na^+ homeostasis in salt-stressed barley *Plant Physiol.* **145** 1714–25

[165] Gavaghan C L, Li J V, Hadfield S T, Hole S, Nicholson J K, Wilson I D, Howe P W, Stanley P D and Holmes E 2011 Application of NMR-based metabolomics to the investigation of salt stress in maize (*Zea mays*) *Phytochem. Anal.* **22** 214–24

[166] Hill C B, Taylor J D, Edwards J, Mather D, Bacic A, Langridge P and Roessner U 2013 Whole-genome mapping of agronomic and metabolic traits to identify novel quantitative trait loci in bread wheat grown in a water-limited environment *Plant Physiol.* **162** 1266–81

[167] Gagneul D, Aïnouche A, Duhazé C, Lugan R, Larher F R and Bouchereau A 2007 A reassessment of the function of the so-called compatible solutes in the halophytic plumbaginaceae *Limonium latifolium Plant Physiol.* **144** 1598–611

[168] Gong Q, Li P, Ma S, Indu Rupassara S and Bohnert H J 2005 Salinity stress adaptation competence in the extremophile *Thellungiella halophila* in comparison with its relative *Arabidopsis thaliana Plant J.* **44** 826–39

[169] Dias D A, Hill C B, Jayasinghe N S, Atieno J, Sutton T and Roessner U 2015 Quantitative profiling of polar primary metabolites of two chickpea cultivars with contrasting responses to salinity *J. Chromatogr.* B **1000** 1–13

[170] Lu Y, Lam H, Pi E, Zhan Q, Tsai S, Wang C, Kwan Y and Ngai S 2013 Comparative metabolomics in *Glycine max* and *Glycine soja* under salt stress to reveal the phenotypes of their offspring *J. Agric. Food Chem.* **61** 8711–21

[171] Zhang J, Zhang Y, Du Y, Chen S and Tang H 2011 Dynamic metabonomic responses of tobacco (*Nicotiana tabacum*) plants to salt stress *J. Proteome Res.* **10** 1904–14

[172] Lim J-H, Park K-J, Kim B-K, Jeong J-W and Kim H-J 2012 Effect of salinity stress on phenolic compounds and carotenoids in buckwheat (*Fagopyrum esculentum* M.) sprout *Food Chem.* **135** 1065–70

[173] Ni J, Yang X, Zhu J, Liu Z, Ni Y, Wu H, Zhang H and Liu T 2015 Salinity-induced metabolic profile changes in *Nitraria tangutorum* Bobr. suspension cells *Plant Cell, Tissue and Organ Culture (PCTOC)* **122** 239–48

[174] Petridis A, Therios I, Samouris G, Koundouras S and Giannakoula A 2012 Effect of water deficit on leaf phenolic composition, gas exchange, oxidative damage and antioxidant activity of four Greek olive (*Olea europaea* L.) cultivars *Plant Physiol. Biochem.* **60** 1–11

[175] Tounekti T, Vadel A M, Ennajeh M, Khemira H and Munné-Bosch S 2011 Ionic interactions and salinity affect monoterpene and phenolic diterpene composition in rosemary (*Rosmarinus officinalis*) *J. Plant Nutr. Soil Sci.* **174** 504–14

[176] Valifard M, Mohsenzadeh S, Kholdebarin B, Rowshan V, Niazi A and Moghadam A 2019 Effect of salt stress on terpenoid biosynthesis in *Salvia mirzayanii*: from gene to metabolite *J. Hort. Sci. Biotechnol.* **94** 389–99

[177] Bendaly A, Messedi D, Smaoui A, Ksouri R, Bouchereau A and Abdelly C 2016 Physiological and leaf metabolome changes in the xerohalophyte species *Atriplex halimus* induced by salinity *Plant Physiol. Biochem.* **103** 208–18

[178] Ren S, Ma K, Lu Z, Chen G, Cui J, Tong P, Wang L, Teng N and Jin B 2019 Transcriptomic and metabolomic analysis of the heat-stress response of *Populus tomentosa* carr *Forests* **10** 383

[179] Liu B, Kong L, Zhang Y and Liao Y 2021 Gene and metabolite integration analysis through transcriptome and metabolome brings new insight into heat stress tolerance in potato (*Solanum tuberosum* L.) *Plants* **10** 103

[180] Kaplan F, Kopka J, Haskell D W, Zhao W, Schiller K C, Gatzke N, Sung D Y and Guy C L 2004 Exploring the temperature-stress metabolome of *Arabidopsis Plant Physiol.* **136** 4159–68

[181] Serrano N, Ling Y, Bahieldin A and Mahfouz M M 2019 Thermopriming reprograms metabolic homeostasis to confer heat tolerance *Sci. Rep.* **9** 1–14

[182] Thomason K, Babar M A, Erickson J E, Mulvaney M, Beecher C and MacDonald G 2018 Comparative physiological and metabolomics analysis of wheat (*Triticum aestivum* L.) following post-anthesis heat stress *PLoS One* **13** e0197919

[183] Templer S E, Ammon A, Pscheidt D, Ciobotea O, Schuy C, McCollum C, Sonnewald U, Hanemann A, Förster J and Ordon F 2017 Metabolite profiling of barley flag leaves under drought and combined heat and drought stress reveals metabolic QTLs for metabolites associated with antioxidant defense *J. Exp. Bot.* **68** 1697–713

[184] Escandón M, Meijón M, Valledor L, Pascual J, Pinto G and Cañal M J 2018 Metabolome integrated analysis of high-temperature response in *Pinus radiata Front. Plant Sci.* **9** 485

[185] Paupière M J, Müller F, Li H, Rieu I, Tikunov Y M, Visser R G and Bovy A G 2017 Untargeted metabolomic analysis of tomato pollen development and heat stress response *Plant Reprod.* **30** 81–94

[186] Cook D, Fowler S, Fiehn O and Thomashow M F 2004 A prominent role for the CBF cold response pathway in configuring the low-temperature metabolome of Arabidopsis *Proc. Natl Acad. Sci.* **101** 15243–8

[187] Zhao X-Q, Wang W-S, Zhang F, Zhang T, Zhao W, Fu B-Y and Li Z-K 2013 Temporal profiling of primary metabolites under chilling stress and its association with seedling chilling tolerance of rice (*Oryza sativa* L.) *Rice* **6** 1–13

[188] Yang J, Duan G, Li C, Liu L, Han G, Zhang Y and Wang C 2019 The crosstalks between jasmonic acid and other plant hormone signaling highlight the involvement of jasmonic acid as a core component in plant response to biotic and abiotic stresses *Front. Plant Sci.* **10** 1349

[189] Cheong B E, Onyemaobi O, Wing Ho Ho W, Biddulph T B, Rupasinghe T W, Roessner U and Dolferus R 2020 Phenotyping the chilling and freezing responses of young microspore stage wheat spikes using targeted metabolome and lipidome profiling *Cells* **9** 1309

[190] Cui P, Li Y, Cui C, Huo Y, Lu G and Yang H 2020 Proteomic and metabolic profile analysis of low-temperature storage responses in *Ipomoea batata* Lam. tuberous roots *BMC Plant Biol.* **20** 1–15

[191] Chai F, Liu W, Xiang Y, Meng X, Sun X, Cheng C, Liu G, Duan L, Xin H and Li S 2019 Comparative metabolic profiling of *Vitis amurensis* and *Vitis vinifera* during cold acclimation *Hort. Res.* **6** 1–12

[192] Sun X, Zhang J, Zhang H, Ni Y, Zhang Q, Chen J and Guan Y 2010 The responses of *Arabidopsis thaliana* to cadmium exposure explored via metabolite profiling *Chemosphere* **78** 840–5

[193] Le Lay P, Isaure M-P, Sarry J-E, Kuhn L, Fayard B, Le Bail J-L, Bastien O, Garin J, Roby C and Bourguignon J 2006 Metabolomic, proteomic and biophysical analyses of *Arabidopsis thaliana* cells exposed to a caesium stress. Influence of potassium supply *Biochimie* **88** 1533–47

[194] Wang Y, Xu L, Shen H, Wang J, Liu W, Zhu X, Wang R, Sun X and Liu L 2015 Metabolomic analysis with GC-MS to reveal potential metabolites and biological pathways involved in Pb & Cd stress response of radish roots *Sci. Rep.* **5** 1–13

[195] Xie Y, Hu L, Du Z, Sun X, Amombo E, Fan J and Fu J 2014 Effects of cadmium exposure on growth and metabolic profile of bermudagrass [*Cynodon dactylon* (L.) Pers.] *PLoS One* **9** e115279

[196] Hédiji H, Djebali W, Belkadhi A, Cabasson C, Moing A, Rolin D, Brouquisse R, Gallusci P and Chaïbi W 2015 Impact of long-term cadmium exposure on mineral content of *Solanum lycopersicum* plants: consequences on fruit production *S. Afr. J. Bot.* **97** 176–81

[197] Bailey N J, Oven M, Holmes E, Nicholson J K and Zenk M H 2003 Metabolomic analysis of the consequences of cadmium exposure in silene cucubalus cell cultures via 1H NMR spectroscopy and chemometrics *Phytochemistry* **62** 851–8

[198] Cheng Y-J, Deng X-P, Kwak S-S, Chen W and Eneji A E 2013 Enhanced tolerance of transgenic potato plants expressing choline oxidase in chloroplasts against water stress *Bot. Stud.* **54** 1–9

[199] Liu M, Li X, Liu Y and Cao B 2013 Regulation of flavanone 3-hydroxylase gene involved in the flavonoid biosynthesis pathway in response to UV-B radiation and drought stress in the desert plant, Reaumuria soongorica *Plant Physiol. Biochem.* **73** 161–7

[200] Abebe T, Guenzi A C, Martin B and Cushman J C 2003 Tolerance of mannitol-accumulating transgenic wheat to water stress and salinity *Plant Physiol.* **131** 1748–55

[201] Wu L, Fan Z, Guo L, Li Y, Zhang W, Qu L-J and Chen Z 2003 Over-expression of an *Arabidopsis* δ-OAT gene enhances salt and drought tolerance in transgenic rice *Chin. Sci. Bull.* **48** 2594–600

[202] You J, Hu H and Xiong L 2012 An ornithine δ-aminotransferase gene OsOAT confers drought and oxidative stress tolerance in rice *Plant S.* **197** 59–69

[203] Cheng C, Hu S, Han Y, Xia D, Huang B-L, Wu W, Hussain J, Zhang X and Huang B 2020 Yellow nutsedge WRI4-like gene improves drought tolerance in *Arabidopsis thaliana* by promoting cuticular wax biosynthesis *BMC Plant Biol.* **20** 1–15

[204] Roy M and Wu R 2001 Arginine decarboxylase transgene expression and analysis of environmental stress tolerance in transgenic rice *Plant Sci.* **160** 869–75

[205] Kumar V, Shriram V, Kishor P K, Jawali N and Shitole M 2010 Enhanced proline accumulation and salt stress tolerance of transgenic indica rice by over-expressing P5CSF129A gene *Plant Biotechnol. Rep.* **4** 37–48

[206] Karthikeyan A, Pandian S K and Ramesh M 2011 Transgenic indica rice cv. ADT 43 expressing a Δ 1-pyrroline-5-carboxylate synthetase (P5CS) gene from *Vigna aconitifolia* demonstrates salt tolerance *Plant Cell, Tissue Organ Culture (PCTOC)* **107** 383–95

[207] Wojas S, Clemens S, Hennig J, Skłodowska A, Kopera E, Schat H, Bal W and Antosiewicz D M 2008 Overexpression of phytochelatin synthase in tobacco: distinctive effects of AtPCS1 and CePCS genes on plant response to cadmium *J. Exp. Bot.* **59** 2205–19

IOP Publishing

Molecular Response and Genetic Engineering for Stress in Plants, Volume 1

Abiotic stress

Pawan Shukla, Anirudh Kumar, Rakesh Kumar and Manish K Pandey

Chapter 3

The molecular basis of mineral toxicity in plants

Abin Sebastian[1]*

[1]*Department of Botany, St. Georges College, Aruvithura 686122, India*

Mineral toxicity limits crop productivity in arable lands. Irrigation, fertiliser application and mining activity increase plant-available mineral ions in the soil. The families of transporter proteins present in the plasma membrane allow influx of mineral ions into roots. Mineral stress causes changes in root architecture and plant growth. The homeostasis of mineral ion concentration in the cells is achieved with the help of plant growth regulators and signalling pathways such as mitogen-activated protein kinase. Plants respond to mineral stress with molecular physiological changes such as adjustment in accumulation of pigments, secretion of phytochemicals, and sequestration of mineral ions in the vacuole. Low-molecular-weight organic acids, phytochelatins, and metal-binding proteins such as metallothionein help crop plants to confront mineral toxicity. The various adaptations of plants against mineral ions offer strategies for crop improvement against mineral stress.

3.1 Introduction

An increase in plant-available ions in soil can cause mineral toxicity in plants. The greater availability of exchangeable ions in the soil depends on the chemical nature of the parent rock material from which the soil is derived. It is well known that calcareous soil is formed as a result of weathering of limestone-containing rocks, and that there is an abundance of calcium ions in calcareous soils which retard plant growth [1]. However, soil physicochemical properties also influence plant-available mineral ions in the soil. Toxicity due to aluminium, iron, and manganese is very common in acidic soil [2]. This kind of soil is common in regions with annual rainfall greater than 60.0–80.0 cm. The high rainfalls cause leaching of cations and increases

* Address for correspondence Email: abin@sgcaruvithura.ac.in.

doi:10.1088/978-0-7503-4921-5ch3

the plant-available H^+ ions in the soil. Secondly, addition of ammonium-based fertilisers decreases soil pH, which in turn creates cation toxicity in plants [3] (figure 3.1). The presence of excess ammonium in the xylem is reported to damage the vascular tissues and retard root growth [4]. Typically, the growth of lateral roots is affected during ammonium stress, with a brown discoloration of the central portion. In addition, the temperature of the soil significantly influences ammonium toxicity in plants, wherein the potential toxicity is increased during chilling periods [4]. This effect is due to the outcome of slow conversion of ammonium to nitrate under low temperature. Soil pHs higher than 8.0 cause immobilisation of metal ions in the soil, which in turn creates nutrient deficiency in plants [1]. The development of a pale green colour is a typical symptom found in plants growing under conditions of alkaline soil [1]. The repeated application of phosphate fertilisers and pesticides not only creates environmental pollution but also causes heavy metal stress in plants. Accumulation of heavy metals such as cadmium in plants is one of the major threats posed by application of phosphate fertilisers. It has been reported that Cd in a phosphate rock range from 20.0–90.0 mg kg^{-1}, along with the application of Cd-contaminated phosphate fertiliser, resulted in an accumulation of Cd in the plant biomass as high as 417.0 mg kg^{-1} [5]. The usage of both ground water and industrial waste water for irrigation also results in contamination of crop produce with toxic metals and metalloids [5]. Accumulation of arsenic in rice grain was a reported outcome of ground water usage in irrigation [6]. Further, water is naturally enriched

Figure 3.1. Mineral toxicity in plants. Plants suffers mineral toxicity because of excess mineral ion release during soil formation and changes in the physicochemical properties of the soil. (a) Laterite soil, which is often subjected to nutrient leaching, results in mineral toxicity in plants because of exposure to aluminium and iron ions. (b) Phosphorous deficiency in an acidic soil, resulting in yellowing of coconut leaves. (c) Magnesium deficiency in pea plants. Note, however, the grass plants growing in the same location did not experience Mg deficiency. This indicates the putative role of metal transporters that help to cope with mineral deficiency stress in plants. (d) Instability in anthocyanin accumulation in basil, which very often occurs because of dynamics in the presence of selenium and boron in the soil.

with fluoride in many regions, and the usage of fluoride-contaminated water results in the accumulation of fluorine in crops [7]. The utilisation of waste water released from mines, pigment industries, and leather tanning factories for agriculture purposes has resulted in contamination of crops with metals such as cadmium, lead, and chromium [8].

The minerals required for plant growth are classified into two main groups: macronutrients and micronutrients [9]. The macronutrients which are important for plant growth are nitrogen, phosphorus, potassium, calcium, sulphur, and magnesium. These elements are mainly associated with synthesis of plant pigments, carbohydrates, and proteins [9]. Elements such as manganese, boron, copper, zinc, molybdenum, and iron are known as micronutrients, with accumulations of less than 0.1% in plant tissues [10]. These elements mainly act as cofactors of metalloproteins. The specificities in uptake of mineral ions are controlled with the help of nutrient transporter proteins in the plasma membrane [11]. There are diverse nutrient transporter protein families in plants. The mineral transporters in the plasma membrane includes ATPases or proton pumps, channel proteins, and cotransporters [11]. Membrane ATPases transport mineral ions against electrical and chemical gradients with the help of adenosine triphosphate (ATP) hydrolysis [12]. On the other hand, proton pumps help to acidify the cytosol or vacuole. But membrane protein channels work to transport mineral ions down the energy gradient. In a recent study, the channel protein was recorded to have the ability to pass more than 10^6 molecules per second [13]. Cotransporters have also been reported to help membrane transport of solute molecules at a rate of 10^4 molecules per second. Recent studies show that cotransporters function as channels, too. It is noteworthy that many of the transition metal transporters allow transport of more than one type of ion. The transport of ammonium was found to be electrogenic, and the ammonium transporter also allows passage of methylammonium at a rate of 30–300 ions per second [14]. Based on the availability of nitrate, low-affinity transport systems and high-affinity transport systems can be found present in plants [15]. The N transporters were found to show more substrate specificity compared to transition metal transporters. The expression of nutrient transporters in plants is under control of signalling cascades involving transcription factors such as the MYB transcription factor, namely PHR1 [16]. Plant roots also grow in search of nutrients, and the process is regulated at the molecular level with the help of genes involved in synthesis of auxin such as *TAA1/TAR* [17]. It is reported that reactive oxygen species (ROS) levels increases during nutrient deprivation, which might be one of the upstream mediators of nutrient deprivation [18].

Plants respond to mineral toxicity with the help of molecular physiological changes [19]. The strategies of mineral tolerance are broadly divided into avoidance and detoxification. The avoidance mechanism against mineral toxicity operates through morphological changes in the roots and secretion of chelators into the rhizosphere [20]. The secretion of low-molecular-weight organic acids is common during metal stress in plants. This kind of response has been noted during Al, Mn, and Fe stress in plants [21, 22]. The secretion of organic acids not only helps to decrease plant-available metal ions but also enhances the solubility of mineral ions such as phosphate. Deposition of compounds such as callose in the cell wall also help to avoid contact with toxic mineral ions [23]. The detoxification of minerals is carried out with the help of biochemical and molecular changes. Storage of mineral ions in the vacuole is the key molecular-level mineral detoxification process found in plants [5]. The mineral ions are chelated to either tricarboxylic acid or phytochelatin prior to transport into the vacuole. The transporters involved in transfer of mineral complexes to the vacuole mainly belong to the ABC family of proteins. Also, mineral ions are detoxified in the cytoplasm with the help of proteins belonging to the heat shock family (HSPs), chaperones, and metallothioneins [5, 24]. Thus, perception and detoxification of mineral ions in plants is a cooperative action involving changes in molecular levels. The following sections describe how plants adapt to mineral toxicity, with a focus on morphological and molecular mechanisms of mineral stress tolerance.

3.2 Plant growth responses in mineral-rich environments

Mineral toxicity affects plant growth and developmental processes [25]. Generally, plants cope with mineral stress via accumulation of pigments such as anthocyanin [1]. However, an increase in the concentration of mineral ions in the soil immediately affects the growth of roots. The uptake of water and remobilisation of photo-assimilates change during mineral toxicity [2]. The alterations in root growth also affect plant growth, promoting microbes in the rhizosphere which ultimately trigger mineral toxicity [5]. Thus, changes to both the rhizosphere environment and uptake of water are the immediate effects of mineral toxicity in plants. However, there are progressive changes in root growth, such as elongation of lateral roots, which have also been reported in the course of mineral deficiency [25].

Root growth is inhibited in the presence of excess aluminium, manganese, and iron [26]. Secondly, the presence of heavy metals such as cadmium, mercury, nickel, copper, and lead also retard root growth [5]. A decrease in root growth in the course of mineral toxicity could be reflected as an inhibition of tap root elongation. The elongation of a root involves cell wall stress relaxation, which decreases turgor pressure and water potential in the root cells. These changes help to create a pressure gradient for water uptake, which in turn makes the cell wall expansion irreversible. However, mineral stress affects glycan networks via rearrangement of glycoprotein networks and sugar residues. For example, Al bound on the cell wall decreases the extensibility of cellulose polymers, which in turn inhibits the elongation of roots [26]. Both initiation and lengthening of lateral roots also decreases during exposure of plants to mineral toxicity. The root growth of plant species such as rice, sorghum, maize, and wheat have been well studied in the course of Al toxicity. The toxicity of zinc ions on the root architecture of plants such as brassica and soybean has also been well studied [27]. Studies on the effect of heavy metal ions on root growth were mainly described with help of the root morphology of tea plants and tomatoes, respectively [28]. The initial symptoms of metal toxicity appeared within 5.0 hr of exposure to metal ions. The main target tissues of metal toxicity were the epidermis, cortex, and vascular tissues, respectively [29]. The cells of these tissues had a greater number of vacuoles and nucleoli in the course of exposure to metal ions [30]. The

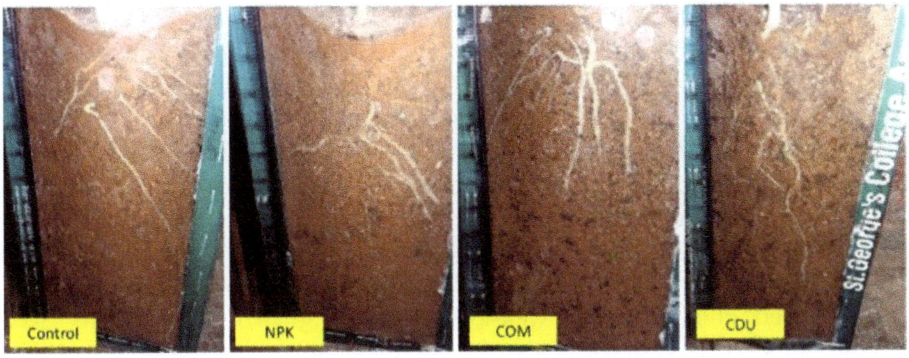

Figure 3.2. Changes in root structure in the course of application of various nitrogen fertilisers. Synthetic fertiliser and compost treatment caused an increase in the girth of roots. However, cow dung, which is rich in metal chelating agents such as humic acids, helped to lengthen roots, which had a significant role in drought tolerance. Abbreviations: Npk: synthetic ammonium fertiliser; COM: vermicompost; CDU: cow dung.

root growth inhibition was very often followed by damage to the roots due to plasmolysis and alterations in the ultrastructure of the root cells.

The architecture of roots is also altered in response to mineral toxicity (figure 3.2). It is reported that heavy metal stress stimulates initiation of lateral roots, although the presence of heavy metals inhibited tap root growth. Stunted roots and poorly developed root hairs are depicted as primary plant responses to accumulation of mineral ions such as Al, Fe, Mn, and Cd. Swollen root apexes, brittle roots, and compact root systems are also reported as primary symptoms of mineral toxicity [28]. Many of these symptoms are the result of changes in the permeability of roots because of lipid peroxidation of biological membranes. A barbed wire appearance of the roots has been observed in the course of heavy metal stress in plants. Roots also develop a brownish coloration during exposure to heavy metals. The change in root colour was noted in particular during nickel stress [31]. Earlier collapse of roots was noticed in the case of Cu and Ni stress. However, increasing concentrations of Mn, Cu, and Zn in micromolar concentrations stimulated formation of lateral roots, and resulted in the development of denser root systems. However, Mn, Al, and Fe become toxic to plants at acidic pHs, wherein the primary symptom was a decrease in root biomass [26]. Excess of ions of macronutrients such as ammonium also inhibit root growth because of the collapse of vascular tissues such as xylem in the root, which ultimately alters the water balance of plants. The exposure of plants to macronutrients such as Ca also altered root architecture because of fluctuations in ionic balance [1]. However, it is noteworthy that an optimum concentration of Ca is necessary for elongation of primary roots in plants.

Mineral toxicity symptoms also appear on the aerial parts of plants. Symptoms of mineral toxicity vary among plant taxa. The most common morphological symptoms of mineral toxicity in plants are a decrease in plant biomass and crop yield [29]. These effects are the outcome of a decrease in photosynthetic activity. Minerals of Fe, Mn, and Al are well known to induce toxic symptoms in plants. Symptoms of toxicity from these elements appear in aerial plant parts as chlorosis because of oxidative stress [26]. The yellowing of leaves mainly appears on the marginal regions of leaves. Necrotic spots and burned regions on leaves are common symptoms of boron toxicity in plants. Plants growing near mine sites very often show these symptoms because of toxicity due to heavy metal cations. Extremely dark green leaves have been noted in plants suffering nitrogen toxicity. Further, the burning of leaves with brown colorations was also noticed in plants exposed to excess nitrogen [18]. However, it should be noted that some plants showed a yellowing of their leaves in response to nitrogen toxicity because of deficiency of other essential elements. It is noteworthy that excess watering aggravates nitrogen toxicity in plants.

Lime-induced chlorosis is a well-known plant growth abnormality seen in calcareous soils [1]. Excess Ca is also found to hinder germination. The yellowing of leaves during Ca stress is the result of inhibition in the uptake of Mn and Fe, respectively. Also, excess Ca salts in the soil can raise the soil pH above 8.0, which will hinder uptake of cations essential for plant growth. The yellowing symptoms start at the margins of leaves and later spread to the inward regions of leaf lamina. The severity of the Ca-stressed condition is indicated by the appearance of dark green areas only around the large veins. Marginal necrosis is a symptom of Mg toxicity in plants, which mimics the symptoms of deficiencies in Ca and K, respectively [32]. Stunted plant growth and the appearance of a coppery colour along the marginal veins are also symptoms of Mg toxicity in plants. Accumulation of excess phosphorus causes necrotic regions along the main veins of the leaves. New leaves have been found to have chlorosis between veins during P toxicity [33]. Exposure of plants to excess potassium caused a reduction in leaf size and stunted plant growth [34]. Plants exposed to high levels of K also tend to accumulate more micronutrients, which in turn creates visual symptoms of micronutrient toxicity such as a decrease in chlorophyll content and dry weight of the plants, while there is an increase in accumulation of plant pigments such as anthocyanins [34].

3.3 Membrane transporters involved in the perception of mineral stress

Membrane transporters help the uptake and transport of mineral nutrients in plants. These transporters can be ion channels, pumps, or carriers. Such transporters help to uptake mineral ions from the soil solution. The expression of mineral nutrient transporters plays a critical role in the regulation of mineral nutrient homeostasis in plants (table 3.1). Molecular analysis of model plants such as *Arabidopsis*, rice and *Chlamydomonas* has revealed numerous membrane transporters that are involved in the transport of mineral nutrients in plants. Studies also showed that many of the mineral nutrient transporters pose high specificity for particular nutrient ions [35]. These transporters may also act as efflux or influx proteins. Secondly, some of the transporters showed polar activity wherein the transport of mineral ions occurs in a single direction. The cooperative action of various mineral transporters is critical for controlling the accumulation of mineral ions. The toxic effect of mineral ions may also be the result of post-translational modifications that alter the structure of membrane transporters, resulting in changes to the transporting activity of ions.

Table 3.1. Summary of families of membrane transporters involved in mineral accumulation.

Membrane transporter family	Mineral ion
AMT	Ammonium
NRT1 (NPF), NRT, CLC, SLAC1/SLAH	Nitrate
KT, HAK, KUP, Trk/HKT, CPA	Potassium
PHT	Phosphorous
MGT, CorA, MRS2	Magnesium
Ca^{2+}ATPase, Ca channels, CaCA, CCX	Calcium
SULTR	Sulphur
NRAMP, YSL, ZIP, CAX, CCX, CDF, MTP, P-type ATPases, VIT	Manganese, copper, zinc, cadmium, molybdenum
HMA3	Cadmium, zinc
ABC	Aluminium, heavy metals
IRT	Iron
ZIP	Copper, cobalt, cadmium, zinc

The nitrogen demands of a plant are achieved with the help of ammonium and nitrate transporters. The transporters of ammonium play an important role in the uptake of ammonium from acidic and waterlogged soils. The low energy expenditure requirement for ammonium uptake makes this mineral a potent source of nitrogen. There are five ammonium transporters reported in *Arabidopsis* roots, namely AtAMT1, AtAMT2, AtAMT3, AtAMT4, and AtAMT5. Among these, both AtAMT2 and AtAMT13 are reported as high-affinity ammonium uptake systems through which about 80.0%–90.0% of ammonium uptake occurs. The function of AtAMT2 is xylem loading of ammonium. The presence of excess ammonium is found to trigger upregulation of usually multiple amino acids move in and out transporters (UMAMITs). These transporters are involved in phloem loading and secretion of amino acids into the external environment [36]. Higher levels of extracellular ammonium are found to inhibit activity of the AMT1 transporter via phosphorylation-dependent pathways [37]. Thus, phosphorylation activity on AMT1 prevents excess accumulation of ammonium in the cell. Further, transporters involved in the uptake and transport of K^+ are reported to be involved in the uptake of ammonium in plants because of the similarity in atomic radius. Nitrate transporters are involved in the uptake of nitrate ions when the soil pH is alkaline. There are four families of nitrate transporters in plants, namely NRT1 (NPF), NRT2, CLC, and SLAC1/SLAH. The dual-affinity transporter NRT1 is found to respond during exposure to higher concentration of nitrate in the rhizosphere. This transporter is also reported as a sensor of nitrate via controlling the expression of NRT2. The above data suggest that plants possess well-developed nitrogen homeostasis mechanisms which allow them to cope with excess nitrogen in the soil.

Membrane transporters belong to families such as the K ($^+$) uptake permeases (KT/HAK/KUP), the K ($^+$) transporter (Trk/HKT) family, and the cation proton antiporters (CPA), which are involved in regulation of potassium accumulation in plants. Transporters in the permeases family are also reported to induce root hairs in response to fluctuations in potassium availability, and these point membrane transporters are able to induce changes in root morphology under varying concentrations of mineral ion in the soil [38]. Potassium channels such as KUP2 are reported to mediate K^+ efflux in roots, which will help to cope with excess potassium accumulation [39]. There are five families of phosphate transporters reported in plants, namely PHT1, PHT2, PHT3, PHT4, and PHT5. The PHT1 transporter is implicated in mediating the uptake of Pi from soil, and also plays an important role during arbuscular-mycorrhizae-mediated Pi uptake [40]. Studies with *Arabidopsis* showed that plants compromise excess Pi in the cytosol with the help of vacuolar P transporters such as VPT1 [41]. Magnesium is a well-known plant nutrient required for the synthesis of chlorophyll. The uptake of Mg is under the control of the MGT, CorA, and MRS2 family of proteins. Among various Mg transporters, expression of *MaMRS2–1* and *MaMRS2–10*, belonging to the MRS2 family, was directly related to excess Mg accumulation in musa plants. It is also reported that expression of *MaMRS2–1* and *MaMRS2–10* downregulates upon low availability of Mg in the rhizosphere [32]. Transporters such as Ca^{2+}-ATPase and Ca^{2+} channels have been reported as key modulators of calcium stress. Expression of these transporters caused vacuolar sequestration of Ca ions, and thus alleviated the toxic effect of Ca in the cytosol. The uptake of sulphur in plants is mediated by sulphate transporters belonging to the SULTR family, which function as a H^+ cotransporter or ABC-type transporter. Sulphur accumulation is mainly regulated with negative regulation of reduced S-containing compounds and positive regulation by O-acetylserine, respectively [42].

The cellular homeostasis of micronutrients is also under the control of transition metal transporters. However, transporters of micronutrients exhibit a broad substrate range. For example, transition metal transporters such as natural resistance associated macrophage protein (NRAMP), yellow-stripe-like (YSL), zinc-regulated transporter/iron-regulated transporter [ZRT/IRT1]-related protein (ZIP), cation exchanger (CAX), calcium cation exchangers (CCXs), cation diffusion facilitator/ metal tolerance protein (CDF/MTP), P-type ATPases, and vacuolar iron transporter (VIT) are implicated in accumulation of Mn, Zn, Fe, Mo, and Cu. Secondly, these transporters also allow accumulation of heavy metals such as Cd, Co, and Ni. The toxicity of micronutrients appears mainly in acidic soil where there are low plant-available macronutrients. It is reported that excess Mn limits the uptake of Ca, Mg, Fe, and P because of the similarity in atomic radius. However, ectopic expression of vacuolar Mn transporters belonging to the CDF family has been reported to increase Mn tolerance in plants [35]. In the case of Zn tolerance in plants, expression of the ZIP19 and ZIP23 transporters was found to be critical in *Arabidospsis halleri* [43]. Also, HMA3, which helped vacuolar sequestration of Zn, was also found to enable Zn tolerance in plants. However, tolerance against toxic metals such as Al and Cd is achieved with the help of metal transporters as well as organic acids. Therefore, operation of transporters involved in both metal and organic acids transport are critical for Al and Cd tolerance in plants. For example, it is reported that aluminium-activated malate transporters (ALMTs) play an important role in Al tolerance [44]. Among various ALMT transporters, ALMT1 was found to localise in the plasma membrane of root cells and take part in the exclusion of Al from the cell in the form of an Al-malate complex. Also, internal detoxification of Al was achieved with the help of AtAlS3, which is an ABC transporter located in the plasma membrane of the epidermal cells of the root cortex and phloem [44]. This transporter is reported to play a key role in decreasing the level of Al accumulation in shoot apexes. Heavy metals such as Cd enter plant cells through transition metal transporters belonging to the ZIP, CDF, and ABC families. Cadmium was also found to cross plasma membranes through transporters involved in the uptake of Fe such as Nramp and IRT1 [45]. Accumulation of Cd in plants is reported to occur through membrane transporters involved in transport of micronutrients. Overexpression of *AtHMA3* not only enhanced vacuolar sequestration of Cd in *Arabidopsis* but also Zn, Pb, and Co. A metal transporter, namely OsZIP1, acted as an efflux transporter to limit the uptake of Cu, Co, Cd, and Zn. The methylation of the histone protein H3K9me2 of OsZIP1 during metal stress has been reported [46]. This result shows that epigenetic control of gene expression also plays a significant role in the exclusion of trace metals from plants.

3.4 Molecular targets of mineral toxicity

Mineral toxicity affects signal transduction and metabolic pathways in the cell. The signalling pathways play a critical role in the perception of mineral stress in plants. Many of the signalling pathways during mineral stress are triggered because of the formation of ROS in the cells. The results of these signalling cascades are often reflected as an increase in chelation activity, ion sequestration in vacuoles, regulation of ion intake by transporters, and intensification of antioxidant activity (figure 3.3). The major cellular signalling pathways that are responsive during mineral stress are calcium signalling, hormone signalling, and mitogen-activated protein kinase (MAPK) signalling [47]. Signalling cascades mediated by microRNAs and transcription factors also play a key role in regulation of mineral stress in plants. The blockage of functional groups and displacement of cofactors are well known to initiate signalling pathways during mineral stress. However, redox-active metals such as iron and copper initiate cellular signalling pathways through changes in redox status [48]. Accumulation of mineral ions is well known to generate ROS in the cell, which is responsible for pioneering of signal transduction pathways in plants under stress. The genes that are regulated in the course of metal stress include genes for metal chelators and transporters.

Calcium signalling plays a key role during heavy metal stress in plants. Expression of glutathione transferase, which initiates antioxidant responses, was found to be related with Ca signalling [49]. However, heavy metals also disturb the flow of Ca through Ca channels. Thus, many of the effects of heavy metal toxicity in plants arise due to disturbance in Ca signalling. It is also reported that cadmium enhances the influx of Ca, which in turn triggers Ca-mediated cellular signalling pathways. Ions of magnesium, potassium, sodium, and zinc are found to bind with

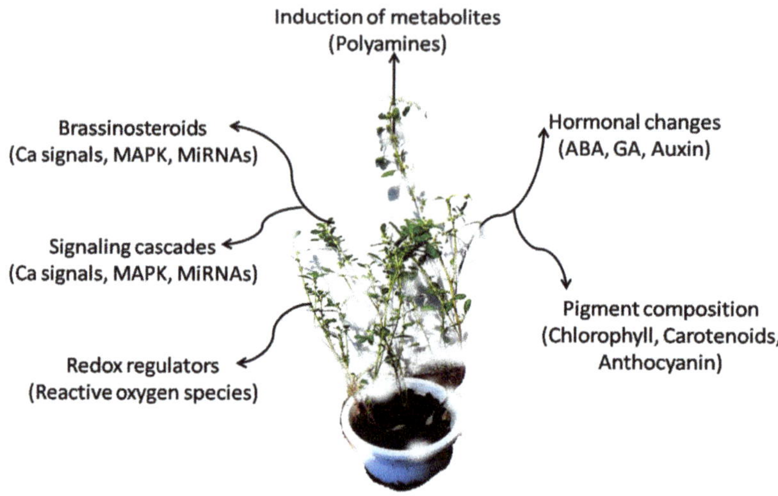

Figure 3.3. Mineral toxicity responses in plants. Metabolites such as polyamines and reactive oxygen species are induced during exposure to mineral ions that initiate signalling pathways. Hormones also play a significant role in triggering various mineral responses in plants. There is an immediate change in pigment composition during mineral stress in plants.

calmodulin, which is a an important protein for Ca-mediated cellular responses. The four binding sites of Ca were found to be occupied by lead, and the binding of Pb with calmodulin stimulated Ca signalling in plants. The nitrous oxide (NO)-signalling cascade in plants is also associated with a Ca spike. Therefore, heavy-metal-induced NO formation in plants is also indirectly related with Ca signalling in plants.

Plant hormones take part in the various stages of development in plants. Among various plant hormones, abscisic acid (ABA) has a critical role in abiotic stress tolerance. It is reported that ABA concentration levels in roots of *Typha latifolia* and *Phragmites australis* increase during Cd stress [50]. Similarly, treatments of mercury, copper, zinc, lead and sodium are found to increase accumulation of ABA in plants. The accumulation of ABA during heavy metal stress is the result of a decrease in water potential, which causes closure of stomata. Reports have also pointed out that the *PYL*, *PP2C*, and *SnRK2* genes take part in ABA signal transduction during heavy metal stress [51]. Exogenous application of ABA is also found to inhibit aerial transport of cadmium and nickel, resulting in the accumulation of metals in roots. Therefore, the application of ABA is a promising method to decrease accumulation of heavy metals such as Ni in leaves and fruits. It is also reported that plants respond to excess ammonium with the help of signal transduction processes involving ABA [50]. A plastid metalloprotease, namely AMOS1/EGY1, has been described as a link integrating ABA into the ammonium signalling response. Almost 90.0% of genes responding to ammonium were regulated by AMOS1/EGY1. It is noteworthy that the ammonium-activated motif also had ABA-responsive elements in the promoter region. Accumulation of macronutrients such as Mg also triggers signalling pathways in plants. It is reported that Mg-mediated signalling pathways operate with the help of protein kinases such as adenylate kinase [52]. Adenylate in the cell also binds with Mg, and thus affects the equilibration of ATP in the cell, which in turn affects various cellular processes depending on ATP. The uptake and accumulation of potassium in plant cells is also reported to trigger signalling pathways. Major responses of dynamics in K^+ ion accumulation in plants were mediated by ROS and Ca ions, respectively. Transcription factors such as MYB59 are also found to modulate K accumulation in plants. Many different proteins modulate the AtAKT1/AtKC1 channel activity involved in uptake of K via phosphorylation and protein–protein interactions [53].

Auxin synthesis is inhibited during cadmium stress, which accounts for a decrease in the size of root apical meristem [54]. The result of these changes is reflected as a decrease in root growth. It is also reported that mineral ions inhibit *PIN* activity, which in turn inhibits the polar transport of auxin [55]. However, Cu stress is found to enhance the activity of auxin in the meristem and elongation zones, respectively [55]. Similarly, an increase in hemicellulose content as a result of NAA:1-naphthaleneacetic acid (NAA) treatment is found to immobilise Cd ions in the cell wall which in turn helps to alleviate the toxic effect of Cd. Auxins also play an important role in stabilising lipid membranes. This hormone is reported to initiate nanoclusturing of membrane proteins, which in turn improves membrane stability. Therefore, a combined treatment of heavy metals such as Pb with auxin was found to avoid alterations in membrane structure [56]. Gibberllic acid as well as salicylic acid is also reported to enable mineral tolerance in plants. These hormones act as ameliorants of ROS in the cell. For example, there was a reported increase in the salicylic acid content in maize plants subjected to Cd stress [45]. Studies have highlighted that phenylalanine and phenylpropanoid pathways, which lead to accumulation of phenolics in plants, also act in the course of salicylic-acid-mediated responses during heavy metal stress. Gibberllic acid is found to trigger activity of the transcription factor *DELLA* during heavy metal stress. The major route of action of *DELLA* was the activation of antioxidant systems [57]. Thus, the role of gibberllic acid during heavy metal stress could be the detoxification of ROS. Gibberllic acid was also found to increase mitotic activity during Cd and Pb stress in *Vicia faba* L. plants. Recent studies have pointed out that the activity of gibberellic acid and salicylic acid are interconnected in the course of mineral accumulation in plants [58]. Cytokinin, ethylene, and jasmonic acid are also reported to act during heavy metal stress in plants. These hormones act by triggering expression of transcription factors such as *DELLA*, AP2m and ERF1 that stimulate activity of antioxidant enzymes and metal-binding proteins in plants [59].

Brassinosteroids are also found to respond during mineral stress in plants. These compounds are poly-hydroxylated steroids that regulate both tissue morphogenesis and senescence. Brassinosteroids have been reported to activate antioxidant enzymes during heavy metal stress. It has been shown that exogenous application of brassinosteroids induces a transient spike in H_2O_2, which activates MAPK-signalling cascades and results in induction of NADPH oxidase [60]. The MAPK-signalling cascade acts by mediating phosphorylation reactions on various protein targets. The initiation of MAPK-mediated signal transduction during mineral stress occurs due to either mineral ion accumulation or ROS. The activation of the MAPK pathway is well depicted in the course of Cd, Cu, and arsenic stress. Both $CdCl_2$ and $CuSO_4$ are found to induce MPK3 and MPK6 in *Arabidopsis*. It is assumed that these MPKs trigger chelation and sequestration of metal ions such as Cd and Cu. It is also reported that there is cooperative action of MAPK cascades and auxin signalling in rice plants under Cd stress [60]. The above reports indicate that distinct MAPK-mediated signalling cascades occur in plants in the course of mineral stress.

3.5 Mineral detoxification in plants

Plants adapt to mineral stress with the help of antioxidants, mineral chelators, and stress-tolerant proteins. Further, activation of metabolic pathways and immobilisation as well as precipitation of mineral ions is well known to confer mineral tolerance in plants (figure 3.4). For example, excess ammonium accumulation in the plant is modulated with the help of enzymes such as glutamine and glutamate synthase [61]. Antioxidant-mediated stress tolerance mechanisms comprise the preliminary steps adopted by a plant to counteract the toxic effect of mineral accumulation. The accumulation of minerals often causes changes in cellular redox status, which modifies cellular metabolic activities. Oxidation-reduction reactions of redox regulators help to adjust the oxidation state of various molecules that are important for the operation of primary metabolism. Glutathione is one of the key redox regulators in the cell. This compound is a thiol peptide formed from L-glutamate, cysteine, and glycine. It is

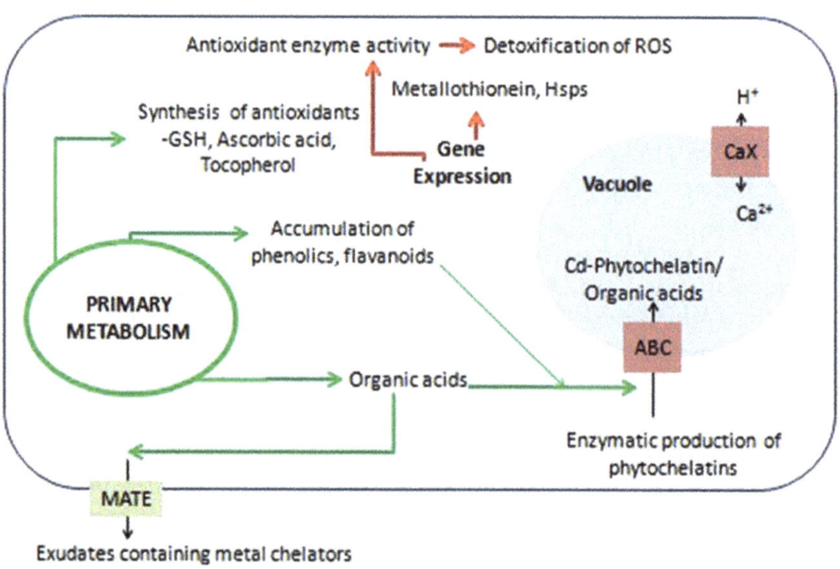

Figure 3.4. Mineral tolerance strategies in plants. Mineral ions are detoxified with the help of intermediates of both primary and secondary metabolism, respectively. Gene expression was involved to initiate antioxidant enzymes and synthesis of stress-tolerant proteins such as heat shock proteins (Hsps) during mineral stress tolerance. Both extracellular and intracellular complexation of ions helped to alleviate toxic effects in the course of mineral ion accumulation in plants.

synthesised by the activity of enzymes such as glutamate–cysteine ligase and glutathione synthetase, respectively. The activity of these enzymes is found to change during mineral stress in plants [62]. This compound acts as both chelator of mineral ions and scavenger of ROS. The presence of the sulfhydril group (–SH) helps glutathione to conjugate with mineral ions. The mineral-glutathione complex is sequestrated in vacuoles. For example, *Brassica chinensis* L. has been found to accumulate Cd-glutathione complex in the vacuole, and an increase in supply of glutathione increased the accumulation of Cd-glutathione conjugate in the vacuole [63].

The antioxidant activity of glutathione is mainly utilised in the ascorbate–glutathione cycle, where the reducing power of glutathione is used to regenerate ascorbic acid in the cell. It is noteworthy that the ascorbate–glutathione cycle plays a key role in the mitigation of heavy-metal-induced oxidative stress in plants [64]. The ascorbate–glutathione cycle pathway operates in a series of enzymatic reactions involving ascorbate peroxidase, monodehydroascorbate reductase, dehydroascorbate reductase, and glutathinone reducatse, respectively. This pathway has been well studied in order to understand photooxidative stress tolerance in plants. However, heavy metals also induce ROS in the cell by altering the activity of photosynthetic electron transport [5]. Also, changes in the activity of membrane-bound NADH—oxidoreductase enzyme also generate ROS during mineral stress. Secondly, accumulation of redox-active metal ions such as iron also causes oxidative stress in plants via taking part in fenton reactions. Furthermore, accumulation of ammonium has been found to increase accumulation of ROS in *Lemna minor,* L. [65]. So, the operation of glutathione-mediated metabolic pathways that act as an antioxidant defence is critical to avoid mineral toxicity in plants. The above reports also highlight that many of the toxic symptoms of mineral stress arise as a result of formation of ROS in plants.

Secondary metabolites belonging to phenolics, flavanoids, and polyamines also act to scavenge toxic effects of mineral accumulation in plants. Phenolics in plants are derived from phenylalanine through the shikimate/phenylpropanoid pathway. These compounds exhibit antioxidant activity, and prevent membrane damage under adverse environmental conditions. It has been reported that plants respond with accumulation of phenolics under cadmium, aluminium, lead, zinc, and copper stress [66]. Amaranth plants respond to higher ammonium levels in the soil with the accumulation of flavonoids and phenolics [67]. Similarly, tomato plants respond to manganese stress with an increase in production of polyphenols [68]. Accumulation of phenolics during mineral stress is the result of an increase in activity of enzymes in the phenylpropanoid pathway such as phenylalanine ammonia-lyase, chalcone synthase, shikimate dehydrogenase, cinnamyl alcohol dehydrogenase, and polyphenol oxidase, respectively. Phenolics such as flavanoids are also known for metal chelation activity [69]. For example, accumulation of anthocyanin increased during magnesium stress in plants. This pigment played a critical role in the localisation of Mg ions in the vacuole. Anthocyanin is also reported to take part in remobilisation of metal ions such as Cd, Pb, Cu, and Ni [70]. Studies with *Azolla imbricate* also showed that anthocyanin acts as both metal chelator and antioxidant [71].

Polyamines are induced during the course of metal accumulation in plants. These compounds mainly act as antioxidant molecules during metal stress. For example, in a study wheat plants responded to Cu and Cd with an increase in accumulation of putrescine and spermine [72]. Polyamines are also reported to enhance metal chelation via promoting synthesis of amino acids that are involved in metal chelation [73]. Polyamines also increase proline synthesis, which in turn acts to regulate osmolytes critical for stress tolerance under mineral toxicity [74].

Metal chelators bind with minerals using functional groups such as the carboxylic and sulfhydril groups. The metal chelator complex is stored in the vacuole, and thus the toxic effect of the mineral ion is nullified (figure 3.4). Also, secretion of metal chelators helps extracellular chelation of metal in the rhizosphere zone. Tricarboxylic acids, such as citric acid, malic acid, and oxalic acid, are found to take part in extracellular chelation of metals such as aluminium [75]. This kind of response is often reported in plants growing in acidic soils. The secretion of organic acids not only helps to chelate metal ions but also helps to acquire essential plant nutrients such as phosphorus. Studies on Al-activated malate transporters (ALMT) have pointed out that the secretion of organic acid into the rhizosphere has multiple benefits such as acquisition of iron from alkaline soils, uptake of nitrate, and vacuolar sequestration of calcium in plants growing in calcareous soil [76]. Storage of mineral ions in the vacuole helps to avoid the toxic effects of ions in the course of hyperaccumulation of metal ions. Apart from organic acids, peptides, namely phytochelatins, also help vacuolar chelation of heavy metals such as cadmium [77]. The metal complexes of phytochelatins are transported across the tonoplast with the help of ABC transporters and ATP hydrolysis, respectively. Phytochelatins are peptides derived from glutathione and are characterised by an abudance of sulphydril groups. Enzymatic synthesis of these peptides occurs with the help of phytochelatin synthase during exposure to metal ions [26]. The enzyme is reported to be activated by several metal ions such as Cd^{2+}, Cu^{2+}, Ag^{+}, Hg^{2+}, and Pb^{2+} and the peptides interact with metal ions through thiolate coordination [78]. Heat shock proteins (Hsps) in plants help to recover the misfolded proteins during metal stress [79]. Molecular-level control of metal ions in plants is also achieved with the help of metallothionein proteins, which also have an abundance of the sulphydril group [26]. These proteins are also able to bind with several metal ions through the formation of mercaptide bonds between the sulphydril group of cysteine residues and the metal. The arrangement of cysteine residues in part controls the metal-binding properties of the metallothionein, and the induction of these proteins help to remove free metal ions present in the cytosol without vacuolar sequestration.

3.6 Conclusions

Minerals are essential for plant growth. However, excess minerals in the environment can cause stress responses in plants. Further, changes in soil physicochemical properties such as pH lead to more plant-available mineral ions in the soil. The root architecture and changes in the coloration of leaves with a decrease in photosynthesis are very often noticed in the course of mineral toxicity. Plants perceive mineral ions with the help of membrane transporters and signal transduction cascades. Some membrane transporters, such as MATE, help to exclude mineral ions. Reactive oxygen species, hormones, and protein kinases play an important role in the perception of mineral stress as well as induction of mineral tolerance. Antioxidants, such as glutathione, metal chelators, and metal-binding proteins all confer mineral tolerance in plants.

Fill in the blanks

1. Excess ammonium causes damage of … tissues in plants.
2. Plants growing in acidic soils secrete … for metal tolerance.
3. Mineral stress is perceived with the help of … protein kinases.
4. Magnesium triggers … mediated signalling pathway.
5. Enzymatic synthesis of … confers metal tolerance in plants.

Answers: 1. vascular tissues; 2. organic acids; 3. mitogen-activated adenylate; 4. adenylate kinases; 5. phytochelatin.

Descriptive-type questions

1. How do plants adapt to mineral-rich environments?
2. What are the plant responses under metal stress?
3. Explain the role of metal transporters in mineral stress.
4. Write a note on signalling pathways involved in mineral stress in plants.
5. Comment on the role of antioxidants involved in mineral tolerance in plants.
6. What is the role of metal chelators in mineral stress?

References

[1] Sebastian A, Nangia A and Prasad M N V 2017 Carbon-bound iron oxide nanoparticles prevent calcium-induced iron deficiency in *Oryza sativa* L. *J. Agric. Food Chem.* **65** 557–64

[2] Chang Y T, Hsi H C, Hseu Z Y and Jheng S L 2013 Chemical stabilization of cadmium in acidic soil using alkaline agronomic and industrial by-products *J. Environ. Sci. Health* **48** 1748–56

[3] Jia K L, Yu H, Feng W Q, Qin Y S, Zhao J, Liao M L, Wang C Q and Tu S H 2009 Effect of different N, P and K fertilizers on soil pH and available Cd under waterlogged conditions *Huan Jing Ke Xue* **30** 3414–21

[4] Liu Y and von Wirén N 2017 Ammonium as a signal for physiological and morphological responses in plants *J. Exp. Bot.* **68** 2581–592

[5] Sebastian A and Prasad M N V 2014 Cadmium minimization in rice: a review *Agron. Sustain. Dev.* **34** 155–73

[6] Roychowdhury T, Tokunaga H, Uchino T and Ando M 2005 Effects of arsenic-contaminated irrigation water on agricultural land soil and plants in West Bengal, India *Chemosphere* **58** 799–810

[7] Gadi B R, Kumar R and Goswami B *et al* 2021 Recent developments in understanding fluoride accumulation, toxicity, and tolerance mechanisms in plants: an overview *J. Soil Sci. Plant Nutr.* **21** 209–228

[8] Tchounwou P B, Yedjou C G, Patlolla A K and Sutton D J 2012 Heavy metal toxicity and the environment *Molecular, Clinical and Environmental ToxicologyExperientia Supplementum* vol 101 (Basel: Springer) pp 133–64

[9] White P J and Brown P H 2010 Plant nutrition for sustainable development and global health *Ann Bot* **105** 1073–80

[10] Conn S and Gilliham M 2010 Comparative physiology of elemental distributions in plants *Ann. Bot.* **105** 1081–102

[11] Hall J L and Williams L E 2003 Transition metal transporters in plants *J. Exp. Bot.* **54** 2601–613

[12] Gadsby D C 2009 Ion channels versus ion pumps: the principal difference, in principle *Nat. Rev. Mol. Cell Biol.* **10** 344–52

[13] Hedrich R 2012 Ion channels in plants *Physiol. Rev.* **92** 1777–811

[14] Neuhäuser B, Dynowski M and Ludewig U 2014 Switching substrate specificity of AMT/MEP/Rh proteins *Channels (Austin)* **8** 496–502

[15] Wang J, Li Y and Zhu F *et al* 2019 Genome-wide analysis of nitrate transporter (NRT/NPF) family in sugarcane *Saccharum spontaneum* L *Tropical Plant Biol.* **12** 133–49

[16] Ambawat S, Sharma P, Yadav N R and Yadav R C 2013 MYB transcription factor genes as regulators for plant responses: an overview *Physiol Mol. Biol. Plants* **19** 307–21

[17] Shao A, Ma W and Zhao X *et al* 2017 The auxin biosynthetic tryptophan aminotransferase related TaTAR2.1–3A increases grain yield of wheat *Plant Physiol.* **174** 2274–88

[18] Shin R, Berg H and Schachtman D P 2005 Reactive oxygen species and root hairs in arabidopsis root response to nitrogen, phosphorus and potassium deficiency *Plant and Cell Physiol.* **46** 1350–357

[19] Jaishankar M, Tseten T, Anbalagan N, Mathew B B and Beeregowda K N 2014 Toxicity, mechanism and health effects of some heavy metals *Interdiscip. Toxicol.* **7** 60–72

[20] Scavo A, Abbate C and Mauromicale G 2019 Plant allelochemicals: agronomic, nutritional and ecological relevance in the soil system *Plant Soil* **442** 23–48

[21] Panda S K, Baluska F and Matsumoto H 2009 Aluminum stress signaling in plants *Plant Signal. Behav.* **4** 592–97

[22] Giehl R F and von Wirén N 2014 Root nutrient foraging *Plant Physiol.* **166** 509–17

[23] De Benedictis M *et al* 2018 The *Arabidopsis thaliana* knockout mutant for phytochelatin synthase1 (cad1–3) is defective in callose deposition, bacterial pathogen defense and auxin content, but shows an increased stem lignification *Front. Plant Sci.* **9** 19

[24] Piano A, Valbonesi P and Fabbri E 2004 Expression of cytoprotective proteins, heat shock protein 70 and metallothioneins, in tissues of *Ostrea edulis* exposed to heat and heavy metals *Chaperones* **9** 134–42

[25] White P J, Broadley M R and Gregory P J 2012 Managing the nutrition of plants and people *Appl. Environ. Soil Sci.* **2012** 104826

[26] Sasaki T, Tsuchiya Y, Ariyoshi M, Ryan P R and Yamamoto Y 2016 A chimeric protein of aluminum-activated malate transporter generated from wheat and *Arabidopsis* shows enhanced response to trivalent cations *Biochim. Biophys. Acta* **1858** 1427–435

[27] Fontes R L F and Cox F R 1998 Iron deficiency and zinc toxicity in soybean grown in nutrient solution with different levels of sulphur *J. Plant Nutr.* **21** 1715–22

[28] Borges K L R, Salvato F, Alcântara B K, Nalin R S, Piotto F Â and Azevedo R A 2018 Temporal dynamic responses of roots in contrasting tomato genotypes to cadmium tolerance *Ecotoxicology* **27** 245–58

[29] Li B, Quan-Wang C, Liu H, Li H X, Yang J, Song W P, Chen L and Zeng M 2014 Effect of Cd^{2+} ions on root anatomical structure of four rice genotypes *J. Environ. Biol.* **35** 751–57

[30] Daud M K *et al* 2009 Cadmium-induced functional and ultrastructural alterations in roots of two transgenic cotton cultivars *J. Hazard. Mater.* **161** 463–73

[31] Ameen N, Amjad M and Murtaza B *et al* 2019 Biogeochemical behavior of nickel under different abiotic stresses: toxicity and detoxification mechanisms in plants *Environ Sci. Pollut. Res.* **26** 10496–0514

[32] Tong M, Liu W, He H, Hu H, Ding Y and Li X *et al* 2020 Identification and functional analysis of the CorA/MGT/MRS2-type magnesium transporter in banana *PLoS One* **15** e0239058

[33] Takagi D, Miyagi A, Tazoe Y, Suganami M, Kawai-Yamada M, Ueda A, Suzuki Y, Noguchi K, Hirotsu N and Makino A 2020 Phosphorus toxicity disrupts Rubisco activation and reactive oxygen species defence systems by phytic acid accumulation in leaves *Plant Cell Environ.* **43** 2033–53

[34] Wang Y Z, Zhang H P, Huang X S, Wang J Z, Cheng R and Chen G D *et al* 2017 Effect of potassium supply on plant potassium distribution and growth and leaf photosynthetic capacity of *Pyrus pyrifolia J. Nanjing Agric Univ* **40** 60–7

[35] Delhaize E, Kataoka T, Hebb D M, White R G and Ryan P R 2003 Genes encoding proteins of the cation diffusion facilitator family that confer manganese tolerance *Plant Cell* **15** 1131–142

[36] Besnard J, Pratelli R, Zhao C, Sonawala U, Collakova E, Pilot G and Okumoto S 2016 UMAMIT14 is an amino acid exporter involved in phloem unloading in *Arabidopsis* roots *J. Exp. Bot.* **67** 6385–397

[37] Lanquar V, Loqué D, Hörmann F, Yuan L, Bohner A, Engelsberger W R, Lalonde S, Schulze W X, von Wirén N and Frommer W B 2009 Feedback inhibition of ammonium uptake by a phospho-dependent allosteric mechanism in *Arabidopsis Plant Cell* **21** 3610–622

[38] Kim E J, Kwak J M, Uozumi N and Schroeder J I 1998 AtKUP1: an *Arabidopsis* gene encoding high-affinity potassium transport activity *Plant Cell* **10** 51–62

[39] Osakabe Y, Arinaga N, Umezawa T, Katsura S, Nagamachi K, Tanaka H, Ohiraki H, Yamada K, Seo S U and Abo M *et al* 2013 Osmotic stress responses and plant growth controlled by potassium transporters in *Arabidopsis Plant Cell* **25** 609–24

[40] Paszkowski U, Kroken S, Roux C and Briggs S P 2002 Rice phosphate transporters include an evolutionarily divergent gene specifically activated in arbuscular mycorrhizal symbiosis *Proc. Natl. Acad. Sci. USA* **99** 13324–3329

[41] Hürlimann H C, Stadler-Waibel M, Werner T P and Freimoser F M 2007 Pho91 is a vacuolar phosphate transporter that regulates phosphate and polyphosphate metabolism in *Saccharomyces cerevisiae Mol. Biol. Cell* **18** 4438–445

[42] Hawkesford M J 2000 Plant responses to sulphur deficiency and the genetic manipulation of sulphate transporters to improve S-utilization efficiency *J. Exp. Bot.* **51** 131–38

[43] Lin Y F, Hassan Z, Talukdar S, Schat H and Aarts M G 2016 Expression of the ZNT1 zinc transporter from the metal hyperaccumulator *Noccaea caerulescens* confers enhanced zinc and cadmium tolerance and accumulation to *Arabidopsis thaliana PLoS One* **11** e0149750

[44] Delhaize E, Gruber B D and Ryan P R 2007 The roles of organic anion permeases in aluminium resistance and mineral nutrition *FEBS Lett.* **581** 2255–262

[45] Thomine S, Wang R, Ward J M, Crawford N M and Schroeder J I 2000 Cadmium and iron transport by members of a plant metal transporter family in *Arabidopsis* with homology to Nramp genes *Proc. Natl. Acad. Sci. USA* **97** 4991–96

[46] Liu X S, Feng S J and Zhang B Q *et al* 2019 OsZIP1 functions as a metal efflux transporter limiting excess zinc, copper and cadmium accumulation in rice *BMC Plant Biol.* **19** 283

[47] Asai T, Tena G, Plotnikova J, Willmann M R, Chiu W L and Gomez-Gomez L *et al* 2002 MAP kinase signalling cascade in *Arabidopsis* innate immunity *Nature* **415** 977–83

[48] Gondor O K, Pál M, Darko E, Janda T and Szalai G 2016 Salicylic acid and sodium salicylate alleviate cadmium toxicity to different extents in maize (*Zea mays* L.) *PLoS One* **11** e0160157

[49] Maksymiec W and Baszyński T 1999 Are calcium and calcium channels involved in the mechanisms of Cu^{2+} toxicity in bean plants? The influence of leaf age *Photosynthetica* **36** 267–78

[50] Li B, Li Q, Xiong L, Kronzucker H J, Krämer U and Shi W 2012 *Arabidopsis* plastid AMOS1/EGY1 integrates abscisic acid signaling to regulate global gene expression response to ammonium stress *Plant Physiol.* **160** 2040–051

[51] Hirayama T and Umezawa T 2010 The PP2C-SnRK2 complex: the central regulator of an abscisic acid signalling pathway *Plant Signal Behav.* **5** 160–63

[52] Tan Y W, Hanson J A and Yang H 2009 Direct Mg^{2+} binding activates adenylate kinase from *Escherichia coli J. Biol. Chem.* **284** 3306–313

[53] Wang Y, Chen Y F and Wu W H 2021 Potassium and phosphorus transport and signalling in plants *J. Integr. Plant Biol.* **63** 34–52

[54] Yuan H and Huang X 2016 Inhibition of root meristem growth by cadmium involves nitric oxide-mediated repression of auxin accumulation and signalling in *Arabidopsis Plant Cell Environ.* **39** 120–35

[55] Bücker-Neto L, Paiva A, Machado R D, Arenhart R A and Margis-Pinheiro M 2017 Interactions between plant hormones and heavy metals responses *Genetics Mol. Biol.* **40** 373–86

[56] Hac-Wydro K, Sroka A and Jablo K 2016 The impact of auxins used in assisted phytoextraction of metals from the contaminated environment on thea alterations caused by lead (II) ions in the organization of model lipid membranes *Colloids Surf. B: Biointerfaces* **143** 124–30

[57] Achard P, Renou J P, Berthomé R, Harberd N P and Genschik P 2008 Plant DELLAs restrain growth and promote survival of adversity by reducing the levels of reactive oxygen species *Curr. Biol.* **18** 656–60

[58] Alonso-Ramírez A, Rodríguez D, Reyes D, Jiménez J A, Nicolás G, López-Climent M, Gómez-Cadenas A and Nicolás C 2009 Cross-talk between gibberellins and salicylic acid in early stress responses in *Arabidopsis thaliana* seeds *Plant Signal Behav.* **4** 50–751

[59] Jalmi S K, Bhagat P K, Verma D, Noryang S, Tayyeba S, Singh K, Sharma D and Sinha A K 2018 Traversing the links between heavy metal stress and plant signalling *Front. Plant Sci.* **9** 12

[60] Jiang Y P, Cheng F, Zhou Y H, Xia X J, Mao W H and Shi K *et al* 2012 Cellular glutathione redox homeostasis plays an important role in the brassinosteroid-induced increase in CO_2 assimilation in *Cucumis sativus New Phytol.* **194** 932–43

[61] Kiraly K A, Pilinszky K, Bittsánszky A, Gyulai G K and Komives T 2013 Importance of ammonia detoxification by plants in phytoremediation and aquaponics *Adria Scientific Workshop* **62** 99–102

[62] Hasanuzzaman M, Nahar K, Anee T I and Fujita M 2017 Glutathione in plants: biosynthesis and physiological role in environmental stress tolerance *Physiol. Mol. Biol. Plants* **23** 249–68

[63] Huang Y, Chen J, Zhang D, Fang B, Jin T Y, Zou J, Chen Y, Su N and Cui J 2021 Enhanced vacuole compartmentalization of cadmium in root cells contributes to glutathione-induced reduction of cadmium translocation from roots to shoots in pakchoi (*Brassica chinensis* L.) *Ecotoxicol. Environ. Saf.* **208** 111616

[64] Bielen A, Remans T, Vangronsveld J and Cu ypers A 2013 The influence of metal stress on the availability and redox state of ascorbate, and possible interference with its cellular functions *Int. J. Mol. Sci.* **14** 6382–413

[65] Huang L, Lu H, Gao X, Du G, Ma X, Liu M, Guo J and Chen Y 2013 Ammonium-induced oxidative stress on plant growth and antioxidative response of duckweed (*Lemna minor* L.) *Ecol. Eng.* **58** 355–62

[66] Sharma A, Shahzad B, Rehman A, Bhardwaj R, Landi M and Zheng B 2019 Response of phenylpropanoid pathway and the role of polyphenols in plants under abiotic stress *Molecules* **24** 2452

[67] Munene R, Changamu E, Korir N and Joseph G 2017 Effects of different nitrogen forms on growth, phenolics, flavonoids and antioxidant activity in amaranth species *Tropical Plant Res.* **4** 81–9

[68] Muzolf-Panek M, Kleiber T and Kaczmarek A 2017 Effect of increasing manganese concentration in nutrient solution on the antioxidant activity, vitamin C, lycopene and polyphenol contents of tomato fruit *Food Addit. Contam.: Part* A **34** 379–89

[69] Handa N, Kohli S K, Sharma A, Thukral A K, Bhardwaj R, Abd_Allah E F, Alqarawi A A and Ahmad P 2019 Selenium modulates dynamics of antioxidative defence expression, photosynthetic attributes and secondary metabolites to mitigate chromium toxicity in *Brassica juncea* L. *Plants Environ. Exp. Bot.* **161** 180–92

[70] Shaked-Sachray L, Weiss D, Reuveni M, Nissim-Levi A and Oren-Shamir M 2002 Increased anthocyanin accumulation in aster flowers at elevated temperatures due to magnesium treatment *Physiol. Plant.* **114** 559–65

[71] Dai L, Dong X and Ma H 2012 Antioxidative and chelating properties of anthocyanins in *Azolla imbricata* induced by cadmium *Pol. J. Environ. Stud.* **21** 837–44

[72] Groppa M D, Tomaro M L and Benavides M P 2007 Polyamines and heavy metal stress: the antioxidant behavior of spermine in cadmium- and copper-treated wheat leaves *Biometals* **20** 185–95

[73] Wen X P, Ban Y, Inoue H, Matsuda N and Moriguchi T 2010 Spermidine levels are implicated in heavy metal tolerance in a spermidine synthase overexpressing transgenic European pear by exerting antioxidant activities *Transgenic Res.* **19** 91–103

[74] Cvikrová M, Gemperlová L, Martincová O and Vanková R 2013 Effect of drought and combined drought and heat stress on polyamine metabolism in proline-over-producing tobacco plants *Plant Physiol. Biochem.* **73** 7–15

[75] Hoffland E, Boogaardn R, Nelemans J and Findenegg G 2006 Biosynthesis and root exudation of citric and malic acids in phosphate-starved rape plants *New Phytol.* **122** 675–80

[76] Tripti S, Ingo D, Leon K and Miguel A P 2016 The ALMT family of organic acid transporters in plants and their involvement in detoxification and nutrient security *Front. Plant Sci.* **7** 1488

[77] Schmöger M E, Oven M and Grill E 2000 Detoxification of arsenic by phytochelatins in plants *Plant Physiol.* **122** 793–801

[78] Vatamaniuk O K, Mari S, Lu Y-P and Rea P A 1999 AtPCS1, a phytochelatin synthase from *Arabidopsis*: isolation and *in vitro* reconstitution *Proc. Natl. Acad. Sci. USA* **96** 7110–115

[79] Kim B M, Rhee J S, Jeong C B, Seo J S, Park G S, Lee Y M and Lee J S 2014 Heavy metals induce oxidative stress and trigger oxidative stress-mediated heat shock protein (hsp) modulation in the intertidal copepod *Tigriopus japonicus Comp. Biochem. Physiol. Part C: Toxicol. Pharmacol.* **166** 65–74

IOP Publishing

Molecular Response and Genetic Engineering for Stress in Plants, Volume 1
Abiotic stress
Pawan Shukla, Anirudh Kumar, Rakesh Kumar and Manish K Pandey

Chapter 4

Mechanistic insight into understanding drought stress response in plants

Vinita Sharma[1], Vinay Sharma[2], Abhishek Bhandawat[1], Ankita Mishra[1], Chibuikem Unamba[3], Joy Roy[1], Vikas Sharma[4], Himanshu Sharma[1*]

[1]*Agri-Biotechnology Division, National Agri-Food Biotechnology Institute, Knowledge City Sector-81, Mohali, Punjab 140306, India*
[2]*International Crops Research Institute for the Semi-Arid Tropics (ICRISAT), Hyderabad 502324, India*
[3]*Department of Plant Science and Biotechnology, Imo State University, Owerri, Nigeria*
[4]*Department of Botany, Sant Baba Bhag Singh University, Khiala, Jalandhar, Punjab 144030, India*

The ever-increasing global food demand has driven scientific efforts toward enhancement of agricultural production. Drought stress affects plant physiological and biochemical processes, and is a serious threat to even existing production rates. In the context of global climate change, development of new crop varieties that can grow under conditions of low tissue water levels is a requisite to ensure food security. Plants adopt different mechanisms to withstand drought stress: understanding these mechanisms will facilitate a way to drought-tolerant crops. Different signaling pathways and their cross-talk regulate drought-responsive genes, resulting in different plant responses to stress. Molecular and genetic analyses have identified many drought-inducible genes in *Arabidopsis*, wheat, rice, and other crops. In this chapter, we summarize different morphological, physiological, and molecular mechanisms involved in drought tolerance in plants. The functions of drought-inducible genes, including functional and regulatory genes, in different plants are discussed. Such functional analysis of drought-inducible genes will strengthen understanding of the drought tolerance strategies adopted by plants, will facilitate genetic manipulation for drought-tolerant crops. This review highlights the opportunities for development of drought-tolerant crops by advanced genetic engineering methods, application of

* Author to whom any correspondence should be addressed Email: himanshusharm@gmail.com.

doi:10.1088/978-0-7503-4921-5ch4

growth regulators, molecular breeding, as well as covering genome-editing tools such as CRISPR/Cas.

4.1 Introduction

The rapidly growing global population has put a lot pressure on the agricultural sector to ensure food for all [1]. The extensive needs for agricultural water, as well as raises in temperature, can hamper water availability [2]. Changes in environmental conditions have adversely affected plant growth and productivity by means of the reduced radiation use efficiency of leaves, leading to decreased harvest index [3].

Often, the term 'stress' is defined physiologically, as a consequence of various conditions. Stress is a term referring to a change in physiological conditions triggered by various events that primarily affect a plant's stability. The constant shift of environmental conditions triggered by periodic daily and seasonal cycles enables plants to adjust to continuous metabolic changes. Drought is a climatic phenomenon defined by insufficient or no rainfall. Drought has an adverse effect on the productivity of most staple crops, such as rice [4] maize [5], and wheat [6, 7]. Therefore, new plant varieties with enhanced productivity in low-water conditions are the need of the hour to ensure food security in the aforementioned context of climate change. The major challenge for the development of drought-tolerant crops is our current understanding of the mechanisms conferring resistance or tolerance to drought stress.

Being sessile in nature, plants have to withstand multiple biotic and abiotic stressors in their environment, which has eventually led to their evolution of several mechanisms to cope with and adapt to these environmental factors [8]. Plants have adapted different strategies to endure under conditions of drought stress, including drought escape, avoidance, and tolerance [9]. Drought escape is an adaptive mechanism enabling a plant to complete its life cycle prior to any coming drought conditions [10]. Changes in leaf size and area, increased stomatal resistance, and changes in leaf anatomy are some of the drought avoidance mechanisms adapted by plants. Whereas, drought tolerance refers to the ability of a plant to maintain its proper functioning under low tissue water potential by increasing its osmotic adjustment capacity [11, 12] or through the presence of rigid cell walls or decreased cell size [13]. These capacities of plants can be measured in terms of physiological traits, including proline content, abscisic acid (ABA) content, sugar content, antioxidant activities, and chlorophyll content of leaves [11]. Therefore, plant physiological traits with a strong correlation with drought are essential to understand the mechanism of dehydration tolerance in water-deficit conditions. Such an understanding of the tolerance mechanism and its associated physiological pathways will facilitate the selection of drought-tolerant crop varieties in future.

Traditional breeding and genetic engineering practices have been employed to develop drought-tolerant crops [14–17]. Many genes related to signaling pathways which control drought tolerance have been identified in different plant species [18–22]. Drought stress, its effect on plants, and plants' responses to drought stress have been discussed in many good reviews focused on either signaling pathways or their related mechanisms [23–25]. In this chapter, we discuss the drought tolerance mechanisms in

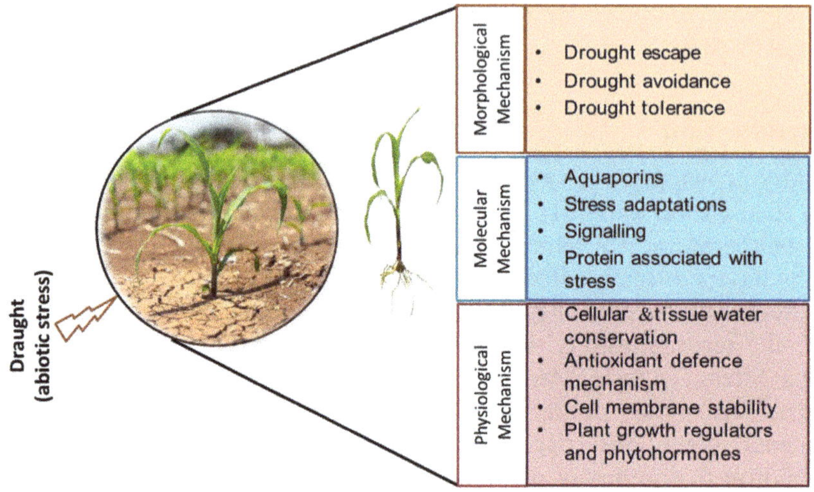

Figure 4.1. A representation of the different mechanisms involved in drought tolerance in plants.

plants, based on previous studies on the morphological, physiological, and biochemical responses to drought and their molecular regulations.

4.2 Mechanisms involved in drought tolerance

Conditions of drought usually occur if there is insufficient water in the soil and constant water loss via evaporation and transpiration. Drought stress is detectable via morphological, physiological, and biochemical responses that include limited water loss by leaf area reduction in stomatal conductance, reduction in cell elongation and growth, stomatal closure, impaired gaseous exchange, and alteration of enzyme-catalyzed reactions. Under extreme drought conditions, photosynthesis and metabolism are severely disrupted, ultimately resulting in the plant's death.

To acclimatize to drought situations, plants activate various response mechanisms that allow them to withstand and probably reduce the adverse impact of limited water, which is discussed below in detail and summarized in figure 4.1.

4.2.1 Morphological mechanisms

Tolerance to drought in plants includes responses at the cellular, molecular, physiological, and genetic stages. Plants usually exhibit certain evolved morphological changes via ontogenetic development, which helps them to adapt in such water-scarce conditions. In this section, we discuss the morphological strategies that plants use to adapt in response to drought conditions.

4.2.1.1 Drought escape

Drought escape involves shortening of the plant life cycle or growth season, which allows them to reproduce before the surrounding conditions get dry [26]. The duration of flowering is critical for drought adaptation, as a short life cycle may result in drought escape [27]. The growth cycle of a plant is genotype and environment dependent. It is important to balance the plant's growth phase with the available soil moisture in order to achieve a higher yield. Thus, early maturity and early blooming benefit plants in avoiding drought stress. Breeding short-duration cultivars has proven to be a useful approach for mitigating loss of yield due to terminal drought, since early maturity permits the crop to skip the upcoming stress phase. Furthermore, yield has often been associated with the duration of the crop under ideal growth conditions, and any shortening of the crop duration below the optimal would result in crop yield reduction [28].

4.2.1.2 Drought avoidance

The primary objective of this approach is to maintain higher water potential in plants. Drought avoidance mechanisms include minimizing water loss through stomatal regulation of transpiration and prolonging absorption of water via a broad and flourishing root system [28]. Root traits such as the biomass of the root, root length, root density, and root depth are important drought avoidance characteristics that are linked with the total yield in terminal drought situations [29]. A wide and dense root system is advantageous for absorption of water from deep levels. Similarly, cuticles and hairy leaves help the plant to maintain higher tissue water potential and are regarded as xeromorphic characteristics for drought avoidance [28].

4.2.1.3 Drought tolerance

Plants which employ a drought tolerance strategy reduce the number and size of their leaves during water scarcity. Nevertheless, this approach results in yield reduction [27]. Additionally, these plants exhibit xeromorphic characteristics such as hairy leaves and trichome development on both the adaxial and abaxial of their leaves. Hairiness reduces the temperature of leaves, and lowers water transpiration loss [30]. Roots are the primary tissue involved in soil water absorption. Thus, root development rate, density, proliferation, and size all play a significant role in determining how plants respond to drought stress. According to previous studies, the primary element affecting plant tolerance is a change in the root system architecture, particularly when tolerance is described as the capacity of a plant to retain its leaf area as well as growth rate over an extended vegetative phase [28]. Other factors enabling drought tolerance in plants include the accumulation of compatible solutes, osmotic adaptation, antioxidant system activation and modifications in metabolic pathways, increase in root/shoot ratio, and the closing of stomata.

4.2.2 Physiological mechanisms

The physiological mechanisms for genetic variability in response to drought are not clearly understood; this seems to be the most complex system. Some of the important components are discussed in detail below.

4.2.2.1 Cellular and tissue water conservation

Water-stressed plants in soil must conserve available water by minimizing transpiration while also fixing sufficient carbon to fulfill their energy requirements. Reduced leaf area and stomatal conductance act quickly to counteract dehydration by reducing transpiration, hence budgeting soil moisture and retaining higher leaf water potential rates. Similarly, solute accumulation in cells under water-scarce conditions for maintaining cell turgor and physiological processes is an important plant adaptation response [31]. Additionally, osmotic adjustment aids in the translocation of pre-anthesis carbohydrate allocation at the time of grain filling, resulting in increased photosynthesis and productivity [32].

4.2.2.2 Antioxidant defense mechanisms

Antioxidant defense responses in plants are composed of both enzymatic and nonenzymatic components. Ascorbate peroxidase, catalase, glutathione reductase, peroxidase, superoxide dismutase, etc, are examples of enzymatic components. Ascorbic acid cysteine and reduced glutathione are some such nonenzymatic components [33]. High antioxidant enzyme activity and a larger proportion of nonenzymatic constituents are important for abiotic stress tolerance, such as drought. Carotenoids, flavonoids, and some other molecules, such as abietane diterpenes, have the ability to reduce lipid peroxy-radicals and singlet oxygen, which limits peroxidation of lipids and production of superoxides in the presence of dehydration stress [34]. Along with its role as an accessory pigment, beta-carotene is a great source of antioxidants and performs a role in maintaining photochemical reactions [35]. Beta-carotene plays a vital role in protecting photosynthetic tissue by directly quenching triplet chlorophyll and limiting the formation of singlet oxygen, thus preventing oxidative stress.

4.2.2.3 Cell membrane stability

Most abiotic stressors first affect the biological membrane. The stability of the cell membrane is proportional to damage to the cell membrane, which is a physiological criterion that is frequently used to evaluate drought tolerance [36]. A previous study reported that the most relevant attribute for screening germplasm for drought resistance was leaf-segment membrane stability [37]. Changes in the cell membrane integrity, regulation of stomata, and root development capacity were shown to be negatively associated with osmotic adjustment [38]. Although the mechanisms by which the membrane is disrupted are not clearly understood, a reduction in cellular volume results in congestion and enhances cytoplasmic constituent viscosity. This enhances the likelihood of molecular interactions that result in denaturation of proteins and fusion of membranes. A wide range of compounds, such as proline, glutamate, glycine betaine, carnitine, etc, responsible for preventing such unfavorable molecular interactions, have been reported for model membranes [28]. Another cause of ion leakage from the cell is temperature stress, which inhibits the function of the membrane-bound enzymes that are involved in maintaining the chemical gradient inside the cell [39]. Another study reported that the leaf membranes of *Arabidopsis* were resistant to water-deficit conditions by maintaining polar lipid levels and compositional stability throughout severe drought [40].

4.2.2.4 Plant growth regulators

Externally given plant growth regulators and internally produced phytohormones are compounds that alter the physiological processes of plants at extremely low levels. Both terms have been used alternately, most commonly in the context of auxin, gibberellin, cytokinin, ethylene, and ABA. Auxins, gibberellins, and cytokinin endogenous concentrations are usually reduced during drought, but ABA and ethylene endogenous concentrations are known to rise [41]. Nonetheless, phytohormones contribute significantly to plants' drought tolerance. Auxins facilitate new root formation by countering the cytokinin-induced root apical dominance. A robust root system is required in response to drought conditions; auxins play a minor but vital role in this process. Drought rhizogenesis is another adaptation response to drought conditions and results in the formation of short, swollen, hairless roots in the *Brassicaceae* and related families. These roots are able to withstand prolonged dehydration, maintain turgor pressure, and, upon rehydration, regenerate a new functioning root system. In a previous study [42], drought rhizogenesis was found to be significantly enhanced in the gibberellic acid biosynthetic mutant *ga5*, showing that gibberellic acids may also participate in this process. Phytohormone ABA plays a central role in response to drought stress. ABA is found in all plants and is universally known as a stress-inducible hormone that controls expression of genes and functions as a marker for the onset of dehydration and other environmental stress adaption mechanisms. It has been hypothesized that ABA and cytokinin have antagonistic functions during drought conditions. Under conditions of water scarcity, an increase in ABA and a decrease in cytokinins promote stomatal closure and minimize transpirational water loss [43]. ABA levels often increase when plants wilt as a consequence of increasing biosynthesis [44]. Accumulation of ABA triggers several changes in developmental and physiological processes [45]. Although ethylene is involved in both environmental growth inhibition and stimulation, it has generally been thought to be a growth inhibitory hormone [46]. Under drought conditions, cereal crops show disrupted leaf functioning and a premature initiation of senescence in older leaves. Ethylene also acts as a regulator of leaf functioning across its life span, determining the beginning of natural senescence and mediating drought-induced senescence [47]. Previous findings reported that ethylene responses frequently involve growth promotion. The role of polyamines in plant growth and development is very well documented. These are cationic in nature, and show interaction with anionic membrane components such as phospholipids, which results in shielding the lipid bilayer from the damaging impact of stress [48]. The function of polyamines in plants' defensive responses to environmental stresses is now receiving more attention, and extensive study has been undertaken over the previous two decades [49]. Similarly, Yang *et al* reported that polyamine synthesis in leaves of rice plants was enhanced during drought stress, particularly the free forms spermine and spermidine, and putrescence as insoluble conjugated forms [50].

4.2.2.5 Osmotic adjustment and accumulation of compatible solutes

The overproduction of various types of compatible solutes is a widely used strategy for stress tolerance in plants [51]. It helps plants adapt to dehydration stress by maintaining turgor and regulating cellular functions. In general, they protect plants in various ways, including by contributing to osmotic adjustments, scavenging reactive oxygen species, and conserving the structures of enzymes and proteins [28]. Osmotic adjustment is a process by which normal water relations are maintained in the presence of osmotic stress. It includes the formation of soluble sugars, proline, organic acids, calcium ions, etc. The cell osmotic potential is lowered during water-stress conditions and as a consequence of solute accumulation, which attracts water into the cell and supports the maintenance of turgor. By maintaining cellular functions, osmotic adjustment enables plants to perform more efficiently in terms of photosynthesis and partitioning of assimilate and plant growth [32]. Indeed, osmotic adjustment has been identified as a key component for reducing the rate of water stress in water scarcity [43].

4.2.3 Molecular mechanisms

Water deficiency in the plant cell can emerge in situations of low soil water content. Gene expression alteration occurs under such situations, which hinder cellular activities. Earlier studies have reported the transcriptional activation of several genes during drought-stress conditions, and the resulting gene products are believed to have a role in response to drought conditions [52]. However, it is commonly known that drought tolerance is a multifaceted phenomenon involving the coordinated action of several genes. Several components are engaged in the molecular processes underlying drought adaptation responses. A few of them are given here.

4.2.3.1 Aquaporins

Aquaporins (AQPs) are intrinsic proteins that form pores in cell membranes to transport water and other uncharged solutes. AQPs are categorized into eight subfamilies in plants, with a few of their members having functional descriptions. The discovery of AQPs in plants led to a paradigm shift in our knowledge of how plants control their water connections [53]. Throughout the plant's life cycle, water passes through its cellular membranes. Water movement across cell membranes is required for various processes such as osmoregulation, expansion of cells, transport of solutes, and generation of turgor potential in leaves [54]. AQPs are thought to be capable of controlling membrane hydraulic conductivity and increasing water permeability by a factor of 10 to 20 [53]. AQP-mediated enhancement in hydrostatic conductivity has been studied extensively in *Oryza sativa* under drought conditions [55]. Several studies on AQPs and plant–water relationships have been undertaken. The cell type-specific interspaced short palindromic repeats-based tissue-specific knockout system (CRISPR-TSKO) technology is helpful for studying AQP gene expression [56]. Recently, the role of many AQP genes in rice has been revealed during drought stress. Another study in transgenic lowland rice reported water channel 3 (RWC3/OsPIP1; 3) protein with a stress-inducible promoter for drought tolerance [57]. Similarly, enhanced hydraulic conductivity and improved survival rates during drought conditions has been seen in transgenic *Arabidopsis* overexpressing the *OsPIP1* and *OsPIP2* genes [58]. According to a recent study [59], *OsPIP2* increases the hydraulic conductivity of the root, stomatal conductance, root character, and concentrations of intercellular CO_2, whereas OsPIP2 RNAi lines decrease the hydraulic conductivity of their root. As a result, AQPs' roles in transporting water molecules throughout the cell membrane and regulating water fluxes in plants during drought-stress conditions have established.

4.2.3.2 Proteins associated with stress

Stress protein biosynthesis is a ubiquitous reaction to persisting stressful situations, like water scarcity. Most stress proteins are soluble in water and help with stress tolerance by hydrating cellular components [60]. Numerous transcription factors (TFs) and stress proteins are synthesized in response to drought tolerance [46]. Dehydration responsive element binding proteins (DREBs) are key TFs that govern tolerance to drought by binding to DRE/CRT promoters and regulating drought-responsive gene expression [61]. DREBs induce the expression of both common and unique regulatory genes encoding late embryogenesis abundant (LEA) proteins, dehydrins, and heat shock proteins (Hsps), etc. LEA proteins and dehydrins play a key role in mitigating dehydration situations through their ability to bind water, sequester ions, protect the structures of proteins, and maintain membrane structure [62, 63]. In addition to their synthesis during the seed's desiccating stage, these proteins accumulate in vegetative tissues during response to a water constraint [64]. There is information which suggests a unique relationship between abiotic stress tolerance and expression of dehydrin genes in plants [65]. As a result, dehydrins have been identified as an important molecular marker for the response of plants to abiotic stress.

Heat shock proteins act as molecular chaperones in sustaining cellular stability in cells in both ideal and adverse growth situations. Hsps facilitate the folding of proteins during cellular metabolism, membrane protein stabilization, and help in protein refolding in plant-stress conditions [66]. Moreover, the role of Hsps/chaperones as sensing and regulatory elements in signaling pathways and transcriptional cascades is not clear, and research into the functional aspects of Hsps/chaperones in response to stress conditions continues. Several recent studies on plants' abiotic stress responses have focused on a family of stress-associated proteins (SAPs) [67, 68]. Around 14 genes encoding SAP proteins have been reported in *A. thaliana* and 18 genes in the *O. sativa* [69]. In rice, the majority of SAP genes are triggered by abiotic stresses such as high temperature and dehydration [68]. SAPs may become a potential focus for stress tolerance improvement in the future, coupled with specific stress-responsive cis-regulatory elements [70]. Additionally, it is necessary to undertake a thorough study to gain knowledge of their function in complex signaling/regulatory cascades in order to fully exploit their possibilities for yield stability under plant-stress conditions.

4.2.3.3 Stress adaptations and signaling

The intricacy of the signaling processes related to stress sensing and triggering of defense and adaptation pathways is thought to be linked various components, including reactive oxygen species (ROS), calcium-dependent components, mitogen-activated protein kinase cascades (MAPKs), and cross-talk between various transcriptional regulators [71]. Wrzaczek and Hirt defined MAPKs as 'essential signaling regulators that control the interaction between environmental stimuli perception and cellular elements' [72]. The Ca^{2+} ion has long been used as a ubiquitous intracellular second messenger in plants. Its functional aspect has been extensively studied in plant cells exposed to dehydration [73] and high-temperature stress [74]. When plants need to communicate external (environmental) and internal (developmental) signals to associated cellular and genetic functions, they use Ca^{2+} ions as auxiliary facilitators [75].

Munir *et al* identified a putative calmodulin-like (*ShCML44*) gene from a wild tomato (*Solanum habrochaites*) that improved drought tolerance in mutant lines [76]. Similarly, in transgenic *Arabidopsis*, a novel *OsMSR2* gene (CML multi-stress-responsive gene 2), from rice confers tolerance to drought in plants via ABA-dependent signaling. Given the above, consideration should be given to investigating the CaM-mediated activation of target proteins in order to gain a better understanding of their function in drought-stress tolerance. Likewise, various chemical signals transmitted during drought stress induce the expression of several genes, resulting in the production of stress-responsive proteins and metabolites that confer stress adaptation to plant species.

4.3 Functions of drought-inducible genes/drought-responsive genes

Drought response is a quantitative trait which involves alterations of the expression of multiple genes. Transcriptome and microarray technologies have assisted significantly in identification of drought-inducible genes. Broadly, there are two basic pathways for drought response: one is ABA dependent and the other is ABA independent [61]. It has been observed that a substantial number of drought-inducible genes are expressed during ABA or salinity stress and, to a lesser extent, during cold, suggesting cross-talk among various abiotic stress pathways [77]. The roles of various drought-responsive genes, including those involved in osmoprotectant biosynthesis, TFs, Hsps, LEAs, and others, is summarized in table 4.1.

4.4 Towards an improved drought tolerance in plants

Water scarcity poses a major threat to global agriculture production. Different strategies are applied to grow crops under conditions of drought stress. Traditional breeding methods have been successfully employed for drought tolerance in different crops. Functional characterization of drought-responsive genes is critical to genetic engineering of drought-tolerant varieties. In-depth understanding of these signaling pathways and cross-talks are a suitable target for the application of advanced genome-editing technologies such as clustered regularly interspaced ahort palindromic repeats (CRISPR)/Cas, zinc finger nucleases (ZFNs), etc. In addition to genetic improvements, use of specific osmoprotectants, plant growth regulators (hormones), and essential nutrients can help to sustain plants under drought stress.

4.4.1 Genome editing for drought-tolerant crops

Advancements in sequencing technologies and the availability of the genome sequences of most of the important crops have opened the way to crop improvement by means of genome editing. Different targeted nucleases have the potential to enable precise and predictable genome editing to accelerate crop improvement [101]. Genome-editing approaches like ZFNs and transcription-activator-like effector nucleases (TALENs) have made it possible for researchers to analyze gene function. However, the involvement of protein engineering steps makes these approaches time consuming and expensive. The advent of CRISPR/Cas technology less than a decade ago, a CRISPR system derived from the bacterium *Streptococcus pyogenes*,

Table 4.1. Summary of drought-responsive genes.

S. No.	Gene	Role in drought and other abiotic stresses	Plant	References
1.	*Arginine decarboxylase* (*ADC2*)	Overexpression of ADC2, a positive regulator of putrescine (a polyamine), enhances drought tolerance.	Rice	[18]
2.	*Spermidine synthase*	Overexpression of spermidine synthase enhances spermidine (a polyamine), enhances tolerance to various stresses including drought.	*Arabidopsis*	[49]
3.	*Galactinol synthase* (*AtGolS2*)	AtGolS2 overexpression resulted in higher accumulation of osmoprotectants like galactinol, and raffinose enhances drought tolerance.	*Arabidopsis*	[78]
4.	*SacB*	Transgenic sugar beet expressing fructan biosynthetic gene *SacB* from *Bacillus subtilis* showed better drought tolerance, and better growth and biomass than wild type.	Sugar beet	[79]
5.	*Mannitol-1-phosphate dehydrogenase* (*mtlD*)	Introduction of bacterial *mtlD* gene enhances drought and salinity tolerance possibly by scavenging hydroxyl radical or stabilizing membrane.	Wheat	[19]
6.	*Myo-inositol O-methyltransferase* (*IMT1*)	Overexpression of *IMT1* increases in D-ononitol accumulation and drought tolerance.	Tobacco	[80]
7.	*D1-pyrroline-5-carboxylate synthase* (*P5CS*)	Antisense suppression of *P5CS* leads to reduced proline accumulation and reduced drought tolerance. Proline is a critical osmolyte that has a multifunctional role in stress management.	*Arabidopsis*	[81]
8.	*9-cis-epoxycarotenoid dioxygenase* (*NCED*)	Involved in ABA biosynthesis; strongly induced by drought stress and confers drought tolerance.	*Arabidopsis*	[82]
9.	*ApGSMT, ApDMT*	Ectopic expression of glycine sarcosine methyltransferase (ApGSMT) and dimethylglycine methyltransferase (ApDMT) synthesizing glycine betain show increased drought, salinity, and chilling tolerance in transgenic plants.	*Arabidopsis*	[83]
10.	*OsPYL3* and *OsPYL9*	Pyrabactin-resistance-like abscisic acid receptor encodes ABA receptor and is found to play a role in drought and cold.	Rice	[84]
11.	*OsSAPK8*	An SnRK2 family gene induced by cold, high salt, and drought stress. Mutant of *OsSAPK8* results in downregulation of other abiotic stress marker genes such as *OsDREB1*, *OsDREB2*, *OsNCED* and *OsRAB21* and plants show reduced tolerance to drought.	Rice	[20]
12.	*MYB* and *MYC*	Transcriptional activators in ABA-dependent pathway involved in drought response.	*Arabidopsis*	[85]
13.	*AREB/ABF*	Master TFs regulating expression of ABA-responsive genes that are also activated by drought. Mutants of *AREB1*, *AREB2*, and *ABF3* show reduced drought tolerance. Overexpression of ABF3 *AREB2/ABF4* leads to ABA hypersensitivity and improved drought tolerance.	*Arabidopsis* and rice	[21, 86]

14.	*OsbZIP46CA1*	A modified active form of *OsbZIP46* significantly improves drought and osmotic stress tolerance.	Rice	[87]
15.	*OsbZIP72*	*OsbZIP72* overexpression lines showed higher expression of ABA-responsive genes such as LEAs and confers better drought tolerance.	Rice	[88]
16.	*AP37*	Overexpression of a rice TF shows increased drought tolerance in vegetative and reproductive tissues.	Rice	[89]
17.	*OsNAC10*	Root-specific expression of *OsNAC10* improves tolerance to drought and yield under drought conditions.	Rice	[90]
18.	*ANAC19, ANAC055, ANAC072*	Drought stress induces the expression of three NAC domain genes, *ANAC19*, *ANAC055*, and *ANAC072*. Overexpression of these genes was found to enhance drought tolerance in transgenics.	*Arabidopsis*	[91]
19.	*ONAC045*	Encodes a stress-responsive NAC TF. Imparts drought and salt tolerance.	Rice	[92]
20.	*SodERF3*	Overexpression of a sugarcane ethylene responsive factor enhances drought and salt tolerance.	Tobacco	[93]
21.	*ABO3/WRKY*	Mutant analysis of *ABO3*, a WRKY TF family member, showed its role in drought tolerance and ABA stress.	*Arabidopsis*	[94]
22.	*OsDIL*	*Oryza sativa* drought-induced LTP (*OsDIL*) encodes a lipid-transfer protein. *OsDIL* overexpression lines showed higher tolerance to drought during vegetative stages and lesser defects during reproductive phases.	Rice	[22]
23.	*NtHSP70*	Overexpression of a heat shock protein gene *NtHSP70–1* from tobacco confers tolerance to water stress.	Tobacco	[95]
24.	*OsCTR1*	A gene-encoding RING Ub E3 ligase found to play an important role in drought tolerance.	*Arabidopsis*	[96]
25.	*CYP707A3*	Mutation in *CYP707A3*, an ABA hydroxylase, leads to reduced transpiration rate thus playing a role in elevating drought tolerance.	*Arabidopsis*	[97]
26.	*HVA1*	ABA-responsive LEA protein involved in drought response.	Wheat	[98]
27.	*ZAT10*	A zinc finger protein involved in abiotic stress tolerance. Improved drought tolerance.	Rice	[99]
28.	*DREB1A/C-repeat binding factor 3* (*CBF3*)	Overexpression of *CBF3* accumulates more galactinol and raffinose and shows improved tolerance to drought and cold.	*Arabidopsis*	[100]

has seen it become the tool of choice of plant molecular biologists for quick and precise genome editing [102].

4.4.2 CRISPR technology: revolutionizing genome editing for crop improvement

CRISPR technology has been adopted for 20 crop species so far including important cereals, maize, rice, and wheat [103]. The technology has played an important role in many transgene-free crop improvement programs by targeting genes related to specific traits. The basic principle and mechanism of the CRISPR/Cas system has been discussed in several good reviews [104, 105]. The protocols for CRISPR/Cas-based targeting either by *Agrobacterium*-mediated or biolistic transformation have been explained in detail in many articles, such as for the examples of rice [106], maize [107, 108], wheat [109, 110], and tobacco [111, 112]. CRISPR/Cas9 has been utilized to confer drought tolerance in different crops; these studies will be discussed below in different subsections for monocots and dicots.

4.4.2.1 Monocots

Monocot crops such as rice, wheat, maize, and barley are susceptible to drought conditions. Drought adversely affects the yield of the aforementioned crops. In the recent past, CRISPR/Cas-based genome editing has been successfully applied in different monocot crops to develop drought tolerance. Examples of these crops are listed in table 4.2.

4.4.2.1.1 Proof-of-concept studies

Among different abiotic stressors, drought significantly affects rice yield. CRISPR/Cas has been utilized to develop rice varieties with improved drought tolerance. *Drought and salt tolerant protein 1* (*OsDST*) is a negative regulator of drought stress. Recently, *OsDST* was knocked out by CRISPR-mediated editing and its mutant lines conferred a moderate-level tolerance to osmotic stress and higher levels of salt tolerance [113]. CRISPR/Cas9-mediated loss-of-function mutants of *osmotic stress/ABA-activated protein kinase 2* (*SAPK2*) resulted in rice with increased drought tolerance, and tolerance to salt as well [114]. The study suggested *SAPK2* as a good candidate for crop improvement.

The auxin-regulated gene involved in organ size (ARGOS) family genes confer drought tolerance by negatively regulating ethylene signaling. The CRISPR-mediated overexpression of *ARGOS8* resulted in higher drought tolerance in maize. The authors inserted a GOS2 promoter in the 5'UTR of *ARGOS8* to develop drought-tolerant maize [115]. These studies provide insights into the elucidation of CRISPR/Cas as a promising technology for stress tolerance in monocots.

4.4.2.2 Dicots

Dicots species like *Arabidopsis* have been utilized for studying the function of different genes for many years. The small and low complexity genome of *Arabidopsis* enabled researchers to demonstrate CRISPR/Cas technology in plants for the first time [116]. CRISPR/Cas has been used successfully for improvement of different

Table 4.2. CRISPR/Cas-mediated genome editing in different crops for drought tolerance.

S. No.	Plant	Target gene	Knockout/ mutation, etc	Delivery approach	References
			Monocots		
1.	Rice	*Drought and salt tolerance* (*OsDST*)	Loss-of-function mutants	*Agrobacterium*-mediated transformation	[118]
2.	Rice	*OsSAPK2*	Loss-of-function mutants	*Agrobacterium*-mediated transformation	[114]
3.	Rice	*Semi-rolled leaf* (*OsSRL1*, *OsSRL2*)	Loss of function	*Agrobacterium*-mediated transformation	[119]
4.	Maize	*ARGOS8*	Overexpression	Biolistic delivery	[115]
			Dicots		
5.	*Arabidopsis*	*Open stomata 2* (*OST2*)	Introduce novel alleles (base editing)	*Agrobacterium*-mediated transformation	[120]
6.	*Arabidopsis*	*AREB1*	Promoter editing	*Agrobacterium*-mediated transformation	[121]
7.	Tomato	*Mitogen-activated protein kinases* (*SlMAPK3*)	Base editing	*Agrobacterium*-mediated transformation	[122]
8.	Tomato	*Nonexpressor of pathogenesis related 1* (*SlNPR1*)	Mutagenesis	*Agrobacterium*-mediated transformation	[123]
9.	Tomato	*Lateral organ boundaries domain* (*SlLBD40*)	Knockout	*Agrobacterium*-mediated transformation	[108]

dicot crop like cotton, soyabean, tomato, potato, etc. Although few studies have targeted drought-related genes, but the availability of efficient transformation methods, genome sequencing will pave a path to successful genome editing for drought tolerance. Some of the studies are discussed below as a proof of concept; they are listed in table 4.2.

4.4.2.2.1 Proof-of-concept studies

Recently, researchers at China Agricultural University developed drought-tolerant tomato by knockout of lateral organ boundaries (LBD) domain-containing genes (*SlLBD40*) by CRISPR/Cas9. All the LBD domain-containing genes are commonly expressed in roots and fruits and are a function protein of lateral organ development. In the same study, they reported severe wilting in the *SlLBD40* overexpression lines [108]. In another study, RNAi and CRISPR-mediated knockout of the UDP-glycosyltransferases UGT79B2 and UGT79B3 increased the susceptibility of the *Arabidopsis* plant to cold, salt, and drought stresses [117].

Thereby, CRISPR technology provides a transgene-free genome-editing approach that is not considered as a genetically modified crop (GMO) in many countries and can be utilized to confer drought tolerance in different crops by precise targeting of stress-related genes.

4.5 Conclusions

A rapidly growing population has emphasized the need for enhanced agricultural produce to achieve global food security. The rise in global temperature and water scarcity have caused serious threats to plants, which in turn leads to yield loss. To overcome such abiotic stressors, plants have developed multiple different mechanisms. A plant's responses to drought stress can be broadly classified into several different categories, such as physiological, morphological, and molecular responses. Plants adopt several physiological mechanisms, namely drought escape, drought tolerance, and drought resistance. Classical breeding and mutational approaches have been used to identify drought-responsive genes and to breed crops for drought tolerance. At the molecular level, different pathways, AQPs, Hsps, numerous TFs, and many drought-inducible genes have been identified. Recent developments in genome-editing tools, particularly CRISPR-Cas technology, have revolutionized crop improvement programs. The CRISPR-Cas technology has been used for genome editing in different monocot and dicot crops to confer drought tolerance and for functional characterization of drought-responsive genes. Genome-editing tools, being precise though labor intensive, hold the keys to the future for crop improvement programs. Understanding of the different factors associated with drought response and their cross-talk in plants will facilitate the development of climate-resilient crops in the ongoing scenario of climate change.

Acknowledgments

The National Agri-Food Biotechnology Institute (NABI), Mohali Punjab, Department of Biotechnology, Govt. of India, is acknowledged for support. DBT-Electronic Library Consortium (DeLCON), Gurugram, India, is acknowledged for access to e-resources.

Multiple choice questions

1. Which of the following mechanisms are associated with plants managing drought stress?
   (a) Drought escape;
   (b) Drought avoidance;
   (c) Drought tolerance;
   (d) All of them.
2. Which of the following genes are involved with ABA biosynthesis and drought tolerance?

   (a) Pectate lyase;
   (b) *NCED*;
   (c) *ZEP*;
   (d) Both (c) and (d).
3. Which of the following genes are involved with ABA signaling and drought tolerance?
   (a) Pyrabactin resistance1 *(PYR1)*;
   (b) *NCED*;
   (c) *ZEP*;
   (d) All of them.
4. Which of the following characteristics could be associated with drought tolerance?
   (a) Thick cuticles on the leaves and bearing thorns on the stem;
   (b) Sunken stomata;
   (c) Reduced stomata;
   (d) All of them.
5. Which of the following proteins are expressed during drought exposure?
   (a) LEA;
   (b) DREB;
   (c) Dehydrin and heat shock proteins;
   (d) All of them.

Answers: 1 (d); 2 (d); 3 (a); 4 (d); 5 (d).

Descriptive-type questions

1. During evolution, terrestrial plants have evolved for drought tolerance by adapting several physiological and morphological changes. Expand on this sentence by providing evidence associated with the morphological and physiological changes in order to combat drought tolerance.
2. Exposure to drought leads to a series of changes at the molecular level. A number of genes encoding for transcription factors, enzymes, and other soluble proteins are expressed in order to provide tolerance. What are those genes and how they are interconnected?
3. What proteins are expressed exclusively during drought exposure to maintain cell physiology and plant phenotype?
4. You are working in a lab that is involved in generating a drought-tolerant rice plant through transgenic and/or CRISPR/Cas method(s). Suggest five genes which can be targeted by using transgenic and/or CRISPR/Cas method (s) by citing the appropriate rationale.
5. The CRISPR/Cas techniques have been significantly used for generating abiotic stress tolerance in crop plants. Write a descriptive note highlighting the drought-tolerant, economically important crops generated through genomic CRISPR/Cas modification.

References

[1] Wheeler T and Braun J V 2013 Climate change impacts on global food security *Science* **80** 341

[2] Dai Z Y and Li Y P 2013 A multistage irrigation water allocation model for agricultural land-use planning under uncertainty *Agric. Water Manag.* **129** 69–79

[3] Earl H J and Davis R F 2003 Effect of drought stress on leaf and whole canopy radiation use efficiency and yield of maize *Agron. J.* **95** 688–96

[4] Bernier J, Atlin G N, Serraj R, Kumar A and Spaner D 2008 Breeding upland rice for drought resistance *J. Sci. Food Agric.* **88** 927–39

[5] Grant R F, Jackson B S, Kiniry J R and Arkin G F 1989 Water deficit timing effects on yield components in maize *Agron. J.* **81** 61–5

[6] Singh N, Mansoori A, Dey D, Kumar R and Kumar A 2021 Potential of metabolomics in plant abiotic stress management ed A Kumar, R Kumar, P Shukla and H K Patel *Omics Technologies for Sustainable Agriculture and Global Food Security* vol II (Singapore: Springer) 193–214

[7] Pradhan G P, Prasad P V V, Fritz A K, Kirkham M B and Gill B S 2012 Effects of drought and high temperature stress on synthetic hexaploid wheat *Funct. Plant Biol.* **39**

[8] Kapoor D, Bhardwaj S, Landi M, Sharma A, Ramakrishnan M and Sharma A 2020 The impact of drought in plant metabolism: how to exploit tolerance mechanisms to increase crop production *Appl. Sci.* **10** 5692

[9] Abobatta W F 2019 Drought adaptive mechanisms of plants—a review *Adv. Agric. Environ. Sci. Open Access.* **2** 42–5

[10] Shavrukov Y, Kurishbayev A, Jatayev S, Shvidchenko V, Zotova L, Koekemoer F, De Groot S, Soole K and Langridge P 2017 Early flowering as a drought escape mechanism in plants: how can it aid wheat production? *Front. Plant Sci.* **8** 1950

[11] Luo L J 2010 Breeding for water-saving and drought-resistance rice (WDR) in China *J. Exp. Bot.* **61** 3509–17

[12] Khan A, Pan X, Najeeb U, Tan D K Y, Fahad S, Zahoor R and Luo H 2018 Coping with drought: stress and adaptive mechanisms, and management through cultural and molecular alternatives in cotton as vital constituents for plant stress resilience and fitness *Biol. Res.* **51** 47

[13] Wilson J, Ludlow M, Fisher M and Schulze E 1980 Adaptation to water stress of the leaf water relations of four tropical forage species *Funct. Plant Biol.* **7** 207–20

[14] Li Y, Ruperao P, Batley J, Edwards D, Khan T, Colmer T D, Pang J, Siddique K H M and Sutton T 2018 Investigating drought tolerance in chickpea using genome-wide association mapping and genomic selection based on whole-genome resequencing data *Front. Plant Sci.* **9** 190

[15] Lopes M S, Royo C, Alvaro F, Sanchez-Garcia M, Ozer E, Ozdemir F, Karaman M, Roustaii M, Jalal-Kamali M R and Pequeno D 2018 Optimizing winter wheat resilience to climate change in rain fed crop systems of Turkey and Iran *Front. Plant Sci.* **9** 563

[16] Bartels D and Sunkar R 2005 Drought and salt tolerance in plants *CRC. Crit. Rev. Plant Sci.* **24** 23–58

[17] Wan J, Griffiths R, Ying J, McCourt P and Huang Y 2009 Development of drought-tolerant canola (Brassica napus L.) through genetic modulation of ABA-mediated stomatal responses *Crop Sci.* **49** 1539–54

[18] Capell T, Bassie L and Christou P 2004 Modulation of the polyamine biosynthetic pathway in transgenic rice confers tolerance to drought stress *Proc. Natl. Acad. Sci. USA* **101** 9909–914

[19] Abebe T, Guenzi A C, Martin B and Cushman J C 2003 Tolerance of mannitol-accumulating transgenic wheat to water stress and salinity *Plant Physiol.* **131** 1748–55

[20] Zhong R, Wang Y, Gai R, Xi D, Mao C and Ming F 2020 Rice SnRK protein kinase OsSAPK8 acts as a positive regulator in abiotic stress responses *Plant Sci.* **292** 110373

[21] Yoshida T, Fujita Y, Sayama H, Kidokoro S, Maruyama K, Mizoi J, Shinozaki K and Yamaguchi-Shinozaki K 2010 AREB1, AREB2, and ABF3 are master transcription factors that cooperatively regulate ABRE-dependent ABA signaling involved in drought stress tolerance and require ABA for full activation *Plant J.* **61** 672–85

[22] Guo C, Ge X and Ma H 2013 The rice OsDIL gene plays a role in drought tolerance at vegetative and reproductive stages *Plant Mol. Biol.* **82** 239–53

[23] Hasanuzzaman M, Al Mahmud J, Anee T I, Islam M T and Nahar K 2018 Drought stress tolerance in wheat: omics approaches in understanding and enhancing antioxidant defense *Abiotic Stress-Mediated Sensing and Signaling in Plants: An Omics Perspective* (Singapore: Springer) pp 267–307

[24] Benlloch R and Lois L M 2018 Sumoylation in plants: mechanistic insights and its role in drought stress *J. Exp. Bot.* **69** 4539–54

[25] Zia R, Nawaz M S, Siddique M J, Hakim S and Imran A 2021 Plant survival under drought stress: implications, adaptive responses, and integrated rhizosphere management strategy for stress mitigation *Microbiol. Res.* **242** 126626

[26] Akhtar I and Nazir N 2013 Effect of waterlogging and drought stress in plants *Int. J. Water Resour. Environ. Sci.* **2** 34–40

[27] Khan M A, Iqbal M, Jameel M, Nazeer W, Shakir S, Aslam M T and Iqbal B 2011 Potentials of molecular based breeding to enhance drought tolerance in wheat (*Triticum aestivum* L.) *African J. Biotechnol.* **10**

[28] Farooq M, Wahid A, Kobayashi N, Fujita D and Basra S M A 2009 Plant drought stress: effects, mechanisms and management *Agron. Sustain. Dev.* **29** 185–212

[29] Comas L H, Becker S R, Cruz V M V, Byrne P F and Dierig D A 2013 Root traits contributing to plant productivity under drought *Front. Plant Sci.* **4** 442

[30] Hamaoka N, Yasui H, Yamagata Y, Inoue Y, Furuya N, Araki T, Ueno O and Yoshimura A 2017 A hairy-leaf gene, BLANKET LEAF, of wild *Oryza nivara* increases photosynthetic water use efficiency in rice *Rice.* **10**

[31] Basu S, Ramegowda V, Kumar A and Pereira A 2016 Plant adaptation to drought stress *F1000Research.* **5** 1554

[32] Subbarao G V, Nam N H, Chauhan Y S and Johansen C 2000 Osmotic adjustment, water relations and carbohydrate remobilization in pigeonpea under water deficits *J. Plant Physiol.* **157** 651–9

[33] Gong H, Zhu X, Chen K, Wang S and Zhang C 2005 Silicon alleviates oxidative damage of wheat plants in pots under drought *Plant Sci.* **169** 313–21

[34] Deltoro V I, Calatayud A, Gimeno C, Abadía A and Barreno E 1998 Changes in chlorophyll a fluorescence, photosynthetic CO_2 assimilation and xanthophyll cycle interconversions during dehydration in desiccation-tolerant and intolerant liverworts *Planta.* **207** 224–8

[35] Havaux M 1998 Carotenoids as membrane stabilizers in chloroplasts *Trends Plant Sci.* **3** 147–51

[36] Premachandra G S, Saneoka H, Kanaya M and Ogata S 1991 Cell membrane stability and leaf surface wax content as affected by increasing water deficits in maize *J. Exp. Bot.* **42** 167–71

[37] Dhanda S S, Sethi G S and Behl R K 2004 Indices of drought tolerance in wheat genotypes at early stages of plant growth *J. Agron. Crop Sci.* **190** 6–12

[38] Villar-Salvador P, Planelles R, Oliet J, Peñuelas-Rubira J L, Jacobs D F and González M 2004 Drought tolerance and transplanting performance of holm oak (*Quercus ilex*) seedlings after drought hardening in the nursery *Tree Physiol.* **24** 1147–55

[39] Reynolds M and Rebetzke G 2010 Application of plant physiology in wheat breeding *Application of Physiology in Wheat Breeding* (Texcoco: CIMMYT) pp 2–10

[40] Gigon A, Matos A R, Laffray D, Zuily-Fodil Y and Pham-Thi A T 2004 Effect of drought stress on lipid metabolism in the leaves of *Arabidopsis thaliana* (Ecotype Columbia) *Ann. Bot.* **94** 345–51

[41] Nilsen E T and Orcutt D M 1996 Physiology of plants under stress *Abiotic factors* (New York: Wiley)

[42] Vartanian N, Marcotte L and Giraudat J 1994 Drought rhizogenesis in *Arabidopsis thaliana*: differential responses of hormonal mutants *Plant Physiol.* **104** 761–67

[43] Bharath P, Gahir S and Raghavendra A S 2021 Abscisic acid-induced stomatal closure: an important component of plant defense against abiotic and biotic stress *Front. Plant Sci.* **12** 615114

[44] Taylor I B, Burbidge A and Thompson A J 2000 Control of abscisic acid synthesis *J. Exp. Bot.* **51** 1563–74

[45] Turner N C, Wright G C and Siddique K H M 2001 Adaptation of grain legumes (pulses) to water-limited environments *Adv. Agron.* **71** 193–231

[46] Joshi R, Wani S H, Singh B, Bohra A, Dar Z A, Lone A A, Pareek A and Singla-Pareek S L 2016 Transcription factors and plants response to drought stress: current understanding and future directions *Front. Plant Sci.* **7** 1029

[47] Young T E, Meeley R B and Gallie D R 2004 ACC synthase expression regulates leaf performance and drought tolerance in maize *Plant J.* **40** 813–25

[48] Bouchereau A, Aziz A, Larher F and Martin-Tanguy J 1999 Polyamines and environmental challenges: recent development *Plant Sci.* **140** 103–25

[49] Kasukabe Y, He L, Nada K, Misawa S, Ihara I and Tachibana S 2004 Overexpression of *Spermidine* synthase enhances tolerance to multiple environmental stresses and up-regulates the expression of various stress-regulated genes in transgenic *Arabidopsis thaliana Plant Cell Physiol.* **45** 712–22

[50] Yang J, Zhang J, Liu K, Wang Z and Liu L 2007 Involvement of polyamines in the drought resistance of rice *J. Exp. Bot.* **58** 1545–55

[51] Serraj R and Sinclair T R 2002 Osmolyte accumulation: can it really help increase crop yield under drought conditions? *Plant, Cell Environ.* **25** 333–41

[52] Kavar T, Maras M, Kidrič M, Šuštar-Vozlič J and Meglič V 2008 Identification of genes involved in the response of leaves of *Phaseolus vulgaris* to drought stress *Mol. Breed.* **21** 159–72

[53] Maurel C and Chrispeels M J 2001 Aquaporins. A molecular entry into plant water relations *Plant Physiol.* **125** 135–8

[54] Du Y, Zhao Q, Chen L, Yao X, Zhang W, Zhang B and Xie F 2020 Effect of drought stress on sugar metabolism in leaves and roots of soybean seedlings *Plant Physiol. Biochem.* **146** 1–12

[55] Gowda V R P, Henry A, Yamauchi A, Shashidhar H E and Serraj R 2011 Root biology and genetic improvement for drought avoidance in rice *F. Crop. Res.* **122** 1–13

[56] Decaestecker W, Buono R A, Pfeiffer M L, Vangheluwe N, Jourquin J, Karimi M, van Isterdael G, Beeckman T, Nowack M K and Jacobs T B 2019 CRISPR-Tsko: s technique for efficient mutagenesis in specific cell types, tissues, or organs in *Arabidopsis Plant Cell.* **31** 2868–87

[57] Lian H L, Yu X, Ye Q, Ding X S, Kitagawa Y, Kwak S S, Su W A and Tang Z C 2004 The role of aquaporin RWC3 in drought avoidance in rice *Plant Cell Physiol.* **45** 481–9

[58] Guo L, Zi Y W, Lin H, Wei E C, Chen J, Liu M, Zhang L C, Li J Q and Gu H 2006 Expression and functional analysis of the rice plasma-membrane intrinsic protein gene family *Cell Res.* **16** 277–86

[59] Ding L, Uehlein N, Kaldenhoff R, Guo S, Zhu Y and Kai L 2019 Aquaporin PIP2;1 affects water transport and root growth in rice (*Oryza sativa* L.) *Plant Physiol. Biochem.* **139** 152–60

[60] Wahid A and Close T J 2007 Expression of dehydrins under heat stress and their relationship with water relations of sugarcane leaves *Biol. Plant.* **51** 104–109

[61] Sakuma Y, Maruyama K, Osakabe Y, Qin F, Seki M, Shinozaki K and Yamaguchi-Shinozaki K 2006 Functional analysis of an *Arabidopsis* transcription factor, DREB2A, involved in drought-responsive gene expression *Plant Cell.* **18** 1292–309

[62] Hanin M, Brini F, Ebel C, Toda Y, Takeda S and Masmoudi K 2011 Plant dehydrins and stress tolerance: versatile proteins for complex mechanisms *Plant Signal. Behav.* **6** 1503–509

[63] Duan J and Cai W 2012 OsLEA3-2, an abiotic stress induced gene of rice plays a key role in salt and drought tolerance *PLoS One* **7** e45117

[64] Eriksson S, Eremina N, Barth A, Danielsson J and Harryson P 2016 Membrane-induced folding of the plant stress dehydrin Lti30 *Plant Physiol.* **171** 932–43

[65] Graether S P and Boddington K F 2014 Disorder and function: a review of the dehydrin protein family *Front. Plant Sci.* **5** 576

[66] Wang J, Sun N, Deng T, Zhang L and Zuo K 2014 Genome-wide cloning, identification, classification and functional analysis of cotton heat shock transcription factors in cotton *BMC Genomics.* **15** 961

[67] Giri J, Dansana P K, Kothari K S, Sharma G, Vij S and Tyagi A K 2013 SAPs as novel regulators of abiotic stress response in plants *BioEssays* **35** 639–48

[68] Dixit A R and Dhankher O P 2011 A novel stress-associated protein 'AtSAP10' from *Arabidopsis thaliana* confers tolerance to nickel, manganese, zinc, and high temperature stress *PLoS One.* **6** e20921

[69] Vij S and Tyagi A K 2006 Genome-wide analysis of the stress associated protein (SAP) gene family containing A20/AN1 zinc-finger(s) in rice and their phylogenetic relationship with *Arabidopsis Mol. Genet. Genomics* **276** 565–75

[70] Ghneim-Herrera T *et al* 2017 Expression of the *Aeluropus littoralis* AlSAP gene enhances rice yield under field drought at the reproductive stage *Front. Plant Sci.* **8** 994

[71] Chon W *et al* 2002 Expression profile matrix of *Arabidopsis* transcription factor genes suggests their putative functions in response to environmental stresses *Plant Cell* **14** 559–74

[72] Wrzaczek M and Hirt H 2001 Plant MAP kinase pathways: how many and what for? *Biol. Cell* **93** 81–7

[73] Zhu J K 2002 Salt and drought stress signal transduction in plants *Annu. Rev. Plant Biol.* **53** 247–73

[74] Goswami S *et al* 2015 Calcium triggers protein kinases-induced signal transduction for augmenting the thermotolerance of developing wheat (*Triticum aestivum*) grain under the heat stress *J. Plant Biochem. Biotechnol.* **24** 441–52

[75] Kudla J, Batistič O and Hashimoto K 2010 Calcium signals: the lead currency of plant information processing *Plant Cell* **22** 541–63
[76] Munir S, Liu H, Xing Y, Hussain S, Ouyang B, Zhang Y, Li H and Ye Z 2016 Overexpression of calmodulin-like (ShCML44) stress-responsive gene from *Solanum habrochaites* enhances tolerance to multiple abiotic stresses *Sci. Rep.* **6** 31772
[77] Seki M *et al* 2002 Monitoring the expression profiles of 7000 *Arabidopsis* genes under drought, cold and high-salinity stresses using a full-length cDNA microarray *Plant J.* **31** 279–92
[78] Taji T, Ohsumi C, Iuchi S, Seki M, Kasuga M, Kobayashi M, Yamaguchi-Shinozaki K and Shinozaki K 2002 Important roles of drought- and cold-inducible genes for galactinol synthase in stress tolerance in *Arabidopsis thaliana Plant J.* **29** 417–26
[79] Pilon-Smits E A H, Terry N, Sears T and Van Dun K 1999 Enhanced drought resistance in fructan-producing sugar beet *Plant Physiol. Biochem.* **37** 313–7
[80] Sheveleva E, Chmara W, Bohnert H J and Jensen R G 1997 Increased salt and drought tolerance by D-ononitol production in transgenic *Nicotiana tabacum* L *Plant Physiol.* **115** 1211–9
[81] Nanjo T, Kobayashi M, Yoshiba Y, Sanada Y, Wada K, Tsukaya H, Kakubari Y, Yamaguchi-Shinozaki K and Shinozaki K 1999 Biological functions of proline in morphogenesis and osmotolerance revealed in antisense transgenic *Arabidopsis thaliana Plant J.* **18** 185–93
[82] Iuchi S, Kobayashi M, Taji T, Naramoto M, Seki M, Kato T, Tabata S, Kakubari Y, Yamaguchi-Shinozaki K and Shinozaki K 2001 Regulation of drought tolerance by gene manipulation of 9-cis-epoxycarotenoid dioxygenase, a key enzyme in abscisic acid biosynthesis in *Arabidopsis Plant J.* **27** 325–33
[83] Waditee R *et al* 2005 Genes for direct methylation of glycine provide high levels of glycinebetaine and abiotic-stress tolerance in *Synechococcus* and *Arabidopsis Proc. Natl. Acad. Sci. USA* **102** 1318–23
[84] Tian X, Wang Z, Li X, Lv T, Liu H, Wang L, Niu H and Bu Q 2015 Characterization and functional analysis of pyrabactin resistance-like abscisic acid receptor family in rice *Rice* **8** 28
[85] Abe H, Urao T, Ito T, Seki M, Shinozaki K and Yamaguchi-Shinozaki K 2003 *Arabidopsis* AtMYC2 (bHLH) and AtMYB2 (MYB) function as transcriptional activators in abscisic acid signaling *Plant Cell* **15** 63–78
[86] Kang J Y, Choi H I, Im M Y and Soo Y K 2002 *Arabidopsis* basic leucine zipper proteins that mediate stress-responsive abscisic acid signaling *Plant Cell* **14** 343–57
[87] Luo R *et al* 2012 SOAPdenovo2: an empirically improved memory-efficient short-read de novo assembler *Gigascience* **1** 2047–217X-1-18
[88] Lu G, Gao C, Zheng X and Han B 2009 Identification of OsbZIP72 as a positive regulator of ABA response and drought tolerance in rice *Planta* **229** 605–15
[89] Kim Y S and Kim J K 2009 Rice transcription factor AP37 involved in grain yield increase under drought stress *Plant Signal. Behav.* **4** 735–6
[90] Jeong J S, Kim Y S, Baek K H, Jung H, Ha S H, Do Choi Y, Kim M, Reuzeau C and Kim J K 2010 Root-specific expression of OsNAC10 improves drought tolerance and grain yield in rice under field drought conditions *Plant Physiol.* **153** 185–97
[91] Tran L S P, Nakashima K, Sakuma Y, Simpson S D, Fujita Y, Maruyama K, Fujita M, Seki M, Shinozaki K and Yamaguchi-Shinozaki K 2004 Isolation and functional analysis of *arabidopsis* stress-inducible NAC transcription factors that bind to a drought-responsive cis-element in the early responsive to dehydration stress 1 promoter *Plant Cell* **16** 2481–98

[92] Zheng X, Chen B, Lu G and Han B 2009 Overexpression of a NAC transcription factor enhances rice drought and salt tolerance *Biochem. Biophys. Res. Commun.* **379** 985–9
[93] Trujillo L E, Sotolongo M, Menéndez C, Ochogavía M E, Coll Y, Hernández I, Borrás-Hidalgo O, Thomma B P H J, Vera P and Hernández L 2008 SodERF3, a novel sugarcane ethylene responsive factor (ERF), enhances salt and drought tolerance when overexpressed in tobacco plants *Plant Cell Physiol.* **49** 512–25
[94] Ren X, Chen Z, Liu Y, Zhang H, Zhang M, Liu Q, Hong X, Zhu J K and Gong Z 2010 ABO3, a WRKY transcription factor, mediates plant responses to abscisic acid and drought tolerance in *Arabidopsis Plant J.* **63** 417–29
[95] Cho E K and Hong C B 2006 Over-expression of tobacco NtHSP70-1 contributes to drought-stress tolerance in plants *Plant Cell Rep.* **25** 349–58
[96] Lim S D, Lee C and Jang C S 2014 The rice RING E3 ligase, OsCTR1, inhibits trafficking to the chloroplasts of OsCP12 and OsRP1, and its overexpression confers drought tolerance in *Arabidopsis Plant, Cell Environ.* **37** 1097–113
[97] Umezawa T, Yoshida R, Maruyama K, Yamaguchi-Shinozaki K and Shinozaki K 2004 SRK2C, a SNF1-related protein kinase 2, improves drought tolerance by controlling stress-responsive gene expression in *Arabidopsis thaliana Proc. Natl. Acad. Sci. USA* **101** 17306–11
[98] Sivamani E, Bahieldin A, Wraith J M, Al-Niemi T, Dyer W E, Ho T H D and Qu R 2000 Improved biomass productivity and water use efficiency under water deficit conditions in transgenic wheat constitutively expressing the barley HVA1 gene *Plant Sci.* **155** 1–9
[99] Xiao B Z, Chen X, Bin Xiang C, Tang N, Zhang Q F and Xiong L Z 2009 Evaluation of seven function-known candidate genes for their effects on improving drought resistance of transgenic rice under field conditions *Mol. Plant* **2** 73–83
[100] Avonce N, Leyman B, Mascorro-Gallardo J O, Van Dijck P, Thevelein J M and Iturriaga G 2004 The *Arabidopsis* trehalose-6-P synthase AtTPS1 gene is a regulator of glucose, abscisic acid, and stress signaling *Plant Physiol.* **136** 3649–59
[101] Bortesi L and Fischer R 2015 The CRISPR/Cas9 system for plant genome editing and beyond *Biotechnol. Adv.* **33** 41–52
[102] Jinek M, Chylinski K, Fonfara I, Hauer M, Doudna J A and Charpentier E 2012 A programmable dual-RNA-guided DNA endonuclease in adaptive bacterial immunity *Science* **337** 816–21
[103] Ricroch A, Clairand P and Harwood W 2017 Use of CRISPR systems in plant genome editing: toward new opportunities in agriculture *Emerg.Top. Life Sci.* **1** 169–82
[104] Hille F and Charpentier E 2016 CRISPR-cas: biology, mechanisms and relevance *Philos. Trans. R. Soc. B Biol. Sci.* **371** 20150496
[105] Jiang F and Doudna J A 2017 CRISPR-Cas9 structures and mechanisms *Annu. Rev. Biophys.* **46** 505–29
[106] Shan Q, Wang Y, Li J and Gao C 2014 Genome editing in rice and wheat using the CRISPR/Cas system *Nat. Protoc.* **9** 2395–410
[107] Lee K, Zhu H, Yang B and Wang K 2019 An *agrobacterium*-mediated CRISPR/Cas9 platform for genome editing in maize *Plant Genome Editing with CRISPR Systems* (Methods in Molecular Biology vol 1917) vol 1917 (New York: Humana) pp 121–43
[108] Liu L, Zhang J, Xu J, Li Y, Guo L, Wang Z, Zhang X, Zhao B, Guo Y D and Zhang N 2020 CRISPR/Cas9 targeted mutagenesis of SlLBD40, a lateral organ boundaries domain transcription factor, enhances drought tolerance in tomato *Plant Sci.* **301** 110683

[109] Bhowmik P K and Islam M T 2020 CRISPR-Cas9-mediated gene editing in wheat: a step-by-step protocol *CRISPR-Cas Methods* (Springer Protocols Handbooks) (New York: Humana) pp 203–22

[110] Bhandawat A, Sharma V, Rishi V and Roy J K 2020 Biolistic delivery of programmable nuclease (CRISPR/Cas9) in bread wheat *Biolistic DNA Delivery in PlantsMethods in Molecular Biology* vol 2124 (New York: Humana) pp 309–29

[111] Jiang W, Zhou H, Bi H, Fromm M, Yang B and Weeks D P 2013 Demonstration of CRISPR/Cas9/sgRNA-mediated targeted gene modification in *Arabidopsis*, tobacco, sorghum and rice *Nucleic Acids Res.* **41** e188

[112] Li H, Fan R, Li L, Wei B, Li G, Gu L, Wang X and Zhang X 2014 Identification and characterization of a novel copper transporter gene family TaCT1 in common wheat *Plant Cell Environ.* **37** 1561–73

[113] Gupta P K, Balyan H S, Sharma S and Kumar R 2020 Genetics of yield, abiotic stress tolerance and biofortification in wheat (*Triticum aestivum* L.) *Theor. Appl. Genet.* **133** 1569–602

[114] Lou D, Wang H, Liang G and Yu D 2017 OsSAPK2 confers abscisic acid sensitivity and tolerance to drought stress in rice *Front. Plant Sci.* **8** 993

[115] Shi J, Gao H, Wang H, Lafitte H R, Archibald R L, Yang M, Hakimi S M, Mo H and Habben J E 2017 ARGOS8 variants generated by CRISPR-Cas9 improve maize grain yield under field drought stress conditions *Plant Biotechnol. J.* **15** 207–16

[116] Feng Z *et al* 2013 Efficient genome editing in plants using a CRISPR/Cas system *Cell Res.* **23** 1229–32

[117] Liang G, Zhang H, Li X, Ai Q and Yu D 2017 BHLH transcription factor bHLH115 regulates iron homeostasis in *Arabidopsis thaliana J. Exp. Bot.* **68** 1743–55

[118] Santosh Kumar V V, Verma R K, Yadav S K, Yadav P, Watts A, Rao M V and Chinnusamy V 2020 CRISPR-Cas9 mediated genome editing of drought and salt tolerance (OsDST) gene in *indica mega* rice cultivar MTU1010 *Physiol. Mol. Biol. Plants.* **26** 1099–110

[119] Liao S, Qin X, Luo L, Han Y, Wang X, Usman B, Nawaz G, Zhao N, Liu Y and Li R 2019 CRISPR/Cas9-induced mutagenesis of semi-rolled Leaf1,2 confers curled leaf phenotype and drought tolerance by influencing protein expression patterns and ROS scavenging in rice (*Oryza sativa* L.) *Agronomy.* **9** 728

[120] Osakabe Y, Watanabe T, Sugano S S, Ueta R, Ishihara R, Shinozaki K and Osakabe K 2016 Optimization of CRISPR/Cas9 genome editing to modify abiotic stress responses in plants *Sci. Rep.* **6** 26685

[121] Roca Paixão J F, Gillet F X, Ribeiro T P, Bournaud C, Lourenço-Tessutti I T, Noriega D D, de Melo B P, de Almeida-Engler J and Grossi-de-Sa M F 2019 Improved drought stress tolerance in *Arabidopsis* by CRISPR/dCas9 fusion with a Histone AcetylTransferase *Sci. Rep.* **9** 8080

[122] Wang L, Chen L, Li R, Zhao R, Yang M, Sheng J and Shen L 2017 Reduced drought tolerance by CRISPR/Cas9-mediated SlMAPK3 mutagenesis in tomato plants *J. Agric. Food Chem.* **65** 8674–82

[123] Li X, Kristiansen K, Rosenqvist E and Liu F 2019 Elevated CO_2 modulates the effects of drought and heat stress on plant water relations and grain yield in wheat *J. Agron. Crop Sci.* **205** 362–71

IOP Publishing

Molecular Response and Genetic Engineering for Stress in Plants, Volume 1
Abiotic stress
Pawan Shukla, Anirudh Kumar, Rakesh Kumar and Manish K Pandey

Chapter 5

Engineering salt tolerance in crops: ion transporters and compatible solutes

Israr Ahmed[1*], Pawan Shukla[2]

[1]*Department of Plant Molecular Biology, University of Delhi South Campus, Benito Juarez Road, Dhaula Kuan, New Delhi 110021, India*

[2]*Seri-Biotech Research Laboratory (SBRL), Carmelram Post, Kodathi, Bangalore 560035, Karnataka, India*

Increasing soil salinity is emerging as one of the serious limiting factors to food security. Salt stress negatively affects plants' growth and development and reduces their yield potential. Plants undergo a series of intracellular changes to respond to the deleterious effects of salt stress. Understanding the perception of salt stress, signaling mechanisms, and adaptive responses is important to tailor crops for salt tolerance. The present chapter describes an overview of plant response mechanisms against salt stress, including signaling mechanisms and transport of Na^+ and K^+ within plant cells to maintain ionic homeostasis. In addition, we also discuss the progress made in developing salt-tolerant crops by genic manipulation of the genes involved in ion transport or synthesis of compatible solutes.

5.1 Introduction

Plants encounter different kinds of abiotic and biotic stresses that severely limit their growth and yield. Abiotic stressors include water limitation, salinity, temperature (high and low), toxic metals, nutrients, light, ozone, etc, whereas biotic stressors include insect pests and pathogens. Together, these constraints significantly reduce crop production and productivity. Soil salinity affects about 800 million hectares of land across the world [1]. Salt-affected soils contain high concentrations of chlorides, sulfates, and carbonates of sodium (Na^+), potassium (K^+), calcium (Ca^{2+}), and magnesium (Mg^{2+}). Soil salinity is expressed in terms of the electrical conductivity

* Author to whom any correspondence should be addressed Email: iahmed67@gmail.com.

doi:10.1088/978-0-7503-4921-5ch5

(EC_e) of the saturation extract of a soil. The saturation extract is the amount of water that is extracted from the soil at its maximum water-holding capacity. Saline soils contain EC_e above 4 dS m^{-1}.

Soil salinity is the outcome of both natural processes (weathering of rocks, sea water intrusions, low precipitation, winds) and human activities (excessive use of chemical fertilizers, continuous use of ground water for irrigation, poor soil-management practices). In line with the scarcity of arable land across the globe, there is a need to increase the global food production by 70% by the year 2050 to fulfill the requirements of a human population of nearly 9.3 billion. Salt-affected soils can be reclaimed by various methods like leaching, surface drainage, or addition of soil amendments (like gypsum). These methods are labor intensive and are not economically suitable. In the present scenario, plant biologists face a challenge to develop new technologies and approaches that will increase global food production under these environmental conditions.

In saline environments, plants are evolved with several adaptive mechanisms to cope with the conditions. Understanding these adaptive responses could play a significant role in engineering or breeding crops that can withstand such saline environment. Plant adaptations to salinity involve a range of changes at the physiological, biochemical, and molecular levels, which contribute toward their survival under the variable environmental conditions. Physiological responses include stomatal closure to reduce the transpiration rate, reduction in leaf area, stunted growth, root growth, leaf abscission, wilting, change in relative water content, and electrolyte leakage. On the other hand, biochemical changes include generation of reactive oxygen species (ROS), accumulation of osmoprotectants like proline, soluble sugars, glycine betaine, etc, changes in chlorophyll, and lipid peroxidation. Molecular responses include signal perception, and activation of signaling cascades to activate transcription factors, which further induce a set of genes. These genes synthesize the compatible molecules or regulate the downstream signaling for signal transduction [2]. Recent research has made efficient use of various omics tools that helped in the identification of complex stress perception and response regulatory networks in model and crop plants [3].

Conventional breeding technologies have been vital in the introgression of quantitative trait loci (QTLs) associated with the desired traits to develop salt-tolerant crops. However, they can bring about certain undesirable agronomic characteristics from the donor parents. Traditional breeding methods fail to introduce the desired genes due to the pollination barrier across species and distant relatives or from non-plant sources. Advancements in plant tissue culture techniques have helped in the development of efficient regeneration systems for various crop plants [4]. In such conditions, insertion of specific genes through genetic engineering provides a potential solution for generation of stress-tolerant crops. Further, genetic manipulation allows for control over the expression of a particular gene in a specific time and tissue for their optimal function. Initially, salt tolerance in crops was obtained through constitutive expression of a stress-induced gene or transcription factor. Later, it was found that, in certain cases, constitutive expression of a transgene is often accompanied with deleterious effects due to continuous accumulation of the gene product or energy consumption. To cope with such problems, stress-inducible promoters were identified and fused with the desired gene to develop stress-tolerant plants [5, 6]. Recently, several attempts were made to co-express two genes to develop salt-tolerant crops [7]. With the complete annotation of several genome sequencing projects, it has become possible to study genes responding to abiotic stress in a comprehensive manner. Comparative transcriptome studies between tolerant and susceptible cultivars have helped in the identification of the stress-related genes and their relationship with the cis-regulatory elements. Stress-induced genes are regulated by complex signaling networks [6].

Broadly, stress-induced proteins are categorized into two groups: (a) proteins with known functions that directly participate in abiotic stress tolerance (e.g. late embryogenesis abundant (LEA) proteins, heat shock proteins (Hsps), transporters, etc), and (b) regulatory proteins involved in intracellular signaling and stress-induced gene expression (e.g. protein kinases, phosphatases, and transcription factors). Single-action genes such as those involved in the synthesis of osmoprotectants, ROS detoxification, and transport, formed the initial targets for plant transformation. Several reports are available in the literature wherein single gene overexpression leads to stress tolerance [6]. Abiotic stresses result in ROS generation in plant cells which need to be detoxified for protection from oxidative damage. Engineering crops with the expression of ROS-detoxifying genes such as catalase (*CAT*), glutathione peroxidase (*GPx*), superoxide dismutase (*SOD*), ascorbate peroxidases (*APX*), dehydroascorbate reductase (*DHAR*), and glutathione reductases (*GR*) alleviates oxidative stress in transgenic plants. Another strategy to engineer salinity tolerance is the reestablishment of ionic homeostasis in cells. This can be achieved either by sequestering the excessive ions inside vacuoles or effluxing out from the roots [5, 6, 8]. Several reports on the overexpression of transporters linked with the sequestration of Na^+ in vacuoles or efflux from root cells conferred transgenic plants which displayed salinity tolerance [5, 8]. Genetic manipulation of genes involved in calcium signaling improved salt stress tolerance in model and crop plants [9–14].

5.2 Physiological effects of salt stress on plant growth and development

Salinity imposes two types of stress via its osmotic and ionic components. Together, they severely limit plant growth. The immediate effect of excess salts is inhibition of water uptake by the roots. Secondly, in the long run, it may accumulate in the tissues at toxic levels. This is a typical effect of excess salt or ions linked to salinity [1, 15]. These two mechanisms distinguish a biphasic growth response in plants.

Phase 1. This is the initial growth response of the plant to salt stress, which arises due to an increase in the osmotic potential of the soil and consequently a reduction in leaf and root growth due to a decrease in turgor pressure. Osmotic stress induces stomatal closure, which results in reduced CO_2 assimilation. These symptoms mainly arise due to water stress [1].

Phase 2. This phase of response includes gradual deposition of Na^+ and Cl^- ions to toxic levels in older leaves. The level of salt present in the cell cytosol is greater than their ability to compartmentalize it into vacuoles. This will alter the K^+/Na^+

ratio and Na^+ and Cl^- ion concentrations within the cytosol, resulting in inhibition of enzymatic activity, metabolic processes, and photosynthesis. Higher levels of salt may deposit in the cell wall and dehydrate the cell [1].

5.3 Halophytes versus glycophytes

Plants are classified into two groups based on their ability to tolerate saline environments. These two groups are halophytes and glycophytes. Halophytes are plants which survive under high salt conditions, normally NaCl, but may also include other types of salts such as Na_2SO_4, $CaSO_4$, $MgSO_4$, KCl, $MgCl_2$, and Na_2CO_3. Glycophytes, on the other hand, lack the mechanism to survive under high salt conditions and are salt-sensitive. Although these are two distinct groups, most plant species follow a continuum in respect of salt tolerance ranging from highly susceptible species, such as rice and chickpeas, to the most tolerant *Tecticornia* species in halophytes. The high salt tolerance adaptive mechanisms of halophytes have been studied in detail and are attributed to differences in physiological and molecular adaptive mechanisms. Several studies have shown that halophyte genes share homology with their glycophytic counterparts, but differ in terms of their expression and function [16, 17]. Halophytes have evolved better capacity to adjust the osmotic potential, ion exclusion and sequestration of excess salts within the cellular organelles, secretion of excess salts through secretary glands through transporters, and maintenance of net K^+ to Na^+ selectivity [18]. Several reports have indicated that stress-responsive gene expression is higher in halophytes compared to glycophytes. Gene homologs from halophytes have been expressed in glycophytes, which in turn displayed improved salt tolerance [5, 8].

5.4 Salt stress sensing and signal transduction components

Plants have developed unique sensory mechanisms to sense hyperosmotic and ion-specific components of salt stress. Understanding these mechanisms of Na^+ entry and translocation to aerial parts and sequestration to intracellular compartments is of vital importance to engineer salt tolerance in plants. Plants have evolved with salt stress sensors that sense the excess salts present in the root growth zone, which transduce stress signals to the intracellular components of signaling to generate stress responses. However, identification of these sensors is not easy due to functional redundancy or lethality of the knockout mutants. The early signaling events with respect to salt stress perception start with a rapid transient rise in $[Ca^{2+}]_{cyt}$. The rise in Ca^{2+} is sensed by various calcium sensor proteins like calmodulin, calmodulin-like proteins (CMLs), calcineurin B-like (CBL) proteins, or calcium-dependent protein kinases (CDPKs). CBLs interact with CBL-interacting protein kinases (CIPKs) and regulate various downstream targets. Hence, the identification of the proteins or transporters that are involved in mediating rapid influx of Ca^{2+} during salt stress will help in discovering the salt stress sensors [15]. Using forward genetic screening and calcium imaging, a few osmotic and ionic sensors have been identified in *Arabidopsis*. For example, reduced hyperosmolality-induced $[Ca^{2+}]$i increase1 (OSCA1) is a plasma membrane-localized osmosensor that is involved in osmotic-stress-induced Ca^{2+} increase in the cytosol [19]. Similarly, another ionic stress sensor, monocation-induced $[Ca^{2+}]_i$ increases 1

(MOCA1), has been identified using forward genetic screening. Phenotypic analysis demonstrated that the moca1 mutant was sensitive to salt stress. MOCA1 encodes for glucuronosyltransferase, which is involved in the synthesis of glycosyl inositol phosphorylceramide (GIPC) sphingolipids. GIPCs directly bind Na^+ and facilitate Ca^{2+} influx into the cytosol. Salt stress causes softening of cell wall damage [20]. Another salt stress sensor, FERONIA, is implicated in maintaining cell wall integrity under salt stress. Loss-of-function mutant *fer* exhibited reduced root growth under salt stress conditions. The mechano-sensitive channel MSL8 senses salt-stress-induced change in the turgor pressure and regulates pollen hydration and germination [21].

Reactive oxygen species such as hydrogen peroxide (H_2O_2), singlet oxygen ($^1O_2$), superoxide anion (O_2^-), and hydroxyl radical (OH^-) are produced in various subcellular compartments such as the apoplast, mitochondria, chloroplast, and peroxisomes [15]. Although higher levels of ROS cause cellular damage, low levels of ROS act as a signaling molecule to activate ROS-responsive genes. It has been demonstrated that *AtRBOHD* and *AtRBOHF* are associated with regulation of Na^+/K^+ homeostasis during salt stress [15]. ROS generated by *RBOHD* mediates propagation of the salt stress signal to distant parts, suggesting their role in systemic signaling.

5.5 Sodium uptake and transport in plants

Na^+ can enter into roots through multiple Na^+ influx pathways comprising nutrient channels and transporters. Typically, plants maintain a low cytosolic Na^+ concentration (1–10 mM) and a negative electrochemical gradient across the plasma membrane. An increase in the extracellular Na^+ ions results in the establishment of a large electrochemical potential gradient and favors passive transport of Na^+ across the plasma membrane [22]. An overview of Na^+ transport mechanisms in plant root cells is shown in figure 5.1.

Inorganic ions enter plant roots via apoplastic and symplastic pathways. The apoplastic pathway constitutes a continuum between the cell wall and extracellular space of the adjacent cells. This pathway generally accounts for less than 1% of the transpirational volume flow in most of species. However, in monocots like rice nearly 50% of the total Na^+ and Cl^- uptake could be facilitated by the apoplastic pathway [22]. In the symplastic pathway, ions enter neighboring cells by crossing the plasma membrane via plasmodesmata. Na^+ crosses the plasma membrane via different ion channels and transporters [22].

Proteins involved in uptake of Na^+ from, and its subsequent transport within, the plant cell are present within the plasma membrane. Under saline conditions, rapid influx of Na ions across the plasma membrane occurs through the non-selective cation channels (NSCCs) and later through the high-affinity K^+ transporter (HKT1) [23].

Plants sense the low water potential outside the rhizosphere through membrane transporters. A low water potential changes the turgidity of the cell. This is perceived by the periplasmic domain of the AtHKT transporter and induces the MAPK signaling cascade to alter the gene expression [23]. Alternatively, cell distortion causes a change in the geometry of mechano-sensitive ion channels that leads to depolarization of the plasma membrane and, finally, increases in the cytoplasmic Ca^{2+} levels to initiate further signaling to respond to the stimulus [24]. As yet no Na^+ specific sensor has been identified in plants.

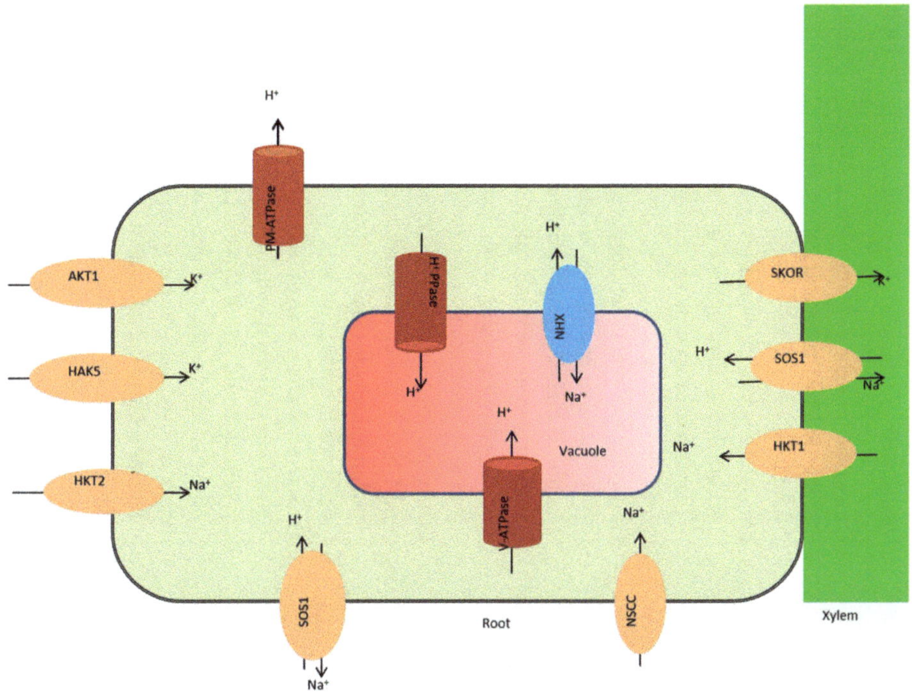

Figure 5.1. Ion transporters involved in maintenance of Na^+ uptake and homeostasis in plant cells.

5.6 Transporters in salt stress

Plants maintain Na^+ ion homeostasis inside the cells through the involvement of ion channels and transporters present across the plasma and endomembrane. In roots, Na^+ enters the plant cell through passive transport due to enormous negative electric potential across the membrane. It may enter through the calcium-permeable NSSCs and glutamate receptor-like (GLR). Most of the Na^+ that enters the root cells is pumped back into the soil through adenosine triphosphate (ATP)-driven Na^+/H^+ antiporters. The intracellular Na^+ is sequestered to vacuoles by tonoplast Na^+/H^+ antiporters. Na^+ reaches the stellar cells by symplastic movement. The net delivery of Na^+ in xylem depends on four components of the transport system: (a) influx of Na^+ into the outer half of the roots from the rhizosphere; (b) efflux back into the soil solution through active transport; (c) efflux of Na^+ into the inner half of the root to xylem; and (d) influx back into these cells from the xylem [1]. An overview of the different transporters associated with the uptake and translocation of Na^+ in plant cells is shown in figure 5.1. Several researchers have reported that the heterologous expression of ion transporters in crop plants resulted in enhanced salt tolerance (table 5.1)

5.6.1 High-affinity potassium transporters

High-affinity potassium transporters (HKTs) belong to the Trk/Ktr/HKT super-family that is present in plants and microorganisms. The first report of involvement

Table 5.1. Ion transporters genetically engineered to improve salt stress tolerance of crop plants.

Gene	Function	Donor species	Recipient species	Description	References
AtNHX1	Na^+/H^+ antiporter	*Arabidopsis*	*Vigna radiata*	Transgenic mungbean plants demonstrated higher K^+/Na^+ ratio and antioxidant activity.	[106]
SbNHXLP	Na^+/H^+ antiporter	*Sorghum bicolor*	*Solanum lycopersicum*	Transgenic plants exhibited lower Na^+ accumulation and higher Ca^{2+} levels. Transgenic tomato plants are tolerant at 200 mM NaCl treatment.	[107]
PpENA1	Na^+ pumping ATPase	*Physcomitrella patens*	*Oryza sativa*	Transgenic rice plants can tolerate 50 mM NaCl stress in hydroponic conditions and exhibited higher biomass production.	[108]
HvNHX2	Vacuolar Na^+/H^+ antiporter	*Hordeum vulgare*	*Solanum tuberosum*	Transgenic plants were salt tolerant and exhibited better root growth, higher biomass and K^+ content compared to wild type (WT).	[109]
StNHX1	Vacuolar Na^+/H^+ antiporter	*Solanum torvum*	*Glycine max*	Transgenic soybean lines were tolerant at 100 mM NaCl. Lower Na^+ and malondialdehyde (MDA) content were observed in transgenic lines.	[110]
TaNHX2	Vacuolar Na+/$^+$ antiporter	*Triticum aestivum*	*Glycine max*	Transgenic plants were salt tolerant at 150 Mm NaCl stress in sand culture. Higher biomass production and no. of flowers per plant was observed in transgenics compared to control.	[111]
SaVHAc1	Vacuolar H^+-ATPase	*Spartina alterniflora*	*Oryza sativa*	Higher K^+/Na^+ ratio in the transgenic plants. Higher relative water content (RWC) due to early closure of stomata and reduced stomatal density.	[62]
*AVP*1	Vacuolar H^+ pyrophosphatase	*Arabidopsis thaliana*	*Hordeum vulgare*	Transgenic barley plants expressing *AVP1* had increased biomass and grain yield in 150 mM NaCl stress in soil conditions.	[112]

(*Continued*)

Table 5.1. (*Continued*)

Gene	Function	Donor species	Recipient species	Description	References
AVP1	Vacuolar H^+ pyrophosphatase	*Arabidopsis*	Cotton	Transgenic cotton plants had vigorous growth, enhanced salt tolerance, and seed fibre yield in green house conditions.	[113]
TsVP	Vacuolar H^+ Pyrophosphatase	*Thellungiella halophila*	*Gossypium hirsutum*	Early seed germination and 14.8% increase in seed cotton yield under saline condition in green house.	[114]
*AVP*1 and *NHX1*	Vacuolar H^+ PyrophosphataseNa^+/H^+ antiporter	*Arabidopsis*	Cotton	Co-expression of *AVP1* and *NHX1* enhanced salt tolerance and fibre yied compared to the non-transformed plants.	[115]
AtAVP1 and *AtNHX1*	Vacuolar H^+-pyrophosphataseNa^+/H^+ antiporter	*A. thaliana*	Soybean	Co-expression of *AtVP1* and *AtNHX1* enhanced salt tolerance of the transgenic plants compared to the transgenic plants either expressing *AtVP1* or *AtNHX1*.	[7]
IbNHX2	Na^+/H^+ antiporter	*Ipomoea batatas* (L.) Lam.	*Ipomoea batatas* (L.) Lam.	Transgenic *Ipomoea* plants exhibited enhanced salt and drought tolerance.	[61]
AtHKT1	High-affinity potassium transporter	*Arabidopsis thaliana*	*Potato*	Overexpression of *AtHKT1* increased salt tolerance by improving K^+/Na^+ ratio, photosynthesis rate, and stomatal conductance.	[116]
AtHKT1;1	High-affinity potassium transporter	*Arabidopsis thaliana*	*Rice*	Tissue-specific expression of *AtHKT1;1* in cortical and epidermal cells increased salt tolerance by retrieving Na^+ from the xylem sap.	[117]
HvHKT2;1	Na^+/K^+ symporters	*Hordeum vulgare*	*Hordeum vulgare*	Overexpression lines exhibited enhanced Na^+ uptake in roots, xylem sap, and translocation to leaves at 50–100 mM NaCl concentration.	[40]

AVP1 and *PgNHX1*	Vacuolar H^+ PyrophosphataseNa^+/H^+ antiporter	*Arabidopsis and Pennisetum*	Tomato	Co-transformants accumulated 1.5 times higher Na^+ than the single transformants accompanied with better osmoregulatory capacity and salt tolerance.	[118]
AtNHX1	Na^+/H^+ antiporter	*Arabidopsis*	*Arachis hypogea*	Overexpression lines of *AtNHX1* exhibited improved tolerance to salt and drought stresses including higher salt and proline content in the leaves.	[119]
OsHAK16	High-affinity potassium transporter	Rice	Rice	*OsHAK16* KO decreases K uptake whereas its overexpression increases K uptake and root to shoot translocation, under low K supply. *OsHAK16* OX lines increased salt tolerance.	[101]
*OsHAK*1	High-affinity potassium transporter	*O. sativa*	*O. sativa*	Loss-of-function mutant *oshak*1 was sensitive to K^+ uptake whereas overexpression lines of OsHAK1 enhanced K^+ uptake and K^+/Na^+ ratio.	[103]
*AtNHX*1 and *TsVP*	Vacuolar Na^+/H^+ antiporter Vacuolar H^+ Pyrophosphatase	*Arabidopsis, Thellungiella halophila*	Cotton	Enhanced seed cotton yield in transgenic plants. Increased seed germination and ionic homeostasis.	[120]

of HKTs in Na^+/K^+ transport came from the characterization of wheat *TaHKT2;1* that mediates Na^+/K^+ transport [23]. Subsequently, these were identified in other crop plants and their role in the transport of Na^+ and K^+ has been extensively investigated. HKTs are grouped into two subclasses based on their cation transport activity [8]. Class I HKTs act as Na^+ uniporters whereas Class II HKTs are reported to act as Na^+/K^+ symporters under potassium limitation or salt stress conditions [8].

HKT1 transporters are the key players in root to shoot Na^+ partitioning [23, 25]. *Arabidopsis HKT1;1* null mutant plants show salt-sensitive phenotype with hyper-accumulation of Na^+ in shoots and reduced accumulation in roots [26, 27]. Tissue-specific expression of *AtHKT1:1* in the stele tissue enhances salt tolerance [28]. *AtHKT1;1* expression has been suggested to stimulate the acquisition of K^+ in the xylem vessels via outward-rectifying K^+ channels, resulting in a high K^+/Na^+ ratio in leaves [29, 30]. QTL analysis of wheat has identified *Nax1*, *Nax2*, and *Kna1* loci that confer salt tolerance [31]. The *Nax1* locus contains the *TaHKT1;4* gene, which was found to contribute to Na^+ removal from xylem in the leaf sheath to protect leaf blades from Na^+ accumulation [32].

HKT transporters are negatively regulated by cytokinins (CKs) via type B response regulators ARR1 and ARR2 [33]. Exogenous application of CKs reduces the expression of HKT transporters. Transcriptome analysis of a salt-tolerant mutant defective CK synthesis showed increased expression of HKT transporters [34]. In addition, transcription factors like ABI4 and AtZIP4 also regulate the expression of *AtHKT1;1*. The loss-of-function mutant *abi4* shows salt tolerance and reduced Na accumulation. Reduced Na^+ accumulation is correlated with enhanced expression of *AtHKT1;1* expression in *abi4* mutants. Overexpression of *ABI4* decreases the expression of *AtHKT1;1* and results in salt-sensitive phenotype [35]. Suppression of AtZIP24 in *Arabidopsis* improves salt tolerance by enhancing the expression of *AtHKT1;1* [36].

Several candidate genes encoding for the HKT1 transporter in crop plants have been identified and functionally characterized during salt stress. Ectopic expression of soybean *GmHKT1;4* and *Sorghum bicolor SbHKT1;4* enhanced salt tolerance through balancing the K^+/Na^+ ratio [37, 38]. In maize, a salt tolerance QTL, namely *Zea mays* Na^+ Content1 (*ZmNC1*), has been identified that encodes the HKT1 transporter. Tissue-specific localization studies revealed that *ZmHKT1* was localized to root stele cells. Loss of function of *ZmHKT1* due to presence of a retrotransposon resulted in increased accumulation of Na^+ in xylem sap and root to shoot Na^+ delivery [39].

Class II HKT transporters are known as Na^+/K^+ symporters and are mainly present in monocots. Several members of the Class II HKT subfamily have been characterized in rice and barley [40, 41]. Transcript profiling revealed a distinct expression profile and ion selectivity among Class II HKT transporters. Class II HKT transporters exhibited differential ion selectivity in salt-tolerant and sensitive plants. Transgenic plants with reduced expression of *TaHKT2;1* resulted in enhanced salt tolerance. Further investigation revealed that enhance salt tolerance of the antisense lines of *TaHKT2;1* was attributed to reduced uptake of Na^+ in the root tissues and reduced translocation to the xylem sap [42]. In contrast to

TaHKT2;1, barley *HvHKT2;1* overexpression led to enhanced Na^+ uptake, enhanced xylem loading, and translocation to leaves during salt stress [40]. This class of transporter has been implicated in the maintenance of the K^+/Na^+ ratio under salt stress conditions [25].

5.6.2 Salt overly sensitive

SOS1 (also known as NHX7) is a plasma membrane-localized Na^+ transporter, which is involved in the efflux of a major portion of the Na^+ ions which enters a cell. Upon entry of excess Na^+ ions inside the cell, a rapid rise in the cytoplasmic calcium takes place. The rise in $[Ca^{2+}]_{cyt}$ is sensed by calcineurin B-like protein (CBL4), which was originally identified as SOS3 [43]. Ca^+–SOS3 interaction leads to dimerization of SOS3 that further interacts with a CBL-interacting protein kinase (CIPK24), originally named SOS2 [44]. The SOS3/SOS2 complex is targeted at the plasma membrane, enabling the phosphorylation and activation of membrane-bound SOS1 [45–47]. This module operates in both root and shoot tissues to reduce cytosolic Na^+ concentration. However, the sensor protein that is involved in sensing Ca^{2+} differs in these two tissues. In roots, it is CBL4 that interacts with SOS2, while in shoots it is the SOS3-like Ca^{2+} binding protein, SCaBP or CBL10, that is involved in binding SOS2 [124]. In normal (unstressed) conditions, SOS2 forms a complex with 14–3–3 and GIGANTEA (GI) proteins that masks its kinase activity. However, under salt stress conditions, the 14–3–3 and GI proteins become degraded, which leads to the release of SOS2 from the complex. Protein phosphatase 2C ABI2 interacts with SOS2 and regulates its function. SOS1 is autoinhibited under normal conditions, but under salt stress its autoinhibition is relieved by phosphorylation of Ser1044 residue at its C-terminal domain [15]. Overexpression of *SOS1* or *SOS2* gene enhanced salt tolerance in transgenic plants [48–51].

Another important strategy to maintain low cytosolic Na^+ is to sequester excess Na^+ ions to intracellular compartments such as vacuoles. Tonoplast-localized Na^+/H^+ exchanger 1 (*NHX1*) overexpression in *Arabidopsis* and crop plants led to enhanced salt tolerance. NHX exchangers are classified into two classes: Class I NHX exchangers contain four members (NHX1 to NHX4). These transporters are localized to the tonoplast. Class II NHX members include NHX5 and NHX6, which are present in the endosomal compartments. NHX5 and NHX6 were reported to be present in the membrane of golgi and trans-golgi network membranes and are associated with transport of endosomal vesicles to the vacuole [52]. Co-overexpression of plasma membrane-localized *AtSOS1* and vacoular *AtNHX1* significantly improved salt tolerance compared to the transformed lines expressing either *AtSOS1* or *AtNHX1* [53].

5.6.3 Proton pumps

In plants three types of proton transport proteins are present in the plant cell: plasma membrane H^+-ATPases or P-type ATPases, vacuolar H^+-ATPases (V-H^+-ATPases), and vacuolar H^+ pyrophosphatases (V-HPPases) [5, 8]. P-type H^+-ATPases are activated to maintain more negative membrane potential across the plasma membrane and are linked to salt tolerance [5, 54]. Plasma membrane H^+-ATPases pump H^+ outside the cell to maintain electrochemical potential gradient across the plasma membrane. These pumps use energy released by hydrolyzing ATP to pump out protons across the plasma membrane. The *Arabidopsis* genome contains 11 genes encoding for H^+-ATPase [55]. Transcripts of three AHA genes (*AHA1*, *AHA2* and *AHA3*) are induced by salt stress. Also, the mRNA levels of H^+-ATPase-coding genes were higher in halophytes compared to glycophytes. Higher activity of H^+-ATPase in halophytes is linked to salt tolerance [54]. H^+-ATPase contains an autoinhibitory C-terminal domain that regulates its activity. Activation of these pumps is mediated by the removal of the C-terminal autoinhibitory domain through phosphorylation and binding of 14–3–3 regulatory proteins. Constitutive expression of *NpPMA4* from *Nicotiana plumbaginifolia* does not alter salt tolerance whereas the expression of truncated *NpPMA4* lacking autoinhibitory domain confers enhanced salt tolerance to genetically transformed plants [56]. In another study, ectopic expression of a plasma membrane H^+-ATPase from the halophyte *Sesuvium portulacastrum* conferred salt tolerance to transformed *Arabidopsis* [57].

5.6.4 Vacuolar H^+-ATPase

Tonoplast-localized H^+-ATPase is associated with transport of protons from cytosol to the vacuolar (V) compartment. V-H^+-ATPases are multi-subunit enzymes consisting of two major domains: V1 and V0. V1 is the membrane peripheral domain responsible for ATP hydrolysis, whereas V0 is the transmembrane domain responsible for transfer of H^+ across the tonoplast membrane [58]. V-H^+-ATPases help in acidification of vacuolar lumen and establish an electrochemical gradient across the tonoplast membrane. This electrochemical gradient helps in the sequestration of Na^+ in the vacuolar compartment. It has been reported that several genes coding for the subunits of V-H^+-ATPase get induced by salt stress [59–61]. The activity of V-H^+-ATPase was also induced during salt stress [60, 61].

Ectopic expression of a gene, *SaVHAc*, coding for the vacuolar H^+-ATPase subunit c1 from the halophyte *Spartina alterniflora* improved salt tolerance in transformed rice plants [62]. Similarly, expression of *ScVHA-B, ScVHA-C,* and *ScVHA-H* encoding V-H^+-ATPase subunits B, C, and H respectively, imparted tolerance to salt and saline alkali stress to genetically transform *Medicago sativa* [61]. The main strategy of halophyte *Suaeda salsa* to survive under high salt conditions is due to upregulated activity of V-H^+-ATPase [63]. This halophyte accumulates more Na^+ ions in the shoots compared to roots under salt stress. Accordingly, activity of H^+-ATPase was higher in the shoots compared to root tissue under NaCl treatment [64].

5.6.5 Vacuolar H^+ pyrophosphatases

These pumps are associated with the translocation of cytosolic H^+ across membranes coupled with hydrolysis of the pyrophosphatases that are released during several anabolic reactions such as synthesis of DNA, RNA, and proteins [65].

Unlike V-H^+-ATPases, which are composed of multi-subunits, V-HPPase is a single protein pump [65]. These pumps are involved in maintaining electrochemical gradient across the tonoplast for the active transport of Na^+ and K^+ in the vacuole. Two phylogenetically distinct classes of V-HPPases exists in plants: Class I is K^+ dependent whereas Class II is K^+ independent [65]. Salt stress induced higher expression of gene encoding for H^+-PPase in salt barley cultivar compared to that of salt-sensitive [66]. Overexpression of the vacuolar H^+-pyrophosphatase 1 (*OVP1*) gene in rice provided greater tolerance to salt stress [67]. Overexpression of *AVP1* not only improved salt tolerance but also improved growth under nitrogen and phosphorous limitations in transgenic tomato, rice, and romaine lettuce [68–70]. Another report observed that the expression of two HPPase-encoding genes (*SeVP1* and *SeVP2*) were induced by salt and nitrogen limitations in the euhalophyte *Salicornia europaea* [71]. Interestingly, transgenic *Arabidopsis* or wheat plants expressing *SeVP1* or *SeVP2* conferred tolerance to the combined stress of salt and low nitrogen [71]. Co-expression of vacoular *AtAVP1* and *AtNHX1* together enhanced salt tolerance of transgenic soybean plants compared to those either expressing *AtAVP1* or *AtNHX1*. The higher salt tolerance of these plants was attributed to enhanced compartmentalization of Na^+ into vacuoles [7].

5.7 Compatible solutes

The synthesis of compatible solutes has been used as one of the many strategies to survive under abiotic stresses. They are small, nontoxic soluble molecules that accumulate at higher concentrations in response to stressors. These compatible solutes include betaine and related compounds; proline; and sugar and sugar alcohols such as trehalose, mannitol, and sorbitol. Genetic transformation with genes linked to generation of the various compatible solutes resulted in improved tolerance to abiotic stresses [5]. Alternatively, exogenous application of glycine betaine (GB) or proline has significantly enhanced salt stress tolerance [72–74].

5.7.1 Glycine betaine

Glycine betaine is a well-known osmoprotectant. Typically, it displays very low expression when plants grow normally without any stress but its accumulation increases during salt, drought, and temperature stresses [75]. Several reports have revealed that GB protects the quaternary structure of proteins and photosynthetic machinery, and maintains membrane integrity against the damaging effects of several environmental cues [75, 76]. Exogenous application of GB confers salinity tolerance to plants by modulating the antioxidative defense mechanism [75]. GB biosynthesis occurs via two distinct pathways. One of the biosynthetic pathways is the two-step oxidation reaction of choline to betaine aldehyde as an intermediate product and finally to GB. This pathway is common in animals, plants, and microorganisms. The first oxidation reaction is catalyzed by choline monooxygenase (CMO) in plants whereas in animals and microorganisms this reaction is catalyzed by choline dehydrogenase (CDH). The second oxidation reaction to convert toxic betaine aldehyde to GB is catalyzed by NAD^+-dependent betaine aldehyde dehydrogenase (BADH) [76, 77]. In soil-borne bacteria *Arthobacter globiformis* and *Arthrobacter pascens*, GB is synthesized directly from choline by a single-step reaction catalyzed by choline oxidase (COD). In some extreme halophytes, GB can be synthesized from glycine by three successive N-methylation reactions catalyzed by glycinesarcosine methyltransferase (GSMT) and sarcosine dimethylglycine methyltransferase (SDMT) [77].

The genes coding for the enzymes involved in the biosynthesis of GB have been cloned and used to engineer plants with enhanced salinity tolerance. Ectopic expression of the *codA* gene encoding for COD enhances salt tolerance in *Arabidopsis*, rice, *Brassica*, and tomato [78–81]. The rice genome lacks a functional gene encoding for CMO but it contains two copies of a gene which encodes for BADH. Ectopic expression of CMO from spinach enhances GB accumulation and confers salt tolerance to genetically transformed rice seedlings [82]. It was observed that halophytes accumulate higher levels of GB under different abiotic stresses. Genes encoding for CMO and BADH from halophytes were expressed in glycophytes to enhance their salt tolerance [83–85].

5.7.2 Proline

Proline is another important osmolyte that accumulates during abiotic stresses including salinity. Several reports have indicated proline accumulation linked with abiotic stress tolerance. Biosynthesis of proline in plants occurs from glutamate and ornithine. Proline synthesis and degradation forms a cycle involving different cellular compartments. Glutamate-derived proline synthesis involves two enzymes, Δ^1-pyrroline-5-carboxylate synthetase (P5CS) and Δ^1-pyrroline-5-carboxylate reductase (P5CR). Proline is catabolized back to glutamate via action of two enzymes, namely proline dehydrogenase (PDH) and Δ^1-pyrroline-5-carboxylate dehydrogenase [86]. P5CS1 is a rate-limiting step in proline biosynthesis whereas reduced expression of *PDH* favors proline accumulation [86]. Comparative transcript profiling genes associated in biosynthesis and catabolism of proline in *A. thaliana* and *T. halophila* revealed differential gene expression of the *PDH* gene in these species [87]. Expression of the *PDH* gene is undetectable in the shoots and accordingly lower PDH enzyme activity was found in *T. halophila* seedlings, which could be the reason for higher proline accumulation in halophytes [87]. Transcripts of *P5CS1* were upregulated by salt and drought stresses [88, 89]. Expression of the *P5CS* gene is regulated by feedback inhibition by free proline accumulation. Substitution of phenyl alanine to alanine at the 129 position of *Vigna aconitifolia* P5CS enzyme (P5CSF129A) removed feedback inhibition in transgenic plants [90, 91]. Foliar application of proline improved salt tolerance by reducing ion toxicity, increasing biological nitrogen fixation, and upregulation of genes associated with abiotic stress [92]. *P5CS1* was engineered in several crop species to increase proline accumulation and provide tolerance to abiotic stress [93–95]. Expression of the *P5CS* gene driven by a stress-inducible promoter exhibited faster growth compared to its constitutive expression in transgenic rice under salt stress [96].

5.8 Potassium transporters in salt tolerance

Potassium constitutes approximately 10% of the total dry matter in plants. It is required for many enzymatic reactions, ionic homeostasis, pH maintenance, and membrane potential. Recently, in addition to the nutritional context, recognition of the role of K^+ as a signaling molecule is gaining importance [22, 97]. Maintenance of K^+ homeostasis is important for several stress adaptive responses including salinity. A number of K^+ transporters involved in transport and maintaining ionic homeostasis have been identified. K^+ enters root cells through two types of transporters: high-affinity K^+ transporters, which operate under low extracellular K availability (μM range), and low-affinity K^+ transporters, which are active under high extracellular K supply (mM range) [22]. High-affinity transporters include members of the KT/HAK/KUP family of transporters whereas low-affinity K^+ transporters include AKT1. Cytosolic K^+ retention is particularly important for determining the salinity tolerance in plants. It has been observed that salt-tolerant varieties have higher cytosolic K^+ retention capacity compared to sensitive ones [98–100]. Salt stress induces efflux of K^+ from the cytosol. Influx of Na^+ depolarizes the plasma membrane, resulting in activation of the guard cell outward-rectifying K^+ channel (GORK). Also, Na^+-induced ROS production activates NSCC channels. Together, the efflux of K^+ from cytosol to apoplast mediates activation of the GORK and NSCC channels. The salt tolerance of barley cultivars depends on cytosolic K^+ retention in the root tissue. Salt-sensitive varieties of wheat and barley exhibited higher salt-induced K^+ efflux compared to salt-tolerant varieties [22]. Plant salinity tolerance is greatly dependent on the transporters involved in uptake and distribution of Na^+ and K^+ to various tissues and cytosolic K^+/Na^+ ratio maintenance [22]. Attempts have been made to genetically manipulate potassium transporters to develop salt-tolerant transgenic plants. *OsHAK16* is induced under salt stress and K^+-deficient conditions. Overexpression of *OsHAK16* increased K^+ uptake and its translocation to the shoot, whereas its disruption diminished K^+ uptake in the root and consequently reduced K^+/Na^+ ratio [101]. Similarly, the functions of other HAK transporters have been genetically characterized for providing tolerance against salt stress [102–104]. Expression of *AKT1* from salt-tolerant *Puccinellia tenuiflora* conferred salt tolerance to transformed *Arabidopsis* plants by increasing K^+ uptake [105]. Recently, several articles have been published wherein the complex molecular mechanisms of many abiotic stress tolerances and responses have been explained in detail [121–123].

5.9 Conclusions

Crop yield is negatively affected by salt stress. Advances in genetic and molecular technologies have helped in dissecting salt tolerance mechanisms in plants, but still we are far from a complete understanding of this complex trait. Comparative studies of stress response at the transcriptomic, proteomic, and metabolic levels have helped in the identification of several genes, proteins, or metabolites that provide stress tolerance to halophytes. Plants maintain ionic homeostasis in the cytosol by controlling the activity of different ion channels and transporters. It has been reported that salt tolerance mechanisms differ with species, cultivars, and populations. Hence, it would be an oversimplification to directly use the information from model plants to engineer salt tolerance in crops. Genetic manipulation of crops for salt tolerance has been attempted either by constitutive expression of single genes or co-expression of two genes. Also, in most cases, studies of genetic manipulation have been limited to model plants and *in vitro* conditions only. Constitutive expression of genes constantly demands energy and is often associated with undesirable effects. Hence, salt-stress-inducible promoters should be used to express the genes of interest. Phenotypic analysis of transgenic plants should be validated not only in model plants but also in crops under field conditions. Another important hurdle in engineering salt tolerance in crops is due to its polygenic inheritance. Multiple genes should be manipulated at a time to engineer stress tolerance in crops. This can be practically achieved by use of improved CRISPR/Cas9 genome-editing tools.

Acknowledgments

I.A. acknowledges University Grants Commission, Govt. of India, for providing a D.S. Kothari Post-doctoral Fellowship.

Multiple choice questions

1. Soil salinity is expressed in terms of ...
   (a) Electrical conductivity;
   (b) Meter;
   (c) Kg/square feet;
   (d) Ml.
2. Saline soil contains sodium salts of ...
   (a) Bicarbonates;
   (b) Carbonates;
   (c) Both (a) and (b);
   (d) None.
3. Salt-affected soils can be reclaimed by ...
   (a) Gypsum;
   (b) Calcium chloride;
   (c) Sodium carbonate;
   (d) Calcium oxide.
4. Which of the following compounds has been produced in transgenic plants to improve tolerance to salt stress?
   (a) Sucrose;
   (b) Mannitol;
   (c) Nicotine;
   (d) Octopine.
5. Which of the following plants is salt tolerant?
   (a) Rice;
   (b) Chickpea;

(c) Mustard;
(d) Tecticornia.

Answers: 1 (a); 2 (c); 3 (a); 4 (b); 5 (d).

Descriptive-type questions

1. How does salt stress affect plant growth and development and result in yield loss in crops?
2. Explain the mechanisms involved in salt stress perception and signaling to reprogram the transcriptional machinery needed to adapt to saline environments.
3. How do ion transporters mediate Na^+ entry into root cells and translocate it to shoot tissues?
4. What is the role of compatible solutes like proline and glycine betaine in salt stress and how does the genetic manipulation of the genes involved in their synthesis affect the salt tolerance of transgenic plants?
5. How are potassium transporters involved in maintaining ion homeostasis under salt stress?

References

[1] Munns R and Tester M 2008 Mechanisms of salinity tolerance *Annu. Rev. Plant Biol.* **59** 651–81

[2] Ahuja I, de Vos R C H, Bones A M and Hall R D 2010 Plant molecular stress responses face climate change *Trends Plant Sci.* **15** 664–74

[3] Ahmed I, Shukla P and Gautam R 2021 *Omics Technologies for Sustainable Agriculture and Global Food Security* vol 1 (Berlin: Springer) pp 169–89

[4] Gautam R, Meena R K, Rohela G K, Singh N K and Shukla P 2021 *Omics Technologies for Sustainable Agriculture and Global Food Security* vol 1 (Berlin: Springer) pp 125–48

[5] Wani S H, Kumar V, Khare T, Guddimalli R, Parveda M, Solymosi K, Suprasanna P and Kishor P K 2020 Engineering salinity tolerance in plants: progress and prospects *Planta* **251** 1–29

[6] Cabello J V, Lodeyro A F and Zurbriggen M D 2014 Novel perspectives for the engineering of abiotic stress tolerance in plants *Curr. Opin. Biotechnol.* **26** 62–70

[7] Nguyen N T, Vu H T, Nguyen T T, Nguyen L-A T, Nguyen M-C D, Hoang K L, Nguyen K T and Quach T N 2019 Co-expression of *Arabidopsis* AtAVP1 and AtNHX1 to improve salt tolerance in soybean *Crop Sci.* **59** 1133–43

[8] Kotula L, Garcia Caparros P, Zörb C, Colmer T D and Flowers T J 2020 Improving crop salt tolerance using transgenic approaches: an update and physiological analysis *Plant, Cell Environ.* **43** 2932–56

[9] Ahmed I, Yadav D, Shukla P, Vineeth T, Sharma P and Kirti P 2017 Constitutive expression of *Brassica juncea* annexin, AnnBj2 confers salt tolerance and glucose and ABA insensitivity in mustard transgenic plants *Plant Sci.* **265** 12–28

[10] Ahmed I, Yadav D, Shukla P and Kirti P 2018 Heterologous expression of *Brassica juncea* annexin, AnnBj2 confers salt tolerance and ABA insensitivity in transgenic tobacco seedlings *Funct. Integrat. Genomics* **18** 569–79

[11] Yadav D, Ahmed I, Shukla P, Boyidi P and Kirti P B 2016 Overexpression of *Arabidopsis* AnnAt8 alleviates abiotic stress in transgenic Arabidopsis and tobacco *Plants* **5** 18
[12] Su Y, Guo A, Huang Y, Wang Y and Hua J 2020 GhCIPK6a increases salt tolerance in transgenic upland cotton by involving in ROS scavenging and MAPK signaling pathways *BMC Plant Biol.* **20** 1–19
[13] Campo S, Baldrich P, Messeguer J, Lalanne E, Coca M and San Segundo B 2014 Overexpression of a calcium-dependent protein kinase confers salt and drought tolerance in rice by preventing membrane lipid peroxidation *Plant Physiol.* **165** 688–704
[14] Singh N K, Shukla P and Kirti P 2020 A CBL-interacting protein kinase AdCIPK5 confers salt and osmotic stress tolerance in transgenic tobacco *Sci. Rep.* **10** 1–14
[15] Zhao C, Zhang H, Song C, Zhu J-K and Shabala S 2020 Mechanisms of plant responses and adaptation to soil salinity *Innovation* **1** 100017
[16] Bartels D and Dinakar C 2013 Balancing salinity stress responses in halophytes and non-halophytes: a comparison between Thellungiella and *Arabidopsis thaliana Funct. Plant Biol.* **40** 819–31
[17] Barkla B J, Castellanos-Cervantes T, Diaz de Leon J L, Matros A, Mock H-P, Perez-Alfocea F, Salekdeh G H, Witzel K and Zörb C 2013 Elucidation of salt stress defense and tolerance mechanisms of crop plants using proteomics—current achievements and perspectives *Proteomics* **13** 1885–900
[18] Shabala S 2013 Learning from halophytes: physiological basis and strategies to improve abiotic stress tolerance in crops *Annal. Bot.* **112** 1209–21
[19] Yuan F, Yang H, Xue Y, Kong D, Ye R, Li C, Zhang J, Theprungsirikul L, Shrift T and Krichilsky B 2014 OSCA1 mediates osmotic-stress-evoked Ca^{2+} increases vital for osmosensing in *Arabidopsis Nature* **514** 367–71
[20] Jiang Z, Zhou X, Tao M, Yuan F, Liu L, Wu F, Wu X, Xiang Y, Niu Y and Liu F 2019 Plant cell-surface GIPC sphingolipids sense salt to trigger Ca^{2+} influx *Nature* **572** 341–6
[21] Hamilton E S, Jensen G S, Maksaev G, Katims A, Sherp A M and Haswell E S 2015 Mechanosensitive channel MSL8 regulates osmotic forces during pollen hydration and germination *Science* **350** 438–41
[22] Isayenkov S V and Maathuis F J 2019 Plant salinity stress: many unanswered questions remain *Front. Plant Sci.* **10** 80
[23] Deinlein U, Stephan A B, Horie T, Luo W, Xu G and Schroeder J I 2014 Plant salt-tolerance mechanisms *Trends Plant Sci.* **19** 371–9
[24] Maathuis F J M 2014 Sodium in plants: perception, signalling, and regulation of sodium fluxes *J. Exp. Bot.* **65** 849–58
[25] Hamamoto S, Horie T, Hauser F, Deinlein U, Schroeder J I and Uozumi N 2015 HKT transporters mediate salt stress resistance in plants: from structure and function to the field *Curr. Opin. Biotechnol.* **32** 113–20
[26] Berthomieu P, Conéjéro G, Nublat A, Brackenbury W J, Lambert C, Savio C, Uozumi N, Oiki S, Yamada K and Cellier F 2003 Functional analysis of AtHKT1 in Arabidopsis shows that Na^+ recirculation by the phloem is crucial for salt tolerance *EMBO* J **22** 2004–14
[27] Rus A, Lee B-H, Muñoz-Mayor A, Sharkhuu A, Miura K, Zhu J-K, Bressan R A and Hasegawa P M 2004 AtHKT1 facilitates Na^+ homeostasis and K^+ nutrition in planta *Plant Physiol.* **136** 2500–11
[28] Moller I S, Gilliham M, Jha D, Mayo G M, Roy S J, Coates J C, Haseloff J and Tester M 2009 Shoot Na^+ exclusion and increased salinity tolerance engineered by cell type-specific alteration of Na^+ transport in *Arabidopsis Plant Cell* **21** 2163–78

[29] Ren Z-H, Gao J-P, Li L-G, Cai X-L, Huang W, Chao D-Y, Zhu M-Z, Wang Z-Y, Luan S and Lin H-X 2005 A rice quantitative trait locus for salt tolerance encodes a sodium transporter *Nat. Genet.* **37** 1141–6

[30] Horie T, Motoda J, Kubo M, Yang H, Yoda K, Horie R, Chan W Y, Leung H Y, Hattori K and Konomi M 2005 Enhanced salt tolerance mediated by AtHKT1 transporter-induced Na^+ unloading from xylem vessels to xylem parenchyma cells *Plant J.* **44** 928–38

[31] Munns R and James R A 2003 Screening methods for salinity tolerance: a case study with tetraploid wheat *Plant Soil* **253** 201–18

[32] Huang S B, Spielmeyer W, Lagudah E S, James R A, Platten J D, Dennis E S and Munns R 2006 A sodium transporter (HKT7) is a candidate for Nax1, a gene for salt tolerance in durum wheat *Plant Physiol.* **142** 1718–27

[33] Mason M G, Jha D, Salt D E, Tester M, Hill K, Kieber J J and Eric Schaller G 2010 Type-B response regulators ARR1 and ARR12 regulate expression of AtHKT1; 1 and accumulation of sodium in *Arabidopsis* shoots *Plant J.* **64** 753–63

[34] Nishiyama R, Le D T, Watanabe Y, Matsui A, Tanaka M, Seki M, Yamaguchi-Shinozaki K, Shinozaki K and Tran L-S P 2012 Transcriptome analyses of a salt-tolerant cytokinin-deficient mutant reveal differential regulation of salt stress response by cytokinin deficiency *PLoS One* **7** e32124

[35] Shkolnik-Inbar D, Adler G and Bar-Zvi D 2013 ABI4 downregulates expression of the sodium transporter HKT1; 1 in *Arabidopsis* roots and affects salt tolerance *Plant J.* **73** 993–1005

[36] Yang O, Popova O V, Suthoff U, Luking I, Dietz K J and Golldack D 2009 The *Arabidopsis* basic leucine zipper transcription factor AtbZIP24 regulates complex transcriptional networks involved in abiotic stress resistance *Gene* **436** 45–55

[37] Wang T T, Ren Z J, Liu Z Q, Feng X, Guo R Q, Li B G, Li L G and Jing H C 2014 SbHKT1; 4, a member of the high-affinity potassium transporter gene family from *Sorghum bicolor*, functions to maintain optimal Na^+/K^+ balance under Na^+ stress *J. Integr. Plant Biol.* **56** 315–32

[38] Chen H, Chen X, Gu H, Wu B, Zhang H, Yuan X and Cui X 2014 GmHKT1; 4, a novel soybean gene regulating Na^+/K^+ ratio in roots enhances salt tolerance in transgenic plants *Plant Growth Regul.* **73** 299–308

[39] Zhang M, Cao Y, Wang Z, Wang Z Q, Shi J, Liang X, Song W, Chen Q, Lai J and Jiang C 2018 A retrotransposon in an HKT1 family sodium transporter causes variation of leaf Na^+ exclusion and salt tolerance in maize *New Phytol.* **217** 1161–76

[40] Mian A, Oomen R J, Isayenkov S, Sentenac H, Maathuis F J and Véry A A 2011 Over-expression of an Na^+-and K^+-permeable HKT transporter in barley improves salt tolerance *Plant J.* **68** 468–79

[41] Horie T, Brodsky D E, Costa A, Kaneko T, Lo Schiavo F, Katsuhara M and Schroeder J I 2011 K^+ transport by the OsHKT2; 4 transporter from rice with atypical Na^+ transport properties and competition in permeation of K^+ over Mg^{2+} and Ca^{2+} ions *Plant Physiol.* **156** 1493–507

[42] Laurie S, Feeney K A, Maathuis F J, Heard P J, Brown S J and Leigh R A 2002 A role for HKT1 in sodium uptake by wheat roots *Plant J.* **32** 139–49

[43] Liu J and Zhu J-K 1998 A calcium sensor homolog required for plant salt tolerance *Science* **280** 1943–5

[44] Liu J, Ishitani M, Halfter U, Kim C-S and Zhu J-K 2000 The *Arabidopsis thaliana* SOS2 gene encodes a protein kinase that is required for salt tolerance *Proc. Natl Acad. Sci.* **97** 3730–4

[45] Quintero F J, Ohta M, Shi H, Zhu J-K and Pardo J M 2002 Reconstitution in yeast of the *Arabidopsis* SOS signaling pathway for Na^+ homeostasis *Proc. Natl Acad. Sci.* **99** 9061–6

[46] Qiu Q-S, Barkla B J, Vera-Estrella R, Zhu J-K and Schumaker K S 2003 Na^+/H^+ exchange activity in the plasma membrane of *Arabidopsis Plant Physiol.* **132** 1041–52

[47] Shi H, Ishitani M, Kim C and Zhu J-K 2000 The *Arabidopsis thaliana* salt tolerance gene SOS1 encodes a putative Na^+/H^+ antiporter *Proc. Natl Acad. Sci.* **97** 6896–901

[48] Huertas R, Olias R, Eljakaoui Z, Gálvez F J, Li J, De Morales P A, Belver A and Rodríguez-Rosales M P 2012 Overexpression of SlSOS2 (SlCIPK24) confers salt tolerance to transgenic tomato *Plant Cell Environ.* **35** 1467–82

[49] Shi H, Lee B-H, Wu S-J and Zhu J-K 2003 Overexpression of a plasma membrane Na^+/H^+ antiporter gene improves salt tolerance in *Arabidopsis thaliana Nat. Biotechnol.* **21** 81–5

[50] El Mahi H, Pérez-Hormaeche J, De Luca A, Villalta I, Espartero J, Gámez-Arjona F, Fernández J L, Bundó M, Mendoza I and Mieulet D 2019 A critical role of sodium flux via the plasma membrane Na^+/H^+ exchanger SOS1 in the salt tolerance of rice *Plant Physiol.* **180** 1046–65

[51] Yang Y, Tang R J, Jiang C M, Li B, Kang T, Liu H, Zhao N, Ma X J, Yang L and Chen S L 2015 Overexpression of the PtSOS2 gene improves tolerance to salt stress in transgenic poplar plants *Plant Biotechnol. J.* **13** 962–73

[52] Bassil E, Tajima H, Liang Y-C, Ohto M-A, Ushijima K, Nakano R, Esumi T, Coku A, Belmonte M and Blumwald E 2011 The *Arabidopsis* Na^+/H^+ antiporters NHX1 and NHX2 control vacuolar pH and K^+ homeostasis to regulate growth, flower development, and reproduction *Plant Cell* **23** 3482–97

[53] Pehlivan N, Sun L, Jarrett P, Yang X, Mishra N, Chen L, Kadioglu A, Shen G and Zhang H 2016 Co-overexpressing a plasma membrane and a vacuolar membrane sodium/proton antiporter significantly improves salt tolerance in transgenic *Arabidopsis* plants *Plant Cell Physiol.* **57** 1069–84

[54] Mansour M M F 2014 The plasma membrane transport systems and adaptation to salinity *J. Plant Physiol.* **171** 1787–800

[55] Falhof J, Pedersen J T, Fuglsang A T and Palmgren M 2016 Plasma membrane H^+-ATPase regulation in the center of plant physiology *Mol. Plant* **9** 323–37

[56] Gévaudant F, Duby G, von Stedingk E, Zhao R, Morsomme P and Boutry M 2007 Expression of a constitutively activated plasma membrane H^+-ATPase alters plant development and increases salt tolerance *Plant Physiol.* **144** 1763–76

[57] Fan Y, Wan S, Jiang Y, Xia Y, Chen X, Gao M, Cao Y, Luo Y, Zhou Y and Jiang X 2018 Over-expression of a plasma membrane H^+-ATPase SpAHA1 conferred salt tolerance to transgenic *Arabidopsis Protoplasma* **255** 1827–37

[58] Cipriano D J, Wang Y, Bond S, Hinton A, Jefferies K C, Qi J and Forgac M 2008 Structure and regulation of the vacuolar ATPases *Biochimica et Biophys. Acta (BBA)-Bioenerg.* **1777** 599–604

[59] Kirsch M, Zhigang A, Viereck R, Löw R and Rausch T 1996 Salt stress induces an increased expression of V-type H^+-ATPase in mature sugar beet leaves *Plant Mole. Biol.* **32** 543–7

[60] Zhang M, Fang Y, Liang Z and Huang L 2012 Enhanced expression of vacuolar H^+-ATPase subunit E in the roots is associated with the adaptation of *Broussonetia papyrifera* to salt stress *PLoS One* **7** e48183

[61] Wang F-W, Chao W, Yao S, Nan W, Li X-W, Dong Y-Y, Na Y, Liu X-M, Huan C and Chen X-F 2016 Overexpression of vacuolar proton pump ATPase (V-H^+-ATPase) subunits B, C and H confers tolerance to salt and saline-alkali stresses in transgenic alfalfa (*Medicago sativa* L.) *J. Integr. Agri.* **15** 2279–89

[62] Baisakh N, RamanaRao M V, Rajasekaran K, Subudhi P, Janda J, Galbraith D, Vanier C and Pereira A 2012 Enhanced salt stress tolerance of rice plants expressing a vacuolar H^+-ATPase subunit c1 (SaVHAc1) gene from the halophyte grass *Spartina alterniflora* Löisel *Plant Biotechnol. J.* **10** 453–64

[63] Wang B, Lüttge U and Ratajczak R 2001 Effects of salt treatment and osmotic stress on V-ATPase and V-PPase in leaves of the halophyte *Suaeda salsa* *J. Exp. Bot.* **52** 2355–65

[64] Yang M F, Song J and Wang B S 2010 Organ-specific responses of vacuolar H^+-ATPase in the shoots and roots of C3 halophyte *Suaeda salsa* to NaCl *J. Integr. Plant Biol.* **52** 308–14

[65] Segami S, Asaoka M, Kinoshita S, Fukuda M, Nakanishi Y and Maeshima M 2018 Biochemical, structural and physiological characteristics of vacuolar H^+-pyrophosphatase *Plant Cell Physiol.* **59** 1300–8

[66] Elsawy H I, Mekawy A M M, Elhity M A, Abdel-Dayem S M, Abdelaziz M N, Assaha D V, Ueda A and Saneoka H 2018 Differential responses of two Egyptian barley (*Hordeum vulgare* L.) cultivars to salt stress *Plant Physiol. Biochem.* **127** 425–35

[67] Kim J-J, Park S-I, Kim Y-H, Park H-M, Kim Y-S and Yoon H-S 2020 Overexpression of a proton pumping gene OVP1 enhances salt stress tolerance, root growth and biomass yield by regulating ion balance in rice (*Oryza sativa* L.) *Environ. Exp. Bot.* **175** 104033

[68] Yang H, Knapp J, Koirala P, Rajagopal D, Peer W A, Silbart L K, Murphy A and Gaxiola R A 2007 Enhanced phosphorus nutrition in monocots and dicots over-expressing a phosphorus-responsive type I H^+-pyrophosphatase *Plant Biotechnol. J.* **5** 735–45

[69] Gaxiola R A, Sanchez C A, Paez-Valencia J, Ayre B G and Elser J J 2012 Genetic manipulation of a 'vacuolar' H^+-PPase: from salt tolerance to yield enhancement under phosphorus-deficient soils *Plant Physiol.* **159** 3–11

[70] Paez-Valencia J, Sanchez-Lares J, Marsh E, Dorneles L T, Santos M P, Sanchez D, Winter A, Murphy S, Cox J and Trzaska M 2013 Enhanced proton translocating pyrophosphatase activity improves nitrogen use efficiency in romaine lettuce *Plant Physiol.* **161** 1557–69

[71] Lv S, Jiang P, Nie L, Chen X, Tai F, Wang D, Fan P, Feng J, Bao H and Wang J 2015 H^+-pyrophosphatase from *Salicornia europaea* confers tolerance to simultaneously occurring salt stress and nitrogen deficiency in *A rabidopsis* and wheat *Plant, Cell Environ.* **38** 2433–49

[72] Heuer B 2003 Influence of exogenous application of proline and glycinebetaine on growth of salt-stressed tomato plants *Plant Sci.* **165** 693–9

[73] Raza S H, Athar H R, Ashraf M and Hameed A 2007 Glycinebetaine-induced modulation of antioxidant enzymes activities and ion accumulation in two wheat cultivars differing in salt tolerance *Environ. Exp. Bot.* **60** 368–76

[74] Malekzadeh P 2015 Influence of exogenous application of glycinebetaine on antioxidative system and growth of salt-stressed soybean seedlings (*Glycine max* L.) *Physiol. Mol. Biol. Plants* **21** 225–32

[75] Ashraf M and Foolad M R 2007 Roles of glycine betaine and proline in improving plant abiotic stress resistance *Environ. Exp. Bot.* **59** 206–16

[76] Hussain Wani S, Brajendra Singh N, Haribhushan A and Iqbal Mir J 2013 Compatible solute engineering in plants for abiotic stress tolerance-role of glycine betaine *Curr. Genomics* **14** 157–65

[77] Chen T H and Murata N 2011 Glycinebetaine protects plants against abiotic stress: mechanisms and biotechnological applications *Plant, Cell Environ.* **34** 1–20

[78] Hayashi H, Mustardy L, Deshnium P, Ida M and Murata N 1997 Transformation of *Arabidopsis thaliana* with the codA gene for choline oxidase; accumulation of glycinebetaine and enhanced tolerance to salt and cold stress *Plant J.* **12** 133–42

[79] Prasad K, Sharmila P, Kumar P and Saradhi P P 2000 Transformation of *Brassica juncea* (L.) Czern with bacterial codA gene enhances its tolerance to salt stress *Mol. Breed.* **6** 489–99

[80] Goel D, Singh A K, Yadav V, Babbar S B, Murata N and Bansal K C 2011 Transformation of tomato with a bacterial codA gene enhances tolerance to salt and water stresses *J. Plant Physiol.* **168** 1286–94

[81] Mohanty A, Kathuria H, Ferjani A, Sakamoto A, Mohanty P, Murata N and Tyagi A 2002 Transgenics of an elite indica rice variety Pusa Basmati 1 harbouring the codA gene are highly tolerant to salt stress *Theor. Appl. Genetics* **106** 51–7

[82] Shirasawa K, Takabe T, Takabe T and Kishitani S 2006 Accumulation of glycinebetaine in rice plants that overexpress choline monooxygenase from spinach and evaluation of their tolerance to abiotic stress *Annal. Bot.* **98** 565–71

[83] Jia G-X, Zhu Z-Q, Chang F-Q and Li Y-X 2002 Transformation of tomato with the BADH gene from Atriplex improves salt tolerance *Plant Cell Rep.* **21** 141–6

[84] Wu W, Su Q, Xia X, Wang Y, Luan Y and An L 2008 The *Suaeda liaotungensis* kitag betaine aldehyde dehydrogenase gene improves salt tolerance of transgenic maize mediated with minimum linear length of DNA fragment *Euphytica* **159** 17–25

[85] Liu Z-H, Zhang H-M, Li G-L, Guo X-L, Chen S-Y, Liu G-B and Zhang Y-M 2011 Enhancement of salt tolerance in alfalfa transformed with the gene encoding for betaine aldehyde dehydrogenase *Euphytica* **178** 363–72

[86] Bhaskara G B, Yang T-H and Verslues P E 2015 Dynamic proline metabolism: importance and regulation in water limited environments *Front. Plant Sci.* **6** 484

[87] Kant S, Kant P, Raveh E and Barak S 2006 Evidence that differential gene expression between the halophyte, *Thellungiella halophila*, and *Arabidopsis thaliana* is responsible for higher levels of the compatible osmolyte proline and tight control of Na^+ uptake in *T. halophila Plant, Cell Environ.* **29** 1220–34

[88] Yoshiba Y, Nanjo T, Miura S, Yamaguchi-Shinozaki K and Shinozaki K 1999 Stress-responsive and developmental regulation of Δ1-pyrroline-5-carboxylate synthetase 1 (P5CS1) gene expression in *Arabidopsis thaliana Biochem. Biophys. Res. Commun.* **261** 766–72

[89] Su M, Li X-F, Ma X-Y, Peng X-J, Zhao A-G, Cheng L-Q, Chen S-Y and Liu G-S 2011 Cloning two P5CS genes from bioenergy sorghum and their expression profiles under abiotic stresses and MeJA treatment *Plant Sci.* **181** 652–9

[90] Zhang C-s, Lu Q and Verma D P S 1995 Removal of feedback inhibition of Δ1-pyrroline-5-carboxylate synthetase, a bifunctional enzyme catalyzing the first two steps of proline biosynthesis in plants *J. Biol. Chem.* **270** 20491–6

[91] Hong Z, Lakkineni K, Zhang Z and Verma D P S 2000 Removal of feedback inhibition of Δ1-pyrroline-5-carboxylate synthetase results in increased proline accumulation and protection of plants from osmotic stress *Plant Physiol.* **122** 1129–36

[92] El Moukhtari A, Cabassa-Hourton C, Farissi M and Savouré A 2020 How does proline treatment promote salt stress tolerance during crop plant development? *Front. Plant Sci.* **11** 1127

[93] Hmida-Sayari A, Gargouri-Bouzid R, Bidani A, Jaoua L, Savouré A and Jaoua S 2005 Overexpression of Δ1-pyrroline-5-carboxylate synthetase increases proline production and confers salt tolerance in transgenic potato plants *Plant Sci.* **169** 746–52

[94] Pavei D, Gonçalves-Vidigal M C, Schuelter A R, Schuster I, Vieira E S N, Vendruscolo E C G and Poletine J P 2016 Response to water stress in transgenic ('p5cs' gene) wheat plants ('*Triticum aestivum*' L.) *Aust. J. Crop Sci.* **10** 776–83

[95] Bhatnagar-Mathur P, Vadez V, Devi M J, Lavanya M, Vani G and Sharma K K 2009 Genetic engineering of chickpea (*Cicer arietinum* L.) with the P5CSF129A gene for osmoregulation with implications on drought tolerance *Mol. Breed.* **23** 591–606

[96] Su J and Wu R 2004 Stress-inducible synthesis of proline in transgenic rice confers faster growth under stress conditions than that with constitutive synthesis *Plant Sci.* **166** 941–8

[97] Wu H, Zhang X, Giraldo J P and Shabala S 2018 It is not all about sodium: revealing tissue specificity and signalling roles of potassium in plant responses to salt stress *Plant Soil* **431** 1–17

[98] Chakraborty K, Bose J, Shabala L and Shabala S 2016 Difference in root K^+ retention ability and reduced sensitivity of K^+-permeable channels to reactive oxygen species confer differential salt tolerance in three *Brassica* species *J. Exp. Bot.* **67** 4611–25

[99] Gao Y, Lu Y, Wu M, Liang E, Li Y, Zhang D, Yin Z, Ren X, Dai Y and Deng D 2016 Ability to remove Na^+ and retain K^+ correlates with salt tolerance in two maize inbred lines seedlings *Front. Plant Sci.* **7** 1716

[100] Wu H, Shabala L, Barry K, Zhou M and Shabala S 2013 Ability of leaf mesophyll to retain potassium correlates with salinity tolerance in wheat and barley *Physiol. Plant.* **149** 515–27

[101] Feng H, Tang Q, Cai J, Xu B, Xu G and Yu L 2019 Rice OsHAK16 functions in potassium uptake and translocation in shoot, maintaining potassium homeostasis and salt tolerance *Planta* **250** 549–61

[102] Zhang H, Xiao W, Yu W, Jiang Y and Li R 2020 Halophytic *Hordeum brevisubulatum* HbHAK1 facilitates potassium retention and contributes to salt tolerance *Int. J. Mol. Sci.* **21** 5292

[103] Chen G, Hu Q, Luo L, Yang T, Zhang S, Hu Y, Yu L and Xu G 2015 Rice potassium transporter OsHAK1 is essential for maintaining potassium-mediated growth and functions in salt tolerance over low and high potassium concentration ranges *Plant Cell Environ.* **38** 2747–65

[104] Shen Y, Shen L, Shen Z, Jing W, Ge H, Zhao J and Zhang W 2015 The potassium transporter OsHAK21 functions in the maintenance of ion homeostasis and tolerance to salt stress in rice *Plant Cell Environ.* **38** 2766–79

[105] Ardie S W, Liu S and Takano T 2010 Expression of the AKT1-type K^+ channel gene from *Puccinellia tenuiflora*, PutAKT1, enhances salt tolerance in *Arabidopsis Plant Cell Rep.* **29** 865–74

[106] Sahoo D P, Kumar S, Mishra S, Kobayashi Y, Panda S K and Sahoo L 2016 Enhanced salinity tolerance in transgenic mungbean overexpressing Arabidopsis antiporter (NHX1) gene *Mol. Breed.* **36** 1–15

[107] Kumari P H, Kumar S A, Sivan P, Katam R, Suravajhala P, Rao K, Varshney R K and Kishor P B K 2017 Overexpression of a plasma membrane bound Na^+/H^+ antiporter-like

protein (SbNHXLP) confers salt tolerance and improves fruit yield in tomato by maintaining ion homeostasis *Front. Plant Sci.* **7** 2027
[108] Jacobs A, Ford K, Kretschmer J and Tester M 2011 Rice plants expressing the moss sodium pumping ATPase PpENA1 maintain greater biomass production under salt stress *Plant Biotechnol. J.* **9** 838–47
[109] Bayat F, Shiran B, Belyaev D, Yur'eva N, Sobol'kova G, Alizadeh H, Khodambashi M and Babakov A 2010 Potato plants bearing a vacuolar Na^+/H^+ antiporter HvNHX2 from barley are characterized by improved salt tolerance *Russian J. Plant Physiol.* **57** 696–706
[110] Chen G-H, Yan W, Yang L-F, Gai J-Y and Zhu Y-L 2014 Overexpression of *StNHX1*, a novel vacuolar Na^+/H^+ antiporter gene from *Solanum torvum*, enhances salt tolerance in transgenic vegetable soybean *Hortic. Environ. Biotechnol.* **55** 213–21
[111] Cao D, Hou W, Liu W, Yao W, Wu C, Liu X and Han T 2011 Overexpression of TaNHX2 enhances salt tolerance of 'composite' and whole transgenic soybean plants *Plant Cell, Tissue Organ Culture (PCTOC)* **107** 541–52
[112] Schilling R K, Marschner P, Shavrukov Y, Berger B, Tester M, Roy S J and Plett D C 2014 Expression of the *A rabidopsis* vacuolar H^+-pyrophosphatase gene (AVP 1) improves the shoot biomass of transgenic barley and increases grain yield in a saline field *Plant Biotechnol. J.* **12** 378–86
[113] Pasapula V, Shen G, Kuppu S, Paez-Valencia J, Mendoza M, Hou P, Chen J, Qiu X, Zhu L and Zhang X 2011 Expression of an *Arabidopsis* vacuolar H^+-pyrophosphatase gene (AVP1) in cotton improves drought-and salt tolerance and increases fibre yield in the field conditions *Plant Biotechnol. J.* **9** 88–99
[114] Zhang K, Song J, Chen X, Yin T, Liu C, Li K and Zhang J 2016 Expression of the *Thellungiella halophila* vacuolar H^+-pyrophosphatase gene (TsVP) in cotton improves salinity tolerance and increases seed cotton yield in a saline field *Euphytica* **211** 231–44
[115] Shen G, Wei J, Qiu X, Hu R, Kuppu S, Auld D, Blumwald E, Gaxiola R, Payton P and Zhang H 2015 Co-overexpression of AVP1 and AtNHX1 in cotton further improves drought and salt tolerance in transgenic cotton plants *Plant Mol. Biol. Report.* **33** 167–77
[116] Wang L, Liu Y, Li D, Feng S, Yang J, Zhang J, Zhang J, Wang D and Gan Y 2019 Improving salt tolerance in potato through overexpression of *AtHKT1* gene *BMC Plant Biol.* **19** 1–15
[117] Plett D, Safwat G, Gilliham M, Skrumsager Møller I, Roy S, Shirley N, Jacobs A, Johnson A and Tester M 2010 Improved salinity tolerance of rice through cell type-specific expression of AtHKT1; 1 *PLoS One* **5** e12571
[118] Bhaskaran S and Savithramma D 2011 Co-expression of *Pennisetum glaucum* vacuolar Na^+/H^+ antiporter and *Arabidopsis* H^+-pyrophosphatase enhances salt tolerance in transgenic tomato *J. Exp. Bot.* **62** 5561–70
[119] Asif M A, Zafar Y, Iqbal J, Iqbal M M, Rashid U, Ali G M, Arif A and Nazir F 2011 Enhanced expression of AtNHX1, in transgenic groundnut (*Arachis hypogaea* L.) improves salt and drought tolerence *Mol. Biotechnol.* **49** 250–6
[120] Cheng C, Zhang Y, Chen X, Song J, Guo Z, Li K and Zhang K 2018 Co-expression of AtNHX1 and TsVP improves the salt tolerance of transgenic cotton and increases seed cotton yield in a saline field *Mol. Breed.* **38** 1–15
[121] Singh N, Mansoori A, Dey D, Kumar R and Kumar A 2021 Potential of metabolomics in plant abiotic stress management ed A Kumar, R Kumar, P Shukla and H K Patel *Omics*

Technologies for Sustainable Agriculture and Global Food Security vol II (Singapore: Springer) pp 193–214

[122] Kumar R *et al* 2021 Understanding omics driven plant improvement and *de novo* crop domestication: some examples *Front Genet.* **12** 637141

[123] Sharma V, Gupta P, Priscilla K, Kumar S, Hangargi B, Veershetty A, Ramrao D, Suresh S, Narasanna R and Naik G *et al* 2021 Metabolomics intervention towards better understanding of plant traits *Cells* **10** 346

[124] Quan R, Lin H, Mendoza I, Zhang Y, Cao W, Yang Y, Shang M, Chen S, Pardo J M and Guo Y 2007 SCABP8/CBL10, a putative calcium sensor, interacts with the protein kinase SOS2 to protect Arabidopsis shoots from salt stress *Plant Cell* **19** 1415–1431

IOP Publishing

Molecular Response and Genetic Engineering for Stress in Plants, Volume 1
Abiotic stress
Pawan Shukla, Anirudh Kumar, Rakesh Kumar and Manish K Pandey

Chapter 6

Cold stress: molecular insight and way forward

Nikhil Malhotra[1*], Shivani Sharma[2]

[1]*ICAR-National Bureau of Plant Genetic Resources Regional Station, Phagli, Shimla 171004, Himachal Pradesh, India*
[2]*ICAR-Indian Agricultural Research Institute Regional Station, Amartara Cottage, Shimla 171004, Himachal Pradesh, India*

Cold stress is an important abiotic and physiological stress that limits the yield and geographical distribution of several important crop plants. Plants have evolved a number of molecular regulatory mechanisms in response to cold stress, notably at the transcriptional level. Substantial progress has been made in recent past years at identifying key genes and their mechanisms associated with cold stress tolerance by understanding regulatory processes. The underlying mechanisms, on the other hand, are complex and little understood. The interpretation of the cold stress tolerance process, especially the signaling network during low-temperature stress, is crucial to improve agro-economical plants, including vegetable crop plants. We here in this chapter outline the important recent key breakthroughs associated with cold stress signaling and answer questions about plant cold stress responses. However, the present area needs to be further explored and more research contributions are required to answer the questions that could lead towards better understanding of the physiological and molecular mechanisms that strengthen plant tolerance to cold stress.

6.1 Introduction

Cold stress is considered as an important abiotic stress that limits the growth, productivity, and geographic spread of plants, including agricultural crops. Every year, abrupt frosts in the autumn, unusually cold/low temperatures in the winter, and the late cold spring events cause severe losses. Many plants increase their cold stress tolerance in response to low temperatures through a process of cold acclimation. This is a multilevel biochemical and cell biological process, which

* Author to whom any correspondence should be addressed Email: nikhil.malhotra86@gmail.com.

doi:10.1088/978-0-7503-4921-5ch6

permits plants to build tolerance against cold stress [1]. These responses are the result of reprogramming of gene expression and accumulation of different metabolites. Molecular responses start with the perception of stress as it occurs, and subsequently information is communicated via a signal transduction pathway. Because cold stress appears to have some similarities to other abiotic stresses such as salinity and drought, it is critical to catalog and separate events that occur as a consequence of cold stress from those that occur as an outcome of other stress responses among several plant species [2].

In recent years, our understanding of the signaling pathways that lead from the perception to end response in plants has grown manifold. The mechanisms related to cold stress are complicated due to differences in signal transduction pathways and metabolisms. An understanding of how plants react to cold stress can provide useful information, and also help in identifying genetic resources for crop breeding for cold stress tolerance. In the recent past, several components of cold stress signaling pathways have been uncovered, including messenger molecules, transcription factors (TFs), protein kinases, and phosphatases. Among them, the ICE–CBF–COR signaling pathway appears to be a well-studied pathway. *Inducer of CBF expression* (ICE), a myelocytomatosis proto-oncogene (MYC)-type basic helix-loop-helix (bHLH) family TF, is a pioneer of cold acclimation [1, 3]. Under cold stress, the *C-repeat binding factor/dehydration-responsive element-binding protein1* (*CBF/DREB1*) genes are rapidly activated, and they play an essential role in cold acclimation of plants [4, 5]. The low temperature induced genes includes Cold-regulated (*COR*) genes, cold inducible (*KIN*), and low temperature induced (*LTI*), are all controlled by cold stress; some encode for osmoprotectants and cryprotective proteins to safeguard the plants from frost bite [6, 7]. The DNA-binding CBF proteins can bind to the *COR* promoters to increase their expression, thereby improving cold tolerance [4, 5]. However, the ICE–CBF–COR-mediated signaling pathway is quite complex and needs further investigation. Here, we describe the recent advances which have encouraged our understanding of the regulatory mechanisms driving cold stress tolerance.

6.2 Sensing of cold stress signals

Understanding how plants detect cold stress signals is crucial in knowing how they avoid damage by low temperatures. The ability of plants to detect such signals has long been a source of fascination for scientists. Many studies have suggested that cold is perceived at multiple sensory levels rather than recognized by a single protein. Plants respond to environmental signals through a sequence of responses, known collectively as signal transduction.

6.2.1 Cell membrane fluidity

The initial line of defense after exposure to low temperature is the loss of cell membrane fluidity, also believed to be one of the cold-sensing mechanisms. The plant cytoskeleton alters when exposed to cold stress, forming bundles of microtubules and filaments [8]. This provides a hint towards the probable role of the cytoskeleton in inducing cold stress signaling. Further, treatment with taxol, an antimicrotubule agent, decreased the expression of gene *BN115* during cold treatment; however, treatment with colchicine, a microtubule depolymerizing agent, enhanced *BN115* expression, according to previous findings by Orvar *et al* [9] and Sangwan *et al* [10]. As a result, depolymerization of the cytoskeleton is required for low-temperature gene activation in plant cells. Furthermore, researchers have developed a technique for measuring membrane fluidity and offer additional experimental proof for such changes in plant cells in response to cold temperatures [11]. The amount of desaturated fatty acids in the plasma membrane affects the fluidity of the plasma membrane. One of the key enzymes, oleate desaturase, which is required for membrane fluidity, is encoded by *fatty acid desaturation2* (*FAD2*). Some physiological responses to cold stress are affected by *FAD2* mutations, including leaf number and hypocotyl length [11]. These findings back the idea that a loss in cell membrane fluidity is a fundamental way to detect cold signals. Given the close relationship between the cytoskeleton and the plasma membrane, it is possible that cytoskeletal alterations are associated with early stages of cold stress signaling.

6.2.2 Calcium (Ca^{2+}) channels

In the majority of animals, transient receptor potential ion channels detect low temperatures [12, 13]. Plants, on the other hand, lack these channels. Instead, they use Ca^{2+} ions as messengers to respond to environmental changes. A rapid increase in cytosolic Ca^{2+} concentration via Ca^{2+} channels upon cold treatment is considered as an initial signaling event in both animals and plants [14–16]. Likewise, the cold-induced gene *COR* is dependent on Ca^{2+} [14]. As a result, it is apparent that Ca^{2+} channels and electrophysiological responses in plants also play a role in low-temperature sensing.

In addition, cyclic nucleotide-gated channels (CNGCs) are a broad class of nonselective cation-conducting channels found mostly in the plasma membranes of plants [17]. Ca^{2+} channels are important for root hair formation and adaptation to low temperatures in cold-stress-tolerant *Arabidopsis* [18]. Moreover, on the basis of phylogeny in rice, *OsCNGCs* are separated into four major groups (I–IV) and two subgroups (IV-A and IV-B) [19]. An abundance of transcripts of *OsCNGCs* from groups I, II, and III are considerably stimulated by cold stress, but the expression of *OsCNGCs* from group IV is suppressed. These recordings confirm that *OsCNGCs* play a key role in plant cold stress responses [19]. Furthermore, in plants, glutamate receptors (GLRs), which mediate Ca^{2+} changes across membranes, lead to production of immense numbers of external and endogenous signals. The GLR channel-like gene *AtGLR3*.4 produces a plasma membrane-bound protein. Cold stress increases the expression of *AtGLR3*.4 in a Ca^{2+}-dependent way, which is linked to a rise in the levels of glutamate and cytosolic Ca^{2+} [20, 21]. The evidence is currently inadequate to infer whether GLRs or CNGCs are the ultimate cold stress sensors in plants [22].

6.2.3 Phytochromes

Phytochromes are photoreceptors which regulate photomorphogenesis. According to two groundbreaking investigations, temperature changes the phytochrome B (phyB) protein state. Thus, it could be involved in temperature sensing, with phyB moving from the active Pfr state to an inactive Pr state. The photoreceptor phyB binds to the promoters of its key target genes in a temperature-dependent manner [23, 24]. These findings show that phyB regulates photomorphogenesis by sensing light and ambient temperatures at different environmental conditions; however, whether phyB is also involved in cold stress sensing requires further investigation.

6.3 Molecular alterations during cold stress

Understanding the fundamental molecular pathways under cold stress necessitates investigating the transcriptional changes in plants during cold acclimation. Many studies have documented a large number of cold-responsive genes and various gene regulatory networks in plants up to this date (table 6.1). ICE–CBF–COR is considered one of the most widely reported routes among these discoveries. Cold stress triggers the ICE–CBF–COR pathway in most plant species, including

Table 6.1. Summary of cold-stress-related genes in crops and their expression under various stress environments.

Gene	Family	Crop species	Stress environment	References
AtCBF3	AP2/ERF	*Arabidopsis thaliana*	Cold stress	[33]
AtGLR1.2		*A. thaliana*	Cold stress	[34]
AtHAP5A		*A. thaliana*	Cold stress	[35]
FDA8	FDA	*A. thaliana*	Cold stress	[36]
PUB25/26		*A. thaliana*	Cold stress	[37]
STCH4		*A. thaliana*	Cold stress	[38]
BpUVR8	UVR	*Betula platyphylla*	ABA responsive and cold stress	[39]
SET		*Brassica rapa*	Cold and heat stress	[40]
CsCPK	CPK	*Camellia sinensis*	Cold stress	[41]
CsLEA	LEA	*C. sinensis*	Cold and dehydration stress	[42]
DlICE1	bHLH	*Dimocarpus longan*	Cold stress	[43]
FtbHLH2	bHLH	*Fagopyrum tataricum*	Cold stress	[44]
GHDREB1	DREB	*Gossypium hirsutum*	Cold stress	[45]
MdMYB108L	MYB	*Malus domestica*	Cold stress	[46]
MdHY5	bZIP	*M. domestica*	Cold stress	[46]
MaPIP2–7	AQP	*Musa acuminata*	Cold, drought and salt stress	[47]
MYBS3	MYB	*Oryza sativa*	Cold stress	[48]
OsGH3–2	GH3	*O. sativa*	Cold and drought stress	[49]
OsMADS57		*O. sativa*	Cold stress	[50]
COR413	COR	*Saussurea involucrata*	Cold and drought stress	[51]
AtGRXS17	GRX	*Solanum lycopersicum*	Cold stress	[33]
Sb08g007310	GST	*Sorghum bicolor*	Cold stress	[52]
TaTPS11		*Triticum aestivum*	Cold stress	[53]
TaSMT1		*T. aestivum*	Cold stress	[54]
VaERF080	AP2/ERF	*Vitis amurensis*	Cold stress	[44]
VvCBF	DREB	*V. vinifera*	Cold stress	[55]

agricultural crops, which then activates the appropriate expression of downstream genes that encode osmoregulatory chemicals [25]. According to reports, *ICE1* is involved in *CBF3* cold induction. *CBF*, also known as *DREB*, functions as a regulatory gene that aids in cold adaptation [26]. CBF is a member of the apetala2/ethylene response factor (AP2/ERF) family that controls *COR* gene expression [27]. The AP2/ERF family is a TF that is defined by the presence of at least one AP2 DNA-binding domain [28]. CBF activates *COR* gene production by binding to a cis-element in the promoter of the gene [26, 29]. CBF is involved in a different abiotic stresses in addition to cold stress [7, 29, 30]. In the promoter regions of several cold-responsive genes there is an abscisic acid (ABA)-responsive element as well as a dehydration-responsive element [31]. Azzeme *et al* have shown that oil palm *EgDREB1* is functionally associated with enhancing cold stress tolerance [32].

Even though research pertaining to cold stress responses in plants has been made in and around studies on model crop systems like *Arabidopsis* and rice, the role of other crops in regulating this phenomenon can not be ruled out. Many studies have indicated that *Capsella bursa-pastoris* can tolerate cold better than *A. thaliana*; however, the reason underlying this improved cold stress tolerance is still unknown and requires further investigation [56]. Similarly, Yan *et al* [57] and Ma *et al* [58] have demonstrated the ability of *Brassica napus* and *B. rapa* to tolerate low temperatures, respectively. Likewise, transcriptome-wide analysis of the cold stress responses in pumpkin [59] and sugarcane [60] have provided new insights into the molecular mechanism linked with low-temperature tolerance. Further, CBFs are known to control cold stress signaling in wheat and barley crops [48]. To improve cold acclimation, winter barleys require the connection of the *vernalization1* (*VRN1*) and *CBF* genes. Spring barley, on the other hand, does not require vernalization, making it susceptible to cold stress. In a study by Londo *et al* [28], grapevine cultivars showed 214 upregulated genes and 884 downregulated genes post low-temperature treatment, similar to spring barley, implying that cold stress promoted the downregulated genes. Besides, multiple TFs that regulate cold stress signaling have been discovered in a variety of crop species [61]. Alternatively, phytohormones have also been associated with the activation of TFs involved in cold stress tolerance [62, 63].

Despite the fact that the ICE–CBF–COR pathway is important for plant cold stress tolerance, research is currently focusing on the role of hormones in cold stress. Key hormones such as ABA, gibberelic acid, jasmonates, brassinosteroids, cytokinin, auxin, melatonin, and polyamines influence *CBFs*. Changes in membrane structure, sugar content, and the creation of toxic proteins are all aided by cold stress, thus resulting in the release of hormones and Ca^{2+}, together with the formation of reactive oxygen species (ROS) (figure 6.1). Overall, these changes increase cold stress tolerance by upregulating the expression of *CBFs* that bind to cis-elements. Furthermore, the circadian clock plays a significant role in cold stress [64]. It is the first time-keeping network, allowing organisms, including plants, to coordinate their physiological activities with environmental changes at specific times of the day. However, further research is required to learn how plant cells and tissues communicate with one another [64]. Similarly, transcriptomes and bioinformatics

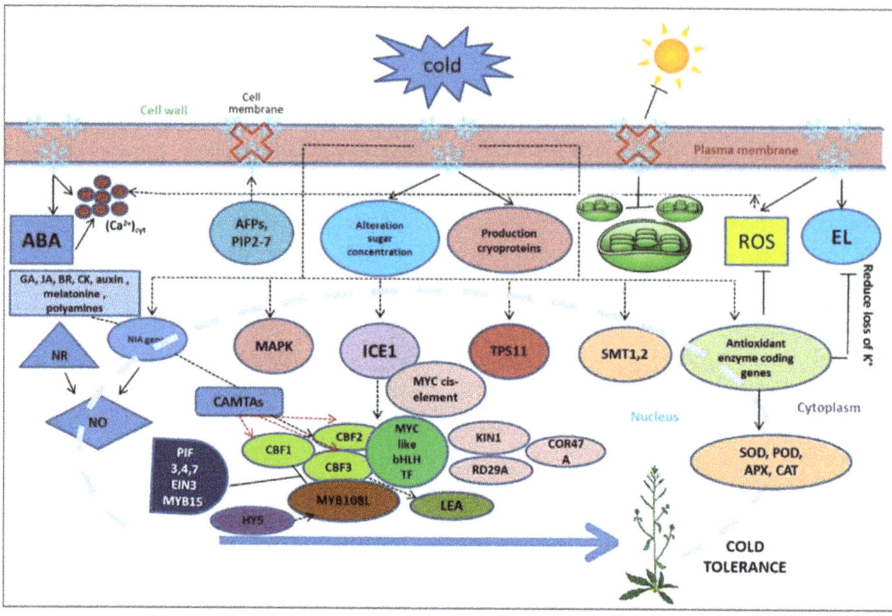

Figure 6.1. Schematic representation of the mechanisms related to cold stress tolerance in plants [76].

have been widely applied to address complicated biological concerns in recent years, leading to in-depth research into cold-stress-related genes [65–69]. Furthermore, a vast variety of mutant lines, including those introduced by genetic engineering and gene-silencing technologies, have been generated for the functional validation of genes in the plant genome [66, 67]. Moreover, clustered regularly interspaced short palindromic repeats (CRISPR)/CRISPR-associated protein 9 (Cas9), a genome-editing technology, has been extensively used as it allows researchers to change an organism's DNA and alter numerous genes in a range of crops [70–74]. Although Jia *et al* demonstrated the use of CRISPR/Cas9 technology to generate mutants for evaluating the role of C-repeat binding factors under cold stress in *Arabidopsis*, the absence of sufficient genomic sequence information in many crop species is a major constraint of this technique [75]. It would be desirable, in order to further enhance the ability of this tool, to elucidate various genes, TFs, and protein functions in plants that are yet to be found and may have functional roles during cold stress. Conclusively, the complexity of different abiotic stress mechanisms has yet to be elucidated to tackle the grave concerns surrounding food security [76–78].

6.4 Conclusions

Cold stress tolerance in crops is a complicated process. Plants have evolved intricate molecular regulatory mechanisms to cope with cold stress, which show similarities and variances among plant species. Despite substantial research having been undertaken to decipher the cold signaling pathways of plants in recent decades, still a clear understanding of the cold signaling process remains obscure. The main reason for this is the extreme similarity with molecular patterns associated with other abiotic stress-induced phenomena.

Even though global temperature is increasing due to climate change, cold extremes and abnormal weather have increased all over the world. It is critical to produce and cultivate cold-resistant crop types in order to expand the available planting area to high latitude and high altitude. The ability of crops to tolerate cold stress could be increased by developing genetically engineered crops using identified cold stress regulators. Furthermore, using high-density mapping populations to identify genome-wide association studies and quantitative trait loci, to discover novel genes associated with plant tolerance to low temperatures, are useful approaches for discovering novel genes linked with plant resistance to low temperatures. More revolutionary research is needed to fully comprehend the methods by which a plant cell's fate is determined. At the molecular level, a better understanding of the balance between plant organogenesis and cold stress adaptation could aid in improving low-temperature tolerance in crops, which could pave the way for developing cold stress-resilient crops.

Multiple choice questions

1. Cold stress leads to …
    - (a) Decrease in the cell membrane permeability;
    - (b) Freezing of cytoplasm;
    - (c) Accumulation of osmoprotectants;
    - (d) All of the above.
2. Which of the following gene(s) could be most important for generating cold stress tolerance through gene-editing techniques?
    - (a) Fatty acid desaturases;
    - (b) Citrate synthase;
    - (c) Tryptophan synthase;
    - (d) All of them.
3. Which of the following gene(s) is involved in cold tolerance?
    - (a) CBF/DREB1;
    - (b) COR;
    - (c) Superoxide dismutase;
    - (d) All of them.
4. Which of the following could be considered as morphological changes suitable for cold tolerance?
    - (a) Increased chlorophyll accumulation;
    - (b) Improved germination;
    - (c) Both (a) and (b);
    - (d) Reduced pollen fertility and seed set.
5. Which of the following gene(s) has been overexpressed for lipid modification to induce cold tolerance?
    - (a) Glycerol-3-phosphate acyl transferase;
    - (b) Superoxide dismutase;

(c) Auxin resistance1;
(d) All of them.

Answers: 1 (d); 2 (a); 3 (d); 4 (c); 5 (d).

Descriptive-type questions

1. How is cold stress tolerance regulated by the ICE–CBF–COR signaling pathway?
2. How is cold stress induced and perceived by the plant cell? Describe the genes which have potential to alter the membrane permeability and enhance cold tolerance.
3. Describe the molecular alterations associated during cold stress.
4. Describe the role of calcium (Ca^{2+}) channels in cold stress signaling.
5. Describe the physiological and molecular mechanisms involved in cold stress.

References

[1] Chinnusamy V, Zhu J and Zhu J-K 2007 Cold stress regulation of gene expression in plants *Trends Plant Sci.* **12** 444–51

[2] Heidarvand L and Amiri R M 2010 What happens in plant molecular responses to cold stress? *Acta Physiol. Plant* **32** 419–31

[3] Wang L, Ma R, Yin Y and Jiao Z 2018 Role of carbon ion beams irradiation in mitigating cold stress in *Arabidopsis thaliana Ecotoxicol. Environ. Saf.* **162** 341–7

[4] Stockinger E J, Gilmour S J and Thomashow M F 1997 *Arabidopsis thaliana CBF1* encodes an AP2 domain-containing transcriptional activator that binds to the C-repeat/DRE, a cis-acting DNA regulatory element that stimulates transcription in response to low temperature and water deficit *Proc. Natl. Acad. Sci. USA* **94** 1035–40

[5] Liu Q, Kasuga M, Sakuma Y, Abe H, Miura S, Yamaguchi-Shinozaki K and Shinozaki K 1998 Two transcription factors, DREB1 and DREB2, with an EREBP/AP2 DNA binding domain separate two cellular signal transduction pathways in drought- and low-temperature-responsive gene expression, respectively, in Arabidopsis *Plant Cell* **10** 1391–406

[6] Yamaguchi-Shinozaki K and Shinozaki K 1994 A novel cis-acting element in an *Arabidopsis* gene is involved in responsiveness to drought, low-temperature, or high-salt stress *Plant Cell* **6** 251–64

[7] Shi Y, Ding Y and Yang S 2018 Molecular regulation of CBF signaling in cold acclimation *Trends Plant Sci.* **23** 623–37

[8] Pokorna J, Schwarzerova K, Zelenkova S, Petrasek J, Janotova I, Capkova V and Opatrny Z 2004 Sites of actin filament initiation and reorganization in cold-treated tobacco cells *Plant Cell Environ.* **27** 641–53

[9] Orvar B L, Sangwan V, Omann F and Dhindsa R S 2000 Early steps in cold sensing by plant cells: the role of actin cytoskeleton and membrane fluidity *Plant J.* **23** 785–94

[10] Sangwan V, Foulds I, Singh J and Dhindsa R S 2001 Cold activation of *Brassica napus BN115* promoter is mediated by structural changes in membranes and cytoskeleton, and requires Ca^{2+} influx *Plant J.* **27** 1–12

[11] Martiniere A, Shvedunova M, Thomson A J, Evans N H, Penfield S, Runions J and McWatters H G 2011 Homeostasis of plasma membrane viscosity in fluctuating temperatures *New Phytol.* **192** 328–37

[12] Bautista D M, Siemens J, Glazer J M, Tsuruda P R, Basbaum A I, Stucky C L, Jordt S E and Julius D 2007 The menthol receptor TRPM8 is the principal detector of environmental cold *Nature* **448** 204–8

[13] Dhaka A, Murray A N, Mathur J, Earley T J, Petrus M J and Patapoutian A 2007 TRPM8 is required for cold sensation in mice *Neuron* **54** 371–8

[14] Knight H, Trewavas A J and Knight M R 1996 Cold calcium signaling in *Arabidopsis* involves two cellular pools and a change in calcium signature after acclimation *Plant Cell* **8** 489–503

[15] Plieth C, Hansen U P, Knight H and Knight M R 1999 Temperature sensing by plants: the primary characteristics of signal perception and calcium response *Plant J.* **18** 491–7

[16] Knight M R and Knight H 2012 Low-temperature perception leading to gene expression and cold tolerance in higher plants *New Phytol.* **195** 737–51

[17] Zelman A K, Dawe A, Gehring C and Berkowitz G A 2012 Evolutionary and structural perspectives of plant cyclic nucleotide-gated cation channels *Front. Plant Sci.* **3** 95

[18] Shao H B, Chu L Y, Shao M A, Li S Q and Yao J C 2008 Bioengineering plant resistance to abiotic stresses by the global calcium signal system *Biotechnol. Adv.* **26** 503–10

[19] Nawaz Z, Kakar K U, Saand M A and Shu Q Y 2014 Cyclic nucleotide-gated ion channel gene family in rice, identification, characterization and experimental analysis of expression response to plant hormones, biotic and abiotic stresses *BMC Genomics* **15** 18

[20] Meyerhoff O, Muller K, Roelfsema M R G, Latz A, Lacombe B, Hedrich R, Dietrich P and Becker D 2005 *AtGLR3.4*, a glutamate receptor channel-like gene is sensitive to touch and cold *Planta* **222** 418–27

[21] Weiland M, Mancuso S and Baluska F 2016 Signalling via glutamate and GLRs in in *Arabidopsis thaliana Funct. Plant Biol.* **43** 1–25

[22] Zhu J K 2016 Abiotic stress signaling and responses in plants *Cell* **167** 313–24

[23] Jung J H *et al* 2016 Phytochromes function as thermosensors in *Arabidopsis Science* **354** 886–9

[24] Legris M, Klose C, Burgie E S, Rojas C C, Neme M, Hiltbrunner A, Wigge P A, Schafer E, Vierstra R D and Casal J J 2016 Phytochrome B integrates light and temperature signals in *Arabidopsis Science* **354** 897–900

[25] Zhang F, Jiang Y, Bai L-P, Zhang L, Chen L-J, Li H G, Yin Y, Yan W-W, Yi Y and Guo Z-F 2011 The ICE-CBF-COR pathway in cold acclimation and AFPs in plants *Middle-East J. Sci. Res.* **8** 493–8

[26] Zhen Y and Ungerer M C 2008 Relaxed selection on the CBF/DREB1 regulatory genes and reduced freezing tolerance in the southern range of *Arabidopsis thaliana Mol. Biol. Evol.* **25** 2547–55

[27] Borba A R, Serra T S, Górska A, Gouveia P, Cordeiro A M, Reyna-Llorens I, Kneˇrová J, Barros P M, Abreu I A and Oliveira M M *et al* 2018 Synergistic binding of bHLH transcription factors to the promoter of the maize NADP-ME gene used in C_4 photosynthesis is based on an ancient code found in the ancestral C_3 state *Mol. Biol. Evol.* **35** 1690–705

[28] Londo J P, Kovaleski A P and Lillis J A 2018 Divergence in the transcriptional landscape between low temperature and freeze shock in cultivated grapevine (*Vitis vinifera*) *Hortic. Res.* **5** 10

[29] Rihan H Z, Al-Issawi M and Fuller M P 2017 Upregulation of CBF/DREB1 and cold tolerance in artificial seeds of cauliflower (*Brassica oleracea* var. *botrytis*) *Sci. Hortic.* **225** 299–309

[30] Lata C and Prasad M 2011 Role of DREBs in regulation of abiotic stress responses in plants *J. Exp. Bot.* **62** 4731–48

[31] Yadav S K 2009 Cold stress tolerance mechanisms in plants *Sustain Agric.* **2** 605–20

[32] Azzeme A M, Abdullah S N A, Aziz M A and Wahab P E M 2017 Oil palm drought inducible *DREB1* induced expression of *DRE/CRT*- and non-*DRE/CRT*-containing genes in lowland transgenic tomato under cold and PEG treatments *Plant Physiol. Biochem.* **112** 129–51

[33] Hu Y, Wu Q, Sprague S A, Park J, Oh M, Rajashekar C B, Koiwa H, Nakata P A, Cheng N and Hirschi K D *et al* 2015 Tomato expressing *Arabidopsis* glutaredoxin gene *AtGRXS17* confers tolerance to chilling stress via modulating cold responsive components *Hortic. Res.* **2** 1–11

[34] Zheng Y, Luo L, Wei J, Chen Q, Yang Y, Hu X and Kong X 2018 The glutamate receptors *AtGLR1.2* and *AtGLR1.3* increase cold tolerance by regulating jasmonate signaling in *Arabidopsis thaliana Biochem. Biophys. Res. Commun.* **506** 895–900

[35] Shi H, Ye T, Zhong B, Liu X, Jin R and Chan Z 2014 *AtHAP5A* modulates freezing stress resistance in *Arabidopsis* through binding to CCAAT motif of *AtXTH21 New Phytol.* **203** 554–67

[36] Kargiotidou A, Deli D, Galanopoulou D, Tsaftaris A and Farmaki T 2008 Low temperature and light regulate delta 12 fatty acid desaturases (*FAD2*) at a transcriptional level in cotton (*Gossypium hirsutum*) *J. Exp. Bot.* **59** 2043–56

[37] Wang S, Huang H, Han R, Liu C, Qiu Z, Liu G, Chen S and Jiang J 2019 Negative feedback loop between *BpAP1* and *BpPI/BpDEF* heterodimer in *Betula platyphylla* × *B. pendula Plant Sci.* **289** 110280

[38] Yu H, Kong X, Huang H, Wu W, Park J, Yun D-J, Lee B-H, Shi H and Zhu J-K 2020 *STCH4/REIL2* confers cold stress tolerance in *Arabidopsis* by promoting rRNA processing and CBF protein translation *Cell Rep.* **30** 229–42

[39] Li X, Ma M, Shao W, Wang H, Fan R, Chen X, Wang X, Zhan Y and Zeng F 2018 Molecular cloning and functional analysis of a UV-B photoreceptor gene, *BpUVR8* (UV Resistance Locus 8), from birch and its role in ABA response *Plant Sci.* **274** 294–308

[40] Liu G, Khan N, Ma X and Hou X 2019 Identification, evolution, and expression profiling of histone lysine methylation moderators in *Brassica rapa Plants* **8** 526

[41] Ding C, Lei L, Yao L, Wang L, Hao X, Li N, Wang Y, Yin P, Guo G and Yang Y 2019 The involvements of calcium-dependent protein kinases and catechins in tea plant [*Camellia sinensis* (L.) O. Kuntze] cold responses *Plant Physiol. Biochem.* **143** 190–202

[42] Wang W, Gao T, Chen J, Yang J, Huang H and Yu Y 2019 The late embryogenesis abundant gene family in tea plant (*Camellia sinensis*): genome-wide characterization and expression analysis in response to cold and dehydration stress *Plant Physiol. Biochem.* **135** 277–86

[43] Yang X, Wang R, Hu Q, Li S, Mao X, Jing H, Zhao J, Hu G, Fu J and Liu C 2019 *DlICE1*, a stress-responsive gene from *Dimocarpus longan*, enhances cold tolerance in transgenic Arabidopsis *Plant Physiol. Biochem.* **142** 490–9

[44] Yao P, Sun Z, Li C, Zhao X, Li M, Deng R, Huang Y, Zhao H, Chen H and Wu Q 2018 Overexpression of *Fagopyrum tataricum FtbHLH2* enhances tolerance to cold stress in transgenic *Arabidopsis Plant Physiol. Biochem.* **125** 85–94

[45] Shan D P, Huang J G, Yang Y T, Guo Y H, Wu C A, Yang G D, Gao Z and Zheng C C 2007 Cotton *GhDREB1* increases plant tolerance to low temperature and is negatively regulated by gibberellic acid *New Phytol.* **176** 70–81

[46] Wang Y, Mao Z, Jiang H, Zhang Z and Chen X 2019 A feedback loop involving *MdMYB108L* and *MdHY5* controls apple cold tolerance *Biochem. Biophys. Res. Commun.* **512** 381–6

[47] Xu Y, Hu W, Liu J, Song S, Hou X, Jia C, Li J, Miao H, Wang Z and Tie W 2020 An aquaporin gene *MaPIP2-7* is involved in tolerance to drought, cold and salt stresses in transgenic banana (*Musa acuminata* L.) *Plant Physiol. Biochem.* **147** 66–76

[48] Mickelbart M V, Hasegawa P M and Bailey-Serres J 2015 Genetic mechanisms of abiotic stress tolerance that translate to crop yield stability *Nat. Rev. Genet.* **16** 237–51

[49] Du H, Wu N, Fu J, Wang S, Li X, Xiao J and Xiong L 2012 A GH3 family member, *OsGH3-2*, modulates auxin and methylation and chromatin patterning abscisic acid levels and differentially affects drought and cold tolerance in rice *J. Exp. Bot.* **63** 6467–80

[50] Chen L, Zhao Y, Xu S, Zhang Z, Xu Y, Zhang J and Chong K 2018 *OsMADS57* together with *OsTB1* coordinates transcription of its target *OsWRKY94* and *D14* to switch its organogenesis to defense for cold adaptation in rice *New Phytol.* **218** 219–31

[51] Guo X, Zhang L, Dong G, Xu Z, Li G, Liu N, Wang A and Zhu J 2019 A novel cold-regulated protein isolated from *Saussurea involucrata* confers cold and drought tolerance in transgenic tobacco (*Nicotiana tabacum*) *Plant Sci.* **289** 110246

[52] Ortiz D, Hu J and Salas Fernandez M G 2017 Genetic architecture of photosynthesis in *Sorghum bicolor* under non-stress and cold stress conditions *J. Exp. Bot.* **68** 4545–57

[53] Liu X, Fu L, Qin P, Sun Y, Liu J and Wang X 2019 Overexpression of the wheat trehalose 6-phosphate synthase 11 gene enhances cold tolerance in *Arabidopsis thaliana Gene* **710** 210–7

[54] Valitova J, Renkova A, Mukhitova F, Dmitrieva S, Beckett R P and Minibayeva F V 2019 Membrane sterols and genes of sterol biosynthesis are involved in the response of *Triticum aestivum* seedlings to cold stress *Plant Physiol. Biochem.* **142** 452–9

[55] Rubio S and Pérez F J 2019 ABA and its signaling pathway are involved in the cold acclimation and deacclimation of grapevine buds *Sci. Hortic.* **256** 108565

[56] Noman A, Kanwal H, Khalid N, Sanaullah T, Tufail A, Masood A, Sabir S, Aqeel M and He S 2017 Perspective research progress in cold responses of *Capsella bursa-pastoris Front. Plant Sci.* **8** 1388

[57] Yan L, Shah T, Cheng Y, Lü Y, Zhang X and Zou X 2019 Physiological and molecular responses to cold stress in rapeseed (*Brassica napus* L.) *J. Integr. Agric.* **18** 2742–52

[58] Ma L *et al* 2019 Transcriptome analysis reveals key cold-stress-responsive genes in winter rapeseed (*Brassica rapa* L.) *Int. J. Mol. Sci.* **20** 1071

[59] Li F, Lu X, Duan P, Liang Y and Cui J 2021 Integrating transcriptome and metabolome analyses of the response to cold stress in pumpkin (*Cucurbita maxima*) *PLoS One* **16** e0249108

[60] Huang X *et al* 2021 Integration of transcriptional and post-transcriptional analysis revealed the early response mechanism of sugarcane to cold stress *Front. Genet.* **11** 581993

[61] Wang S, Yang C, Zhao X, Chen S and Qu G-Z 2018 Complete chloroplast genome sequence of *Betula platyphylla*: gene organization, RNA editing, and comparative and phylogenetic analyses *BMC Genomics* **19** 950

[62] Wang D-Z, Jin Y-N, Ding X-H, Wang W-J, Zhai S-S, Bai L-P and Guo Z-F 2017 Gene regulation and signal transduction in the ICE–CBF–COR signaling pathway during cold stress in plants *Biochemistry* **82** 1103–17

[63] Yang T and Huang X S 2018 Deep sequencing-based characterization of transcriptome of *Pyrus ussuriensis* in response to cold stress *Gene* **661** 109–18

[64] Nohales M A and Kay S A 2016 Molecular mechanisms at the core of the plant circadian oscillator *Nat. Struct. Mol. Biol.* **23** 1061–9

[65] Chen S, Lin X, Zhang D, Li Q, Zhao X and Chen S 2019 Genome-wide analysis of *NAC* Gene family in *Betula pendula Forests* **10** 741

[66] Gang H, Li R, Zhao Y, Liu G, Chen S and Jiang J 2019a Loss of GLK1 transcription factor function reveals new insights in chlorophyll biosynthesis and chloroplast development *J. Exp. Bot.* **70** 3125–38

[67] Gang H, Liu G, Zhang M, Zhao Y, Jiang J and Chen S 2019b Comprehensive characterization of T-DNA integration induced chromosomal rearrangement in a birch T-DNA mutant *BMC Genomics* **20** 311

[68] Wang X, Zeng W, Ding Y, Wang Y, Niu L, Yao J-L, Pan L, Lu Z, Cui G and Li G *et al* 2019 Peach ethylene response factor *PpeERF2* represses the expression of ABA biosynthesis and cell wall degradation genes during fruit ripening *Plant Sci.* **283** 116–26

[69] Wang F, Chen S, Liang D, Qu G-Z, Chen S and Zhao X 2020 Transcriptomic analyses of *Pinus koraiensis* under different cold stresses *BMC Genomics* **21** 1–14

[70] Ma X, Zhu Q, Chen Y and Liu Y-G 2016 CRISPR/Cas9 platforms for genome editing in plants: developments and applications *Mol. Plant* **9** 961–74

[71] Abdelrahman M, Al-Sadi A M, Pour-Aboughadareh A, Burritt D J and Tran L-S P 2018 Genome editing using CRISPR/Cas9–targeted mutagenesis: an opportunity for yield improvements of crop plants grown under environmental stresses *Plant Physiol. Biochem.* **131** 31–6

[72] Jyoti A, Kaushik S, Srivastava V K, Datta M, Kumar S, Yugandhar P, Kothari S L, Rai V and Jain A 2019 The potential application of genome editing by using CRISPR/Cas9, and its engineered and ortholog variants for studying the transcription factors involved in the maintenance of phosphate homeostasis in model plants *Semin. Cell Dev. Biol.* **96** 77–90

[73] Yubing H, Min Z, Lihao W, Junhua W, Qiaoyan W, Rongchen W and Yunde Z 2019 Improvements of TKC technology accelerate isolation of transgene-free CRISPR/Cas9-edited rice plants *Rice Sci.* **26** 109–17

[74] Razzaq M K, Aleem M, Mansoor S, Khan M A, Rauf S, Iqbal S and Siddique K H M 2021 Omics and CRISPR-Cas9 approaches for molecular insight, functional gene analysis, and stress tolerance development in crops *Int. J. Mol. Sci.* **22** 1292

[75] Jia Y, Ding Y, Shi Y, Zhang X, Gong Z and Yang S 2016 The *cbfs* triple mutants reveal the essential functions of *CBFs* in cold acclimation and allow the definition of *CBF* regulons in Arabidopsis *New Phytol.* **212** 345–53

[76] Ritonga F N and Chen S 2020 Physiological and molecular mechanism involved in cold stress tolerance in plants *Plants* **9** 560

[77] Kumar R *et al* 2021 Understanding omics driven plant improvement and *de novo* crop domestication: some examples *Front Genet.* **12** 637141

[78] Sharma V, Gupta P, Priscilla K, Kumar S, Hangargi B, Veershetty A, Ramrao D, Suresh S, Narasanna R and Naik G *et al* 2021 Metabolomics intervention towards better understanding of plant traits *Cells* **10** 346

IOP Publishing

Molecular Response and Genetic Engineering for Stress in Plants, Volume 1

Abiotic stress

Pawan Shukla, Anirudh Kumar, Rakesh Kumar and Manish K Pandey

Chapter 7

Unraveling the molecular and genetic basis of plant responses to heat stress

Ashish Gautam[1*], Vinay Sharma[2], Charupriya Chauhan[1], Ravindra K Panwar[1], Salil K Tewari[1], S K Verma[1]

[1]*G.B. Pant University of Agriculture and Technology, Pantnagar 263145, Uttarakhand, India*

[2]*International Crops Research Institute for the Semi-Arid Tropics (ICRISAT), Hyderabad 502324, India*

Heat stress (HS) imposes a significant limitation on crop productivity, which is projected to worsen with global warming. Thorough knowledge of the molecular mechanisms of heat tolerance (HT) is anticipated to design and develop novel high-yield heat-tolerant varieties. HS stimulates reactive oxygen species (ROS) production, which subsequently leads to protein denaturation and aggregation, as well as degradation of nucleic acids, lipids, and other biomolecules through ROS generated in plant cells. Plants have evolved certain sophisticated mechanisms, such as accumulation of compatible osmolytes, enhanced expression of heat shock proteins, secondary metabolites and antioxidants, to detoxify ROS produced under HS. Conventional models to improve plant resilience to HS rely on screening germplasm to identify genes for HT and incorporating them into crop plants using various biotechnological and breeding approaches. Recent breakthroughs in omics techniques and the huge amount of data collected by them have added further genes, proteins, and metabolites to existing models, and even offer novel signaling pathways for enhancing plant thermotolerance. This chapter summarizes the current understanding of crop plant tolerance to increased temperature and explores approaches for the development of crop varieties with increased HT.

7.1 Introduction

Climate emergency, crisis, or breakdown have been accelerated over the past several decades due to changes in our vegetation and agricultural systems [1]. Climate

* Author to whom an correspondence should be addressed Email: gautam.ashish801@gmail.com.

doi:10.1088/978-0-7503-4921-5ch7

change brings serious social, economic, and environmental challenges, and it is predicted to have an notable impact on all aspects of human activity. Crop production is already one such crucial sector that has been negatively impacted by the strong and consistent trends of increasing temperatures worldwide [2]. In order to thrive, all living species must adapt to changing environmental stressors, such as cold shock, heat stress, oxidative and moisture stress, salinity stress, drought stress, pressure and osmotic stress. These are catastrophic for most species; adaptation to environmental stress is critical. Stress is an induced physiological state in plants triggered by unfavorable environmental changes, leading to an altered metabolic and adaptive pattern in the plant. Because plants are stationary, they are frequently exposed to multiple abiotic stresses, which often occur simultaneously. To cope with such environmental stresses, plants adapt by undergoing appropriate physiological, developmental, and biochemical changes. These typically impair cellular equilibrium and, in the worst-case scenario, result in the plant's death [3]. HS is a primary concern because of its irreversible abnormalities in crop plants [4]. Temperature is one such environmental element that considerably influences the growth of the plant and geographical distribution. Elevated temperatures will result in detrimental effects on agricultural productivity in the future, with the most catastrophic consequences coming from brief bursts of very high temperatures. The rise of HS over the last 200 years may be attributed to a number of causes. Human activity has resulted in an increase in greenhouse gas emissions, chlorofluorocarbons, and CO_2 assimilation, resulting in global warming up to 1.0 °C beyond pre-industrial levels [5]. If present trends continue, temperatures could reach 4.0 °C by the end of the 21st century due to global warming [6]. Reduced agricultural production is one of the effects of global warming [7, 8]. Wheat yields would be reduced by 6% for every degree centigrade rise in the world mean temperature, rice yields by 3%, maize yields by 7.4%, and soybean yields by 3.1% [8]. As per forecasts, the frequency and duration of extreme heat episodes will rise by 50% in 2050 and 90% in 2100, leading to substantial yield losses [9]. For this reason, it is critical to investigate the relationship between crop phenology and temperature in order to gain a better perception of ecosystem evolution, biological activities, as well as underlying plant pathways and responses to climatic fluctuations [10]. The rise in global temperature is one of the most serious challenges [11]. HS may be explained as a temperature rise that exceeds a minimum threshold for an extended duration and is intense enough to permanently disrupt growth and development [12]. High-temperature injuries result in direct impacts like protein denaturation and aggregation, which affect the activation of mitochondrial or chloroplast enzymes or spindle fiber instability [13]. Plants modify their proteome, transcriptome, lipidome, and metabolome processes to compensate for HS. The cellular adaptations are designed to manage protein stability, maintain structural integrity, and mitigate the negative effects of hyperaccumulating molecules such as reactive oxygen species (ROS). Plants resist HS by a rapidly activating unfolded protein response (UPR) process, which activates heat shock proteins (Hsps) and antioxidant defense mechanisms, as well as enzymes such as catalases and superoxide dismutases, as well as osmoprotectant synthesis [14, 15]. Additional tools have been introduced by advanced molecular breeding and genetic engineering techniques that could be used to design crops that are better able to withstand high temperatures. However, attaining this objective will necessitate the collaboration of molecular biologists, plant physiologists, and plant breeders working together.

7.2 Plant responses to heat stress

HS has a significant influence on the growth and development of a plant. Indeed, the practices by which plants adapt to HS are of significant importance. The plant heat stress response (HSR) is governed by an intricate network of transcription factors (TFs) from several families that are implicated in the expression of HSR genes (figures 7.1 and 7.2). The following types of responses to HS exist.

7.2.1 Morphological responses

Rising temperatures and excessive radiation are two of the most critical factors restricting the growth and development of plants, consequently reducing crop productivity. Aside from pre- and post-harvest losses due to fruit injury and reduced yield, high temperatures induce leaf and twig blazing, leaf fall, shoot and root growth retardation, and fruit discoloration [16]. It is a prominent problem in temperate zones, as well, decreasing productivity and limiting dry biomass

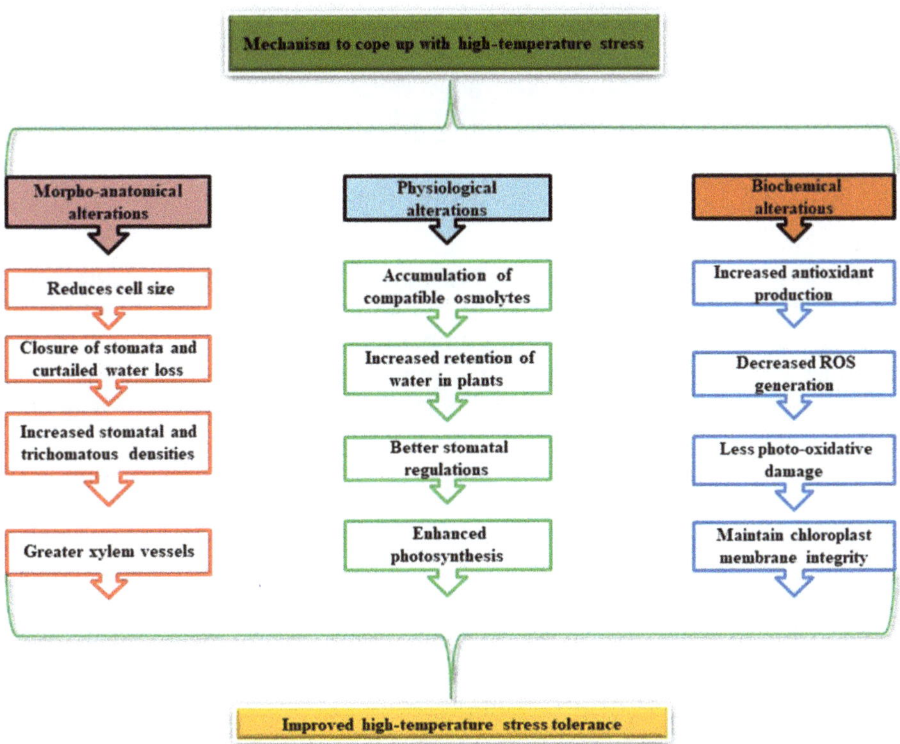

Figure 7.1. Schematic flow diagram of physiological, biochemical, and morphological adaptations of plants to deal with heat stress.

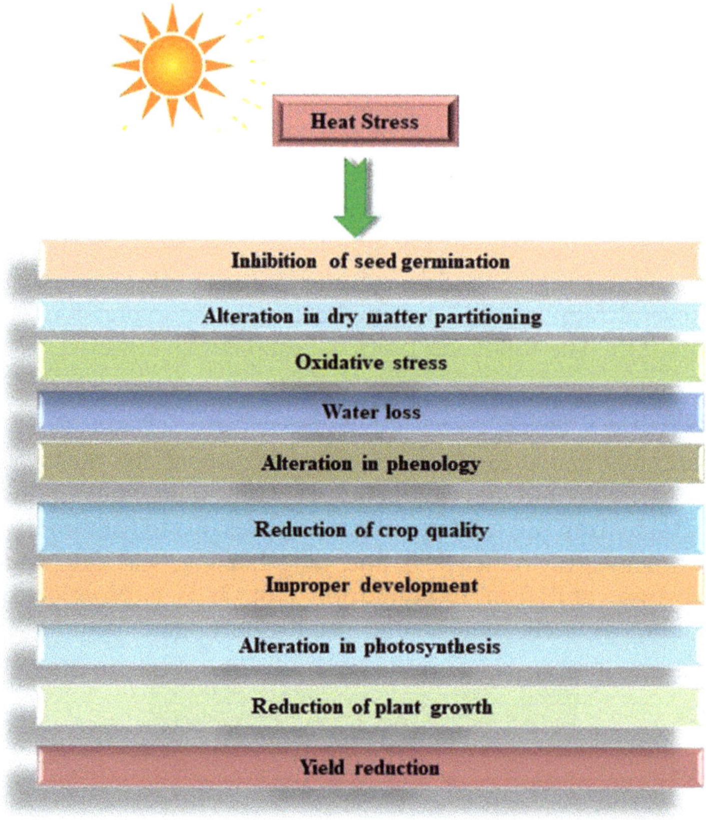

Figure 7.2. Effects of high temperature on plants.

deposition in a variety of commercial crops, particularly maize [17]. Coleoptile and seedling development in maize slow at 40 °C and seize at 45 °C [18]. In wheat, HS at the time of maturity reduces the grain number per spike, affecting both kernel mass and number [19]. Heat susceptibility in common beans and groundnuts results in yield loss in temperate and tropical lowlands [20, 21]. Increased temperatures induced a significant drop in growth, shoot dry biomass, and absorption rate in maize, pearl millet, and sugarcane, but had no impact on leaf expansion [22, 23]. Temperature increase, on the other hand, has a deleterious influence on shoot growth by shortening the length of the first internode and triggering plant death [24]. Sugarcane, for example, exhibited shorter internodes, early senescence, higher tillering, and lower overall plant biomass when exposed to high temperatures [25]. High temperature, alone or with a combination to prolonged drought, is a significant limitation in several temperate cereals during the anther initiation and grain-filling phases [26]. Heat, for example, causes a longer grain-filling phase with less grain growth, resulting in a nearly 7% decrease in the kernel weight of wheat [27]. Similarly, exposure to heat also reduces the carbohydrate, protein, and oil content of maize kernels [28], and the quality attributes of grain in other cereals [29, 30].

Moreover, it also causes the development of an exerted style in the tomato reproductive system [31]. The decline of carbohydrates in sink tissues is associated with poor fruit quality during temperatures rises [23, 32]. According to studies conducted in controlled growth chambers and greenhouses, high temperatures are highly detrimental during the blooming phase, and this vulnerability lasts for around 10–15 days. In many plants, the most vulnerable stages to high temperatures are gamete formation (8–9 days before anthesis) and fertilization [33]. Furthermore, it has a significant impact on the pre- and post-fertilization processes, resulting in a lower yield of crops [31] (figures 7.1 and 7.2).

7.2.2 Anatomical responses

The morphological impacts of elevated temperature on the agricultural plant have been less investigated and are often viewed as those of drought stress. Plant anatomy is disrupted at all cellular levels by suboptimal temperatures, impacting plant growth and yield [34]. Temperature stress (TS) affects cell size, stomatal closure, and results in higher stomatal density, and bigger xylem vessels in the roots and shoots of plants [35, 36]. It causes severe mesophyll cell destruction and increased plasma membrane permeability in grapes. It also reduces grana stacking and causes inflammation in photosynthetic membranes. Stroma lamellae were shown to swell, vacuole contents clump, cristae structure be disturbed, and mitochondria empty in grape plants [37] (figures 7.1 and 7.2).

7.2.3 Phenological responses

HS is a significant element determining the pace of plant growth, which may accelerate up to a certain point and then slow down [13, 38, 39]. It hampers several physiological processes. Excessive daytime temperatures during the vegetative stage, for example, might impair leaf gas exchange abilities. Even a brief spell of this stress during the reproduction phase may accelerate the floral buds and open flower abortion. In high-temperature settings, early heading helps retain more green leaves during anthesis, reducing yield loss [40]. In wheat, temperature rises speed up phenological development, i.e. reducing by days emergence, anthesis, and maturity. As a result, grain-filling time is also reduced (figures 7.1 and 7.2).

7.2.3.1 Photosynthesis

Heat stress in general reduces the rate of photosynthesis, and ends up contracting the plant life cycle and decreasing overall production [41]. Photosynthesis is one of the most temperature-sensitive physiological activities. Photochemical reactions in thylakoid lamellae and carbon metabolism in the chloroplast stroma may be hampered by this [42, 43]. It causes thylakoid membranes to rupture, limiting the activity of electron transporters and enzymes linked to the membrane and, as a result, lowering the overall photosynthesis rate. Since the photosystem II complex is extremely temperature-sensitive, its activity is declines considerably or even ceases under HS conditions [44]. Furthermore, it affects chloroplast structure and the thermostability of elements of the photosynthetic system, affecting Rubisco activity, photosynthetic pigment levels, and the carbon fixation capability [45, 46] (figures 7.1 and 7.2). These are the primary factors contributing to the decrease in photosynthetic efficiency under HS. Thus, a basic knowledge of photosynthetic physiology is important in studying plant thermostability and the impacts of heat on agricultural productivity [47].

7.2.3.2 Thermostability of cell membranes

The primary physiological effect of HS on plants is membrane disruption. Under severe HS, biomolecules' enhanced kinetic energy and mobility across membranes cause chemical bonds to weaken, resulting in membrane lipid breakdown and increased membrane fluidity [48]. It promotes membrane permeability and electrolyte loss, limiting cellular activity and lowering thermotolerance [41]. Furthermore, it facilitates the accumulation of ROS, which leads to membrane disruption, thereby making cells intolerant [49]. Briefly stated, membrane thermostability is crucial for the development of HS tolerance in plant species.

7.2.3.3 Oxidative damage

Exposure of plants to HS produces ROS-singlet oxygen ($^1O_2$), hydrogen peroxide (H_2O_2), superoxide radicals (O_2), and hydroxyl radicals (OH), culminating in oxidative stress [50]. The ROS are mostly formed in photosystem I and photosystem II. Because of the excess energy, chlorophylls in photosystem II form a triplet state, which transfers excitation energy to O_2 to produce singlet oxygen. Excessive degradation of photosystem I generates superoxide anion, which ultimately increases H_2O_2 production [44]. ROS such as O_2^- and H_2O_2 reduce the plant cell viability by inducing oxidative stress via degrading proteins, inactivating enzymes, and altering membrane properties [51]. Free radical injury to the cell membrane under such conditions induces lipid peroxidation [42]. Many plants, such as sorghum, exhibit considerably increased levels of malondialdehyde biomarker when exposed to HS [52]. Under such conditions, ROS can also promote programmed cell death. Plants, on the other hand, have evolved methods of detoxifying ROS and increasing their HT. Plants improve their thermotolerance by employing antioxidant enzymes such as ascorbate peroxidase and superoxide dismutase [53].

7.2.3.4 Other physiological responses

Under fluctuating temperatures, plant water status is often unstable [41]. HS dehydrates plants and impairs their growth and development. When exposed to stress, water potential and relative water content are greatly reduced, lowering photosynthetic efficiency [54]. Plants, on the other hand, modulate respiration rate and transpiration under moderate HS to balance water loss and heat dissipation. Plants also regulate the level of soluble sugars and proteins during these conditions in order to maintain osmotic pressure inside the cell [55]. This significantly diminishes the production of crop species such as cereals, legumes, and oilseeds [56].

7.3 Adaptation mechanisms of plants to heat stress

There are two types of heat adaptation mechanism, heat avoidance and heat tolerance.

7.3.1 Heat avoidance

Plant genotypes have the potential to redirect radiation energy, preventing a rise in plant temperature as a result of stress. The primary mechanism of energy dissipation is transpiration. Transpirational cooling is the link between dehydration and heat avoidance becoming inseparable from those of dehydration avoidance. The reflective properties of leaves determine the proportion of incident solar radiation absorbed by the leaves. Therefore, leaf surface attributes like pubescence, glaucousness, etc, are a component of heat avoidance. In addition, bark may reduce organ and twig temperatures by insulating these organs from heat (figure 7.1).

7.3.2 Heat tolerance

Heat tolerance is the ability of some genotypes to perform better than others when internal temperatures are comparable and in the realm of HS. HT is largely associated with cellular and subcellular components, and its expression is highly dependent on heat hardening or heat acclimation. HT may be defined as an improved capability of a genotype to withstand a period of high temperatures as a consequence of earlier exposure to high temperatures for a given period. The exposure time may vary from 20 s at 50 °C to 4–6 hr, with <1 hr being the most common. In general, the lower the hardening temperature, the lower the level and slower the process of hardening. When heat-hardened plants are subjected to lower temperatures, they lose their hardening; this is called de-hardening. Generally, de-hardening is slower with higher hardening temperatures and with lower de-hardening temperatures. Hsps most likely form an important component of heat hardening induced by a heat shock. The hardening induced by a prolonged (3 days) exposure to moderately high temperatures, however, most likely involves some other mechanism. Membrane stability, reduced heat response of photosystem II, photosynthetic translocation, stem-reserve mobilization, and osmoregulation are all involved in HT. All of these effects are incorporated into stress-induced growth and recovery (figure 7.1).

7.4 Molecular basis of the heat stress response

7.4.1 Oxidative stress and antioxidants

When the temperature rises, the loss of water causes oxidative stress. Under such conditions, various ROS are formed, including singlet oxygen ($^1O2$), hydrogen peroxide (H_2O_2), hydroxyl radicals (OH), and superoxide radicals (O^-) [14]. ROS are used to change membrane permeability by catalyzing the oxidation of membrane lipids and the degradation of pigments [57]. Superoxide radicals are continuously generated in the chloroplast and mitochondrion within the cell [58, 59]. Superoxide radicals and hydrogen peroxides are relatively less toxic than hydroxyl radicals formed during the Haber–Weiss reaction. OH radicals may degrade protein, lipids, DNA and chlorophyll, affecting plant metabolic efficiency, growth, and eventually yield [60]. Plants' ability to withstand oxidative stress is critical for their survival in hot climates [61]. Plants have developed distinct detoxification mechanisms to avoid oxidative damage [60]. Increased ascorbate and glutathione production in turfgrass under heat and cold stress reduces the production of ROS [57]. Also, signaling chemicals like melatonin and abscisic acid may increase the antioxidant capacity of cells during cold stress [62].

7.4.2 Heat shock proteins

Temperature fluctuations induce increased Hsps production in plants [63, 64]. Hsps synthesis is lowered during various developmental phases, such as embryo development, sprouting, pollen production, and fruit development [65]. Most plants synthesize Hsps during the developmental phase, and many of these Hsps are remarkably similar across different species [37]. Hsp100 is a cytoplasmic protein whose molecular mass ranges from 100–114 kDa. Hsp60 has a size range of 10–60 kDa and is found in chloroplasts and mitochondria. Hsp70 is a 69–71 kDa protein found in the cytoplasm, endoplasmic reticulum, and mitochondria. Hsp90 is a cytoplasmic and endoplasmic reticulum-associated protein with a molecular weight of 80–90 kDa. smHsps, with a protein size of 15–30 kDa, is found in the cytoplasm, ER, chloroplast, and mitochondria. Plants have six nuclear gene families for Hsps, each of which encodes to a protein found in particular cell organelles such as the chloroplast, cytosol, mitochondria, ER, and membranes. Hsps are connected with specific cellular structures such as cell walls, ribosomes, mitochondria, and chloroplasts according to immunolocalization studies [66–69]. Corn possesses five low-molecular-weight (LMW) Hsps (28, 23, 22, 20, and 19 kDa proteins) in its mitochondria, which aid in its ability to resist high temperatures (42 °C). Wheat and rye, on the other hand, only have one (20 kDa). High-quality chloroplast Hsps in heat-tolerant maize and tomato have been detected [70]. It was discovered that Hsp101 was frequently increased in response to this in various vegetative and reproductive parts [71]. Elevated temperature causes the synthesis of a 45 kDa Hsps in maize, which has been proven to be related to HS tolerance [66] (table 7.1).

Plants need Hsps to regulate their metabolic systems as well as HT. Plants lengthen and strengthen their roots in order to overcome stress [72]. Some researchers claim that thermal tolerance is linearly associated with Hsps production [73]. Hsps are molecular strategies employed for the regulation of fundamental conformation and activity of cell proteins. For example, Hsps help new proteins acquire the topologies required for normal operations including translation and translocation, proteolysis, chaperoning, and deaggregation, as well as reactivating denatured proteins [37]. LMW Hsps are structural proteins that help maintain cell membrane integrity. In chloroplasts, LMW Hsps preserve the photosystem II and enhance photosynthetic electron transport efficiency during such adverse conditions [67, 74] (table 7.1). In the tomato plant, the LMW Hsp21 protein serves two purposes: it protects the photosystem II from oxidative damage and regulates fruit color during storage under chilling stress [75] (table 7.1).

Table 7.1. Recent heat stress tolerance studies utilizing omics approaches.

Plant species	Primary stress	Secondary stresses	Sample tissue	Conditions	Profiling	References
Oryza sativa	Heat		Leaf and seed	30 °C	qRT–PCR	[90]
			Anther	Up to 37 °C in	iTRAQ LC-MS/MS	[91]
		Low temperature	Suspension cell	44 °C	LC-MS/MS	[92]
		Field condition	Leaf and grain spikelet (milky stage)	42/32 °C, 38 °C, 28 °C	MALDI TOF/TOF MS/MS	[93, 94]
		Variable temperature	Seedling	35 °C, 40 °C and 45 °C, 38 °C	MALDI TOF MS/MS	[95, 96]
			Anther	38 °C	LC-MS/MS	[97]
			Flowering stage	38 °C for 1 day	RNA-seq	[98]
			Seed	Germination at 30 °C (control) and 42 °C (stress)	RNA-seq	[99]
		Oxidative stress	10-day-old seedling	42 ± 1 °C for 30 min with 10 mM H_2O_2	60mer rice 44k array, qRT–PCR	[100]
			Leaf	45 °C	RNA-seq	[101]
			Seed	38 °C	RNA-seq	[102]
		Drought	Leaf	40 °C	Target lipidomics	[103]
		Exogenous brassinosteroids	Leaf	47 °C	Target lipidomics	[104]
Triticum aestivum	Heat		Filling stage	37 °C for 4 hr	Metabolomics, RNA-seq	[105]
		Alteration of lipids	Flowering stage	Up to 35 °C	Metabolomics	[106]
			Seedling	Up to 42 °C	RNA-seq	[107]
			Seedling	35 °C	MALDI TOF/TOF MS/MS	[108]
			Flag leaf, stems, and spikes	38 °C, 37 °C	iTRAQ LC-MS/MS	[109, 110]
		Field condition	Seed	40 °C	MALDI TOF/TOF MS/MS	[111]
		Low nitrogen	Leaf	32 °C	MALDI TOF/TOF MS/MS	[112]
		Drought	Post-anthesis	35 °C	MALDI TOF/TOF MS/MS	[113]
			Leaf	35 °C	Proteomics	[105]
			Leaf of heat-susceptible cultivate and HT cultivar	40 °C for 1 hr with and without heat acclimation (34 °C, 3 hr)	RNA-seq	[114]
			Flag leaf, seedlings	37 °C /17 °C for 3 days in the anthesis stage	Proteomics	[109]
			Spikelet post-anthesis	35 °C	LC/HRMS	[115]

(*Continued*)

Table 7.1. (*Continued*)

Plant species	Primary stress	Secondary stresses	Sample tissue	Conditions	Profiling	References
		Drought	Seedling	40 °C, 20 % (w/v) PEG-6000	RNA-seq	[116]
			Leaf	35 °C	Lipidomics ICP-MS	[106]
			Pollen	35 °C	Lipidomics ICP-MS	[117]
			Leaf	35 °C	Lipidomics ICP-MS/MS	[118]
Zea mays	Heat	Drought	Chloroplast	42 °C	MALDI TOF/TOF MS/MS	[66]
		Drought	Leaf	42 °C	iTRAQ LC-MS/MS	[66, 119]
			Seedling	42 °C	RNA-seq	[120]
			Seedling	32 °C–38 °C	RNA-seq	[121]
		High CO2	Leaf	32 °C	GC-MS	[122]
Hordeum vulgare	Heat	Drought	Plant leaf	26 °C	IC-MS/MS	[123]
			Anthesis	42 °C	Affymetrix	[124]
		Drought	Flag leaf	30 °C+50% FC	RNA-seq	[125]
Cicer arietinum	Heat	Oxidative	Leaf	37 °C	Targeted metabolomics	[126]
			Developmental-stage plant	Up to 42 °C	LC-MS	[127]
Lens culinaris	Heat		Leaf and pods	30 °C–50 °C	Enzyme activities	[128]
			Pods	Up to 33 °C	Enzyme activities	[129]
		Drought	Leaf and pods	30 °C	Enzyme activities	[130]
Arachis hypogaea	Heat		Seedling	40 °C and 30 °C	Physiological measurements, metabolomics	[131]
			Leaf	45 °C	RNA-seq (qRT–PCR), physiological measurements	[132]
Brassica napus	Heat		Emerging seed	Up to 35 °C	95k EST microarray	[133]
			Flowering	Up to 35 °C	Association studies	[134]
Glycine max	Heat	High humidity	Pre-harvest seed	40 °C	MALDI TOF/TOF MS/MS	[135]
			3- to 4-week-old plants	45 °C for 3, 6, 24 hr	RNA-seq	[136]
		High CO2	Leaf	36 °C	GC-MS	[137]
			Root	40 °C	LC-MS/MS; RNA-seq	[138]
			Dry, imbibed, or germinated seeds from HT and heat-sensitive variety	Field conditions	RNA-seq	[139]

Helianthus annuus	Heat	Light	Leaf and immature seed	37 °C	RNA-seq	[140]
Solanum lycopersicum	Heat	Oxidative	Flower buds	36 °C/25 °C for 3 months	LC-MS/MS	[141]
		Lipid antioxidant and galactolipid remodeling	5- to 6-week-old plants	38 °C	MS	[142]
			Pollen	38 °C	Proteomics, LC/MS	[143]
		Submergence	Leaf	39 °C	MALDI TOF/TOF MS/MS	[144]
			Anther	32 °C for 2, 6, 16 or 30 hr	RNA-seq (cDNA-AFLP; 90 K Custom TomatoArray 1.0)	[145]
			Flower bud	43 °C–45 °C for 2 hr	RNA-seq	[146]
			Week-old plants	1 hr at 39 °C	MACE	[147]
			Pollen	38 °C	LC-QTOF-MS + LTQ Orbitrap XL, mass spectrometer	[148]
		N/A	Pollen tube	37 °C	Targeted metabolomics	[149]
		Low temperature	Leaf	38 °C, 20 °C, 10 °C	Untargeted lipidomics LC/MS	[142]
Solanum tuberosum	Heat		Leaf, tubers	Up to 30 °C	GC/MS, microarray	[150]
			Leaf	35 °C	RNA-seq	[151]
			Skin, tubers, and phelloderm	33 °C–35 °C	RNA-seq (qRT–PCR)	[152]
Vitis vinifera	Heat		Leaf	43 °C	iTRAQ LC-MS/MS	[153]
		Variable temperatures	Leaf	35 °C, 40 °C, and 45 °C	iTRAQ LC-MS/MS	[154]
			Leaf	45 °C	RNA-seq (Affymetrix gene chip)	[155]

qRT–PCR: Real-time PCR; iTRAQ LC-MS/MS: Isobaric tags for relative and absolute quantitation (iTRAQ) in combination with the liquid chromatography-tandem mass spectrometry (LC-MS/MS); MALDI TOF/TOF MS/MS : Matrix assisted laser desorption/ionisation-time-of-flight mass spectrometry; LC-MS/MS: liquid chromatography-tandem mass spectrometry; RNA-seq: RNA sequencing; LC/HRMS: Liquid chromatography-high resolution mass spectrometry; Lipidomics ICP-MS : Lipidomics Inductively coupled plasma mass spectrometry; GC-MS: Gas chromatography–mass spectrometry; LC-MS: Liquid chromatography–mass spectrometry; 95k EST microarray: 95k Expressed Sequence Tag microarray; MACE: Massive analysis of cDNA ends; LC-QTOF-MS + LTQ Orbitrap XL: Liquid chromatography quadrupole time-of-flight mass spectrometry + LTQ Orbitrap XL is a hybrid FT mass spectrometer combines a linear ion trap MS and the Orbitrap mass analyzer.

7.4.3 Other heat-activated proteins

Other heat-stress-responsive proteins include ubiquitin [76], cytosolic Cu/Zn-SOD [77], and Mn-POD [78]. For example, in *Prosopis chilensis* and soybeans, ubiquitin and conjugated-ubiquitin synthesis appear to be a significant mechanism of HT during the first 30 min of HS [79]. According to several studies, heat shock stimulates Mn peroxidase, a key enzyme in preventing oxidative damage [80]. In another report, when leaf proteins extracted from the thylakoid with stromal fractions of *Chenopodium murale* were exposed to high temperature, Cu/Zn-SOD from the stromal fraction was much more heat resistant than Cu/Zn-SOD from the thylakoid fraction [81]. It has been shown that Pir proteins (heat and N-induced osmotin-like proteins) are overexpressed in yeast cells, imparting resistance to tobacco osmotin, which is an antifungal [82]. Some proteins, such as late embryogenesis abundant (LEA) proteins, can be used to safeguard the citrate synthase against harsh environmental conditions such as high temperatures and dehydration; recent studies [83, 84] reported increased expression of 25 LEA proteins in hexaploid wheat during the grain-filling stage. Three LMW dehydrins with enhanced HSR have been recently discovered in sugarcane leaves [85]. Involved in this activity, these proteins seem to be connected to protein degradation, reducing dehydration and oxidative damage [86]. Stress protein expression, in a nutshell, is a plant adaptation to HS tolerance. The expression of LMW and high-molecular-weight Hsps in different plants is the most important mechanism. These proteins exhibit organ- and tissue-specific expression and have putative functions such as being chaperones, packaging and unpackaging of cellular proteins, and safeguarding of efficient sites from the detrimental effects of HS. Along with other stress proteins, it has been shown that ubiquitin, LEA, dehydrins, and Pir proteins are also expressed under such conditions. These proteins seem to protect the cellular and subcellular structures against the oxidative and dehydration effects of free radicals.

7.5 Functions of heat-inducible genes

Heat shock proteins are recognized for their participation in a plant's capability to withstand periods of oppressive heat [87]. Numerous experimental studies on soybean crops have revealed a significant correlation between Hsps concentration and seedling HT [88] (tables 7.2 and 7.3). There has been much less research on the specific role of Hsps in legume heat resistance. In a recent study on cowpeas, five quantitative trait locii (QTL) were identified to be positively correlated with the number of pod sets per peduncle under increased TS. Synthetic evaluation of these five QTLs with the soybean genome revealed Hsps and heat shock transcription genes that were exclusive to these QTLs [89] (table 7.1). Unlike the vegetative phases, reproductive stages are more susceptible to elevated temperatures, as already mentioned previously. This may be attributed to pollen's inability to produce Hsps as an adaptation to hot weather [87] (tables 7.2 and 7.3).

Table 7.2. Various transgenics studies for plants traits' improvement under heat stress.

Crop/Plant	Target gene or protein	Promoter	Function	References
Oryza sativa	HSP70/rice	CaMV 35S	Increased tolerance to HT	[156]
	Athsp101/*Arabidopsis*	CaMV 35S		[157]
	OsRab7	CaMV 35S		[158]
	TaMBF1c/wheat	Maize ubiquitin 1		[159]
Triticum aestivum	Hsf6A/wheat	Barley HVA1s promoter	Increased tolerance to HS	[160]
	EF-Tu/maize	Maize ubiquitin 1 promoter		[161]
	TaHsfA6f/wheat	HVA1s		[162]
	TaFER-5B/wheat	Maize ubiquitin 1		[163]
	TaHsfC2a/wheat	—		[164]
	TaGASR1/wheat	—	Increased tolerance to HS and OS	[165]
Zea mays	OsMYB55/rice	Maize ubiquitin 1	Enhanced tolerance to drought and high-temperature stress	[166]
	ZmNF-YB2/maize	Rice actin 1	Improved drought tolerance and enhanced photosynthesis activity	[167]
Solanum tuberosum	HSc70 allelic variant/potato	HSc70 native promoter	Enhanced tolerance to HS	[168]
Glycine max	P5CR/*Arabidopsis*	—	Improved tolerance to increased temperature	[169]
Nicotiana tabacum	HSP70–1/tobacco	CaMV 35S	Improved tolerance to increased temperature	[170]
	HSP70–1/brassica	—		[51]
Solanum lycopersicum	HSP21/tomato	CaMV 35S	Tolerance from heat-induced oxidative stress	[75]

Table 7.3. List of identified genes for improvement of traits under heat stress.

Crop	Target gene	Stress or trait	References
Oryza sativa	*OsDEP1, OsMPK2*	Yield traits	[171]
	GS3, Gn1a	Yield traits	[172]
	OsMPK2, OsPDS, OsBADH2	Abiotic stress tolerance	[173]
	GW5, GW2, TGW6	Grain weight	[137]
	DEP1, Gn1a, GS3	Grain size and number, erect panicles	[62]
Triticum aestivum	*TaERF3, TaDREB2*	Abiotic stress tolerance	[174]
Zea mays	*ARGOS8*	Yield attributing traits under stress	[175]

7.5.1 Activation of signaling pathway

HSR in plants is extensively conserved and includes a number of different pathways and regulatory mechanisms. A minimum of four potential sensors have recently been identified for inducing HSR in plants. A plasma membrane pathway that drives calcium flux inward, a nucleus-localized histone sensor, and two unfolded protein sensors in the endoplasmic reticulum (ER) and cytosol are among them. These potential sensors stimulate a series of HSR genes, resulting in increased HT, but the connection between different signaling cascades and their hierarchical order is unknown.

7.6 Conclusions

Elevated temperatures have profound repercussions on the growth, morphology, and function of plants. HS also results in oxidative stress, restricted photosynthesis, and membrane permeability. Comprehending the molecular and physiological mechanisms underlying HT is critical to developing plants that can sustain high yields under HT. Such HT is a polygenically governed trait and, owing to its complex nature, can only be addressed with a holistic approach that combines traditional and contemporary molecular genetic technologies, resulting in plants with increased thermotolerances. Identifying genetic resources with strong tolerance against HS is a critical step in breeding plants with improved HT. Molecular biology has received a lot of attention as the latest agricultural research technique for detecting molecular markers linked to stress-related attributes and associated candidate genes for stress tolerance. Transgenic pyramiding, which involves combining several beneficial genes into a commercial variety, is also anticipated to be a crucial avenue to crop improvement. These transgenic techniques should be supplemented with attempts in marker-assisted breeding programmes for stress-related genes and QTLs for effective HT. Recent approaches, such as omics technologies, can provide a wealth of data and shed light on novel aspects of HT. The adoption of such high-throughput technologies will aid in acquiring a denser genome-wide marker saturation, speed up QTL identification and transfer via genomic selection, and enhance the flexibility of phenotypic selection and evaluation.

Multiple choice questions

1. Characters involved in heat-tolerant plants typically exhibit which of the following?
   - (a) Monogenic control;
   - (b) Oligogeneic and modifying gene control;
   - (c) Polygenic control;
   - (d) Polygenic + plasmagenic control.
2. High temperatures induce?
   - (a) Leaf and twig blazing;
   - (b) Leaf fall;
   - (c) Shoot and root growth retardation, and fruit discoloration;
   - (d) All of the above.
3. What happens in plants at the biochemical level under severe heat stress?
   - (a) Increase in kinetic energy of biomolecules;
   - (b) Increased membrane fluidity;
   - (c) Decrease in kinetic energy of biomolecules;
   - (d) Both (a) and (b).
4. Exposure of plants to heat stress produce?
   - (a) ROS-singlet oxygen ($^1O_2$);
   - (b) Superoxide radicals (O_2);
   - (c) Hydrogen peroxide (H_2O_2) and hydroxyl radicals (OH);
   - (d) All of the above.
5. When plants are exposed to heat stress …
   - (a) The water potential and relative water content are significantly reduced;
   - (b) Photosynthetic efficiency is increased;
   - (c) The water potential and relative water content are significantly increased;
   - (d) All of the above.

Answers: 1 (c); 2 (d); 3 (a); 4 (d); 5 (a).

Descriptive-type questions

1. Write short notes on …
   - (a) Heat stress;
   - (b) Heat avoidance;
   - (c) Heat tolerance;
   - (d) Reactive oxygen species;
   - (e) Heat shock proteins;
   - (f) Membrane stability under stress.
2. Give a schematic representation of the morphological, physiological, and biochemical adaptations of plants to deal with heat stress?
3. Write in brief about the molecular basis of heat stress? Give a schematic representation of the effects of high temperature on plants?

4. Briefly describe some recent examples of molecular approaches for heat stress and their outcomes?
5. Enlist various transgenic studies for plant trait improvement under heat stress and the genes identified for improvement of traits under heat stress?

References

[1] Rosenzweig C, Elliott J, Deryng D, Ruane A C, Müller C and Arneth A *et al* 2014 Assessing agricultural risks of climate change in the 21st century in a global gridded crop model intercomparison *Proc. Natl. Acad. Sci. USA* **111** 3268–73

[2] Wheeler T and Von Braun J 2013 Climate change impacts on global food security *Science* **341** 508–13

[3] Mittler R 2006 Abiotic stress, the field environment and stress combination *Trends Plant Sci.* **11** 15–9

[4] Hall A E 2001 *Crop Responses to Environment* (Boca Raton, FL: CRC Press)

[5] IPCC 2022 Special report: Global warming of 1.5 ºCI

[6] Betts R A, Collins M, Hemming D L, Jones C D, Lowe J A and Sanderson M G 2011 When could global warming reach 4 °C? *Phil. Trans. R Soc.* A **69** 67–84

[7] Lobell D B, Schlenker W and Costa-Roberts J 2011 Climate trends and global crop production since 1980 *Science* **333** 616–20

[8] Zhao J, Lu Z, Wang L and Jin B 2021 Plant responses to heat stress: physiology, transcription, noncoding RNAs, and epigenetics *Int. J. Mol. Sci.* **22** 117

[9] Handmer J and Dovers S 2013 *Handbook of Disaster Policies and Institutions: Improving Emergency Management and Climate Change Adaptation* (Abingdon: Routledge)

[10] Ding Y, Chen H, Wang L, Yang C Y, He Y and Yang W *et al* 2007 Heat transfer intensification using nanofluids *KONA Pow. Part J.* **25** 23–38

[11] Teixeira E I, Fischer G, Van Velthuizen H, Walter C and Ewert F 2013 Global hot-spots of heat stress on agricultural crops due to climate change *Agri. Forest Meteorol.* **170** 206–15

[12] Li Q, Wang W, Wang W, Zhang G, Liu Y, Wang Y and Wang W 2018 Wheat F-box protein gene TaFBA1 is involved in plant tolerance to heat stress *Front. Plant Sci.* **24** 521

[13] Howarth C J, Ashraf M and Harris P J C 2005 Genetic improvements of tolerance to high temperature *Abiotic Stresses: Plant Resistance through Breeding and Molecular Approaches* (New York: Haworth Press) pp 277–300

[14] Liu X and Huang B 2000 Heat stress injury in relation to membrane lipid peroxidation in creeping bentgrass *Crop Sci.* **40** 503–10

[15] Moreno A A and Orellana A 2011 The physiological role of the unfolded protein response in plants *Biol. Res.* **44** 75–80

[16] Vollenweider P and Günthardt-Goerg M S 2005 Diagnosis of abiotic and biotic stress factors using the visible symptoms in foliage *Environ. Pollut.* **137** 455–65

[17] Giaveno C and Ferrero J 2003 Introduction of tropical maize genotypes to increase silage production in the central area of Santa Fe, Argentina *Crop. Breed. Appl. Biotechnol.* **3** 89–94

[18] Zhao D Q, Li T T, Hao Z J, Cheng M L and Tao J 2019 Exogenous trehalose confers high temperature stress tolerance to herbaceous peony by enhancing antioxidant systems, activating photosynthesis, and protecting cell structure *Cell Stress Chapero.* **24** 247–57

[19] Ferris R, Ellis R H, Wheeler T R and Hadley P 1998 Effect of high temperature stress at anthesis on grain yield and biomass of field-grown crops of wheat *Ann. Bot.* **82** 631–39

[20] Rainey K M and Griffiths P D 2005 Inheritance of heat tolerance during reproductive development in snap bean (*Phaseolus vulgaris* L.) *J. Ame. Soc. Horti. Sci.* **130** 700–6
[21] Vara Prasad P V, Craufurd P Q, Summerfield R J and Wheeler T R 2000 Effects of short episodes of heat stress on flower production and fruit-set of groundnut (*Arachis hypogaea* L.) *J. Exp. Bot.* **51** 777–84
[22] Ashraf M and Hafeez M 2004 Thermotolerance of pearl millet and maize at early growth stages: growth and nutrient relations *Biol. Plant* **48** 81–6
[23] Wahid A, Gelani S, Ashraf M and Foolad M R 2007 Heat tolerance in plants: an overview *Environ. Exp. Bot.* **61** 199–223
[24] Ahmad B, Zaid A, Sadiq Y, Bashir S and Wani S H 2019 Role of selective exogenous elicitors in plant responses to abiotic stress tolerance *Plant Abiotic Stress Tolerance* (Cham: Springer) pp 273–90
[25] Ebrahim M K, Vogg G, Osman M N and Komor E 1998 Photosynthetic performance and adaptation of sugarcane at suboptimal temperatures *J. Plant Physiol.* **153** 587–92
[26] Asseng S, Ewert F, Martre P, Rötter R P, Lobell D B and Cammarano D *et al* 2015 Rising temperatures reduce global wheat production *Nat. Clim. Change.* **5** 143–7
[27] Guilioni L, Wéry J and Lecoeur J 2003 High temperature and water deficit may reduce seed number in field pea purely by decreasing plant growth rate *Funct. Plant Biol.* **30** 1151–64
[28] Wilhelm E P, Mullen R E, Keeling P L and Singletary G W 1999 Heat stress during grain filling in maize: effects on kernel growth and metabolism *Crop Sci.* **39** 1733–41
[29] Maestri E, Klueva N, Perrotta C, Gulli M, Nguyen H T and Marmiroli N 2002 Molecular genetics of heat tolerance and heat shock proteins in cereals *Plant Mol. Biol.* **48** 667–81
[30] Fahad S, Bajwa A A, Nazir U, Anjum S A, Farooq A and Zohaib A *et al* 2017 Crop production under drought and heat stress: plant responses and management options *Front. Plant Sci.* **8** 1147
[31] Ashraf M and Harris P J C 2005 *Abiotic Stresses: Plant Resistance through Breeding and Molecular Approaches* (New York: Haworth Press)
[32] Wien H C 1997 Lettuce *The Physiology of Vegetable Crops* ed H C Wien (London: CAB International) pp 479–509
[33] Zhao L, Oppenheimer M, Zhu Q, Baldwin J W, Ebi K L and Bou-Zeid E *et al* 2018 Interactions between urban heat islands and heat waves *Environ. Res. Lett.* **13** 034003
[34] Praba M L, Cairns J E, Babu R C and Lafitte H R 2009 Identification of physiological traits underlying cultivar differences in drought tolerance in rice and wheat *J. Agron. Crop. Sci.* **195** 30–46
[35] Bañon S, Fernandez J A, Franco J A, Torrecillas A, Alarcón J J and Sánchez-Blanco M J Effects of water stress and night temperature preconditioning on water relations and morphological and anatomical changes of *Lotus creticus* plants *Sci. Hortic.* **101** 333–42
[36] Tombesi S, Nardini A, Frioni T, Soccolini M, Zadra C and Farinelli D *et al* 2015 Stomatal closure is induced by hydraulic signals and maintained by ABA in drought-stressed grapevine *Sci. Rep.* **5** 1–2
[37] Zhang J, Huang W, Pan Q and Liu Y 2005 Improvement of chilling tolerance and accumulation of heat shock proteins in grape berries (*Vitis vinifera* cv. *Jingxiu*) by heat pretreatment *Postharvest Biol. Technol.* **38** 80–90
[38] Hall A E 1992 Breeding for heat tolerance *Plant Breed. Rev.* **10** 129–68
[39] Marcum K B 1998 Cell membrane thermostability and whole-plant heat tolerance of Kentucky bluegrass *Crop Sci.* **38** 1214–8

[40] Tewolde H, Fernandez C J and Erickson C A 2006 Wheat cultivars adapted to post heading high temperature stress *J. Agron. Crop Sci.* **192** 111–20

[41] Xalxo R, Yadu B, Chandra J, Chandrakar V and Keshavkant S 2020 Alteration in carbohydrate metabolism modulates thermotolerance of plant under heat stress *Heat Stress Tolerance in Plants: Physiological, Molecular and Genetic Perspectives* (New York: Wiley) pp 77–115

[42] Hasanuzzaman M, Nahar K, Alam M, Roychowdhury R and Fujita M 2013 Physiological, biochemical, and molecular mechanisms of heat stress tolerance in plants *Int. J. Mol. Sci.* **14** 9643–84

[43] Hu S, Ding Y and Zhu C 2020 Sensitivity and responses of chloroplasts to heat stress in plants *Front. Plant Sci.* **11** 375

[44] Renata Szyman´ska Lesak I, Orzechowska A and Kruk J 2017 Physiological and biochemical responses to high light and temperature stress in plants *Environ. Exp. Bot.* **139** 165–77

[45] Song Y, Chen Q, Ci D, Shao X and Zhang D 2014 Effects of high temperature on photosynthesis and related gene expression in poplar *BMC Plant Biol.* **14** 111

[46] Perdomo J A, Capó-Bauçà S, Carmo-Silva E and Galmés J 2017 Rubisco and rubisco activase play an important role in the biochemical limitations of photosynthesis in rice, wheat, and maize under high temperature and water deficit *Front. Plant Sci.* **13** 490

[47] Bita C and Gerats T 2013 Plant tolerance to high temperature in a changing environment: scientific fundamentals and production of heat stress-tolerant crops *Front. Plant Sci.* **4** 273

[48] Dhanda S S and Munjal R 2012 Heat tolerance in relation to acquired thermotolerance for membrane lipids in bread wheat *Field Crops Res.* **135** 30–7

[49] Mohammed A R and Tarpley L 2009 Impact of high nighttime temperature on respiration, membrane stability, antioxidant capacity, and yield of rice plants *Crop Sci.* **49** 313–22

[50] Nosaka Y and Nosaka A Y 2017 Generation and detection of reactive oxygen species in photocatalysis *Chem. Rev.* **117** 11302–36

[51] Wang D, Heckathorn S A, Mainali K and Tripathee R 2016 Timing effects of heat-stress on plant ecophysiological characteristics and growth *Front. Plant Sci.* **7** 1629

[52] Djanaguiraman M and Prasad P V 2010 Ethylene production under high temperature stress causes premature leaf senescence in soybean *Funct. Plant Biol.* **37** 1071–84

[53] Jespersen D 2020 Heat shock induced stress tolerance in plants: physiological, biochemical, and molecular mechanisms of acquired tolerance *Priming-Mediated Stress and Cross-Stress Tolerance in Crop Plants* (New York: Academic) pp 161–74

[54] Liu X H, Lyu Y S, Yang W, Yang Z T, Lu S J and Liu J X 2020 A membrane associated NAC transcription factor OsNTL3 is involved in thermotolerance in rice *Plant Biotechnol. J.* **18** 1317–29

[55] Wang X, Zhang H, Xie Q, Liu Y, Lv H and Bai R *et al* 2020 SlSNAT interacts with HSP40, a molecular chaperone, to regulate melatonin biosynthesis and promote thermotolerance in tomato *Plant Cell Physiol.* **61** 909–21

[56] Zhang C and Hsieh T F 2013 Heritable epigenetic variation and its potential applications for crop improvement *Plant Breed. Biotechnol.* **1** 307–19

[57] Xu S, Li J, Zhang X, Wei H and Cui L 2006 Effects of heat acclimation pretreatment on changes of membrane lipid peroxidation, antioxidant metabolites, and ultrastructure of chloroplasts in two cool-season turfgrass species under heat stress *Environ. Exp. Bot.* **56** 274–85

[58] Linley E, Denyer S P, McDonnell G, Simons C and Maillard J Y 2012 Use of hydrogen peroxide as a biocide: new consideration of its mechanisms of biocidal action *J. Antimicrob. Chemother.* **67** 1589–96

[59] Sharma D K, Andersen S B, Ottosen C O and Rosenqvist E 2012 Phenotyping of wheat cultivars for heat tolerance using chlorophyll a fluorescence *Funct. Plant Biol.* **39** 936–47

[60] Sairam R K and Tyagi A 2004 Physiology and molecular biology of salinity stress tolerance in plants *Curr. Sci.* **86** 407–21

[61] Larkindale J and Knight M R 2002 Protection against heat stress-induced oxidative damage in *Arabidopsis* involves calcium, abscisic acid, ethylene, and salicylic acid *Plant Physiol.* **128** 682–95

[62] Li H, Ahammed G J, Zhou G, Xia X, Zhou J and Shi K *et al* 2016 Unraveling main limiting sites of photosynthesis under below-and above-ground heat stress in cucumber and the alleviatory role of luffa rootstock *Front. Plant Sci.* **7** 746

[63] Mitra R and Bhatia C R 2008 Bioenergetic cost of heat tolerance in wheat crop *Curr. Sci.* **94** 1049–53

[64] Pessarakli M 2016 Heat shock proteins and acquisition of thermotolerance in plants *Handbook of Plant and Crop Stress* (Boca Raton, FL: CRC Press) pp 549–64

[65] Prasinos C, Krampis K, Samakovli D and Hatzopoulos P 2005 Tight regulation of expression of two *Arabidopsis* cytosolic Hsp90 genes during embryo development *J. Exp. Bot.* **56** 633–44

[66] Hu Z, Song N, Zheng M, Liu X, Liu Z and Xing J *et al* 2015 Histone acetyltransferase GCN 5 is essential for heat stress-responsive gene activation and thermotolerance in *Arabidopsis Plant J.* **84** 1178–91

[67] Kotak S, Larkindale J, Lee U, von Koskull-Döring P, Vierling E and Scharf K D 2007 Complexity of the heat stress response in plants *Curr. Opin. Plant Biol.* **10** 310–6

[68] Mittler R, Finka A and Goloubinoff P 2012 How do plants feel the heat? *Trends Biochem. Sci.* **37** 118–25

[69] Yang J D, Yun J Y, Zhang T H and Zhao H L 2006 Presoaking with nitric oxide donor SNP alleviates heat shock damages in mung bean leaf discs *Bot. Stud.* **47** 129–36

[70] Moriarty T, West R, Small G, Rao D and Ristic Z 2002 Heterologous expression of maize chloroplast protein synthesis elongation factor (EF-Tu) enhances *Escherichia coli* viability under heat stress *Plant Sci.* **163** 1075–82

[71] Young T E, Ling J, Geisler-Lee C J, Tanguay R L, Caldwell C and Gallie D R 2001 Developmental and thermal regulation of the maize heat shock protein, HSP101 *Plant Physiol.* **127** 777–91

[72] Nieto-Sotelo J, Martínez L M, Ponce G, Cassab G I, Alagón A and Meeley R B *et al* 2002 Maize HSP101 plays important roles in both induced and basal thermotolerance and primary root growth *Plant Cell* **14** 1621–33

[73] Chao L M, Liu Y Q, Chen D Y, Xue X Y, Mao Y B and Chen X Y 2017 *Arabidopsis* transcription factors SPL1 and SPL12 confer plant thermotolerance at reproductive stage *Mol. Plant* **10** 735–48

[74] Barua D, Downs C A and Heckathorn S A 2003 Variation in chloroplast small heat-shock protein function is a major determinant of variation in thermotolerance of photosynthetic electron transport among ecotypes of *Chenopodium album Funct. Plant Biol.* **30** 1071–9

[75] Neta-Sharir I, Isaacson T, Lurie S and Weiss D 2005 Dual role for tomato heat shock protein 21: protecting photosystem II from oxidative stress and promoting color changes during fruit maturation *Plant Cell* **17** 1829–38
[76] Sun C W and Callis J 1997 Independent modulation of *Arabidopsis thaliana* polyubiquitin mRNAs in different organs and in response to environmental changes *Plant J.* **11** 1017–27
[77] Herouart D V, Van Montagu M and Inze D 1994 Developmental and environmental regulation of the *Nicotiana plumbaginifolia* cytosolic Cu/Zn-superoxide dismutase promoter in transgenic tobacco *Plant Physiol.* **104** 873–80
[78] Brown J A, Li D, Alic M and Gold M H 1993 Heat shock induction of manganese peroxidase gene transcription in *Phanerochaete chrysosporium Appl. Environ. Microbiol.* **59** 4295–9
[79] Ortiz C and Cardemil L 2001 Heat-shock responses in two leguminous plants: a comparative study *J. Exp. Bot.* **52** 1711–9
[80] Iba K 2002 Acclimative response to temperature stress in higher plants: approaches of gene engineering for temperature tolerance *Annu. Rev. Plant Biol.* **53** 225–45
[81] Khanna-Chopra R and Sabarinath S 2004 Heat-stable chloroplastic Cu/Zn superoxide dismutase in *Chenopodium murale Biochem. Biophys. Res. Commun.* **320** 1187–92
[82] Yun D J, Zhao Y, Pardo J M, Narasimhan M L, Damsz B and Lee H *et al* 1997 Stress proteins on the yeast cell surface determine resistance to osmotin, a plant antifungal protein *Proc. Natl. Acad. Sci. USA* **94** 7082–7
[83] Goyal K, Walton L J and Tunnacliffe A 2005 LEA proteins prevent protein aggregation due to water stress *Biochem. J.* **388** 151–7
[84] Majoul T, Bancel E, Triboï E, Ben Hamida J and Branlard G 2003 Proteomic analysis of the effect of heat stress on hexaploid wheat grain: characterization of heat responsive proteins from total endosperm *Proteomics* **3** 175–83
[85] Wahid A and Close T J 2007 Expression of dehydrins under heat stress and their relationship with water relations of sugarcane leaves *Biol. Plant.* **51** 104–9
[86] Schöffl F, Prandl R and Reindl A 1999 Molecular responses to heat stress *Molecular Responses to Cold, Drought, Heat and Salt Stress in Higher Plants* 83 (Austin, TX: R. G. Landes) pp 81–98
[87] Ohama N, Sato H, Shinozaki K and Yamaguchi-Shinozaki K 2017 Transcriptional regulatory network of plant heat stress response *Trends Plant Sci.* **22** 53–65
[88] Zhang J, Li Y, Jia H X, Li J B, Huang J, Lu M Z and Hu J J 2015 The heat shock factor gene family in *Salix suchowensis*: a genome-wide survey and expression profiling during development and abiotic stresses *Front Plant Sci.* **6** 748
[89] Lucas M R, Ehlers J D, Huynh B L, Diop N N, Roberts P A and Close T J 2013 Markers for breeding heat-tolerant cowpea *Mol. Breed.* **31** 529–36
[90] Phan T T, Ishibashi Y, Miyazaki M, Tran H T, Okamura K and Tanaka S *et al* 2013 High temperature-induced repression of the rice sucrose transporter (*Os SUT 1*) and starch synthesis-related genes in sink and source organs at milky ripening stage causes chalky grains *J. Agron. Crop Sci.* **199** 178–88
[91] Mu Q, Zhang W, Zhang Y, Yan H, Liu K and Matsui T *et al* 2017 iTRAQ-based quantitative proteomics analysis on rice anther responding to high temperature *Int. J. Mol. Sci.* **18** 1811

[92] Gammulla C G, Pascovici D, Atwell B J and Haynes P A 2010 Differential metabolic response of cultured rice (*Oryza sativa*) cells exposed to high and low temperature stress *Proteomics* **10** 3001–19

[93] Liao J L, Zhou H W, Peng Q, Zhong P A, Zhang H Y and He C *et al* 2015 Transcriptome changes in rice (*Oryza sativa* L.) in response to high night temperature stress at the early milky stage *BMC Genom.* **16** 1–4

[94] Das S, Krishnan P, Mishra V, Kumar R, Ramakrishnan B and Singh N K 2015 Proteomic changes in rice leaves grown under open field high temperature stress conditions *Mol. Biol. Rep.* **42** 1545–58

[95] Han F, Chen H, Li X J, Yang M F, Liu G S and Shen S H 2009 A comparative proteomic analysis of rice seedlings under various high-temperature stresses *BBA Prot. Proteo.* **1794** 1625–34

[96] Jagadish S V, Muthurajan R, Oane R, Wheeler T R, Heuer S and Bennett J *et al* 2010 Physiological and proteomic approaches to address heat tolerance during anthesis in rice (*Oryza sativa* L.) *J. Exp. Bot.* **61** 143–56

[97] Kim M, Kim H, Lee W, Lee Y, Kwon S W and Lee J 2015 Quantitative shotgun proteomics analysis of rice anther proteins after exposure to high temperature *Int. J. Genom.* **2015** 1–9

[98] González-Schain N, Dreni L, Lawas L M, Galbiati M, Colombo L and Heuer S *et al* 2016 Genome-wide transcriptome analysis during anthesis reveals new insights into the molecular basis of heat stress responses in tolerant and sensitive rice varieties *Plant Cell Physiol.* **57** 57–68

[99] Mangrauthia S K, Agarwal S, Sailaja B, Sarla N and Voleti S R 2016 Transcriptome analysis of *Oryza sativa* (rice) seed germination at high temperature shows dynamics of genome expression associated with hormones signalling and abiotic stress pathways *Trop. Plant Biol.* **9** 215–28

[100] Mittal D, Madhyastha D A and Grover A 2012 Gene expression analysis in response to low and high temperature and oxidative stresses in rice: combination of stresses evokes different transcriptional changes as against stresses applied individually *Plant Sci.* **197** 102–13

[101] Fang C, Dou L, Liu Y, Yu J and Tu J 2018 Heat stress-responsive transcriptome analysis in heat susceptible and tolerant rice by high-throughput sequencing *Ecol. Genet. Geno.* **6** 33–40

[102] Liao J L, Zhou H W, Peng Q, Zhong P A, Zhang H Y, He C and Huang Y J 2015 Transcriptome changes in rice (*Oryza sativa* L.) in response to high night temperature stress at the early milky stage *BMC Genom.* **16** 1–4

[103] Sharma L, Dalal M, Verma R K, Kumar S V, Yadav S K and Pushkar S *et al* 2018 Auxin protects spikelet fertility and grain yield under drought and heat stresses in rice *Environ. Exp. Bot.* **150** 9–24

[104] Thussagunpanit J, Jutamanee K, Kaveeta L, Chai-arree W, Pankean P and Homvisasevongsa S *et al* 2015 Comparative effects of brassinosteroid and brassinosteroid mimic on improving photosynthesis, lipid peroxidation, and rice seed set under heat stress *J Plant Growth Reg.* **34** 320–31

[105] Wang X, Dinler B S, Vignjevic M, Jacobsen S and Wollenweber B 2015 Physiological and proteome studies of responses to heat stress during grain filling in contrasting wheat cultivars *Plant Sci.* **230** 33–50

[106] Narayanan S, Prasad P V and Welti R 2016 Wheat leaf lipids during heat stress: II. Lipids experiencing coordinated metabolism are detected by analysis of lipid co-occurrence *Plant Cell Environ.* **39** 608–17

[107] Comastri A, Janni M, Simmonds J, Uauy C, Pignone D and Nguyen H T *et al* 2018 Heat in wheat: exploit reverse genetic techniques to discover new alleles within the *Triticum durum* sHsp26 family *Front Plant Sci.* **9** 1337

[108] Gupta O P, Mishra V, Singh N K, Tiwari R, Sharma P and Gupta R K *et al* 2015 Deciphering the dynamics of changing proteins of tolerant and intolerant wheat seedlings subjected to heat stress *Mol. Biol. Rep.* **42** 43–51

[109] Lu Y, Li R, Wang R, Wang X, Zheng W and Sun Q *et al* 2017 Comparative proteomic analysis of flag leaves reveals new insight into wheat heat adaptation *Front. Plant Sci.* **8** 1086

[110] Kumar R R, Singh K, Ahuja S, Tasleem M, Singh I and Kumar S *et al* 2019 Quantitative proteomic analysis reveals novel stress-associated active proteins (SAAPs) and pathways involved in modulating tolerance of wheat under terminal heat *Funct. Integr. Genom.* **19** 329–48

[111] Wang X, Xu Y, Hu Z and Xu C 2018 Genomic selection methods for crop improvement: current status and prospects *Crop J.* **6** 330–40

[112] Yousuf P Y, Abd_Allah E F, Nauman M, Asif A, Hashem A and Alqarawi A A *et al* 2017 Responsive proteins in wheat cultivars with contrasting nitrogen efficiencies under the combined stress of high temperature and low nitrogen *Genes* **8** 356

[113] Zhang X, Högy P, Wu X, Schmid I, Wang X and Schulze W X *et al* 2018 Physiological and proteomic evidence for the interactive effects of post-anthesis heat stress and elevated CO_2 on wheat *Proteomics* **18** 1800262

[114] Qin D, Wu H, Peng H and Yao Y 2008 Heat stress-responsive transcriptome analysis in heat susceptible and tolerant wheat (*Triticum aestivum* L.) by using Wheat Genome Array *BMC Genom.* **9** 432

[115] Thomason K, Babar M A, Erickson J E, Mulvaney M, Beecher C and MacDonald G Comparative physiological and metabolomics analysis of wheat (*Triticum aestivum* L.) following post-anthesis heat stress *PLoS One* **13** e0197919

[116] Liu Z, Xin M, Qin J, Peng H, Ni Z and Yao Y *et al* 2015 Temporal transcriptome profiling reveals expression partitioning of homeologous genes contributing to heat and drought acclimation in wheat (*Triticum aestivum* L.) *BMC Plant Biol.* **15** 1–20

[117] Narayanan S, Prasad P V and Welti R 2018 Alterations in wheat pollen lipidome during high day and night temperature stress *Plant Cell Environ.* **41** 1749–61

[118] Djanaguiraman M, Boyle D L, Welti R, Jagadish S V and Prasad P V 2018 Decreased photosynthetic rate under high temperature in wheat is due to lipid desaturation, oxidation, acylation, and damage of organelles *BMC Plant Biol.* **18** 1–7

[119] Zhao F, Zhang D, Zhao Y, Wang W, Yang H and Tai F *et al* 2016 The difference of physiological and proteomic changes in maize leaves adaptation to drought, heat, and combined both stresses *Front. Plant Sci.* **7** 1471

[120] Shi J, Yan B, Lou X, Ma H and Ruan S 2017b Comparative transcriptome analysis reveals the transcriptional alterations in heat-resistant and heat-sensitive sweet maize (*Zea mays* L.) varieties under heat stress *BMC Plant Biol.* **17** 1–0

[121] Frey F P, Urbany C, Hüttel B, Reinhardt R and Stich B 2015 Genome-wide expression profiling and phenotypic evaluation of European maize inbreds at seedling stage in response to heat stress *BMC Genom.* **16** 1–5

[122] Qu M, Chen G, Bunce J A, Zhu X and Sicher R C 2018 Systematic biology analysis on photosynthetic carbon metabolism of maize leaf following sudden heat shock under elevated CO_2 *Sci. Rep.* **8** 1–1

[123] Templer S E, Ammon A, Pscheidt D, Ciobotea O, Schuy C and McCollum C *et al* 2017 Metabolite profiling of barley flag leaves under drought and combined heat and drought stress reveals metabolic QTLs for metabolites associated with antioxidant defense *J. Exp. Bot.* **68** 1697–713

[124] Mangelsen E, Kilian J, Harter K, Jansson C, Wanke D and Sundberg E 2011 Transcriptome analysis of high-temperature stress in developing barley caryopses: early stress responses and effects on storage compound biosynthesis *Mol. Plant* **4** 97–115

[125] Cantalapiedra C P, García-Pereira M J, Gracia M P, Igartua E, Casas A M and Contreras-Moreira B 2017 Large differences in gene expression responses to drought and heat stress between elite barley cultivar Scarlett and a Spanish landrace *Front. Plant Sci.* **8** 647

[126] Salvi P, Kamble N U and Majee M 2018 Stress-inducible galactinol synthase of chickpea (CaGolS) is implicated in heat and oxidative stress tolerance through reducing stress-induced excessive reactive oxygen species accumulation *Plant Cell Physiol.* **59** 155–66

[127] Parankusam S, Bhatnagar-Mathur P and Sharma K K 2017 Heat responsive proteome changes reveal molecular mechanisms underlying heat tolerance in chickpea *Environ. Exp. Bot.* **141** 132–44

[128] Chakraborty U and Pradhan D 2011 High temperature-induced oxidative stress in *Lens culinaris*, role of antioxidants and amelioration of stress by chemical pre-treatments *J. Plant Interact.* **6** 43–52

[129] Sita K, Sehgal A, Bhandari K, Kumar J, Kumar S and Singh S *et al* 2018 Impact of heat stress during seed filling on seed quality and seed yield in lentil (*Lens culinaris Medikus*) genotypes *J. Sci. Food Agric.* **98** 5134–41

[130] Sehgal A, Sita K, Kumar J, Kumar S, Singh S and Siddique K H *et al* 2017 Effects of drought, heat and their interaction on the growth, yield and photosynthetic function of lentil (*Lens culinaris Medikus*) genotypes varying in heat and drought sensitivity *Front. Plant Sci.* **8** 1776

[131] Singh D, Balota M, Collakova E, Isleib T G, Welbaum G E and Tallury S P Heat stress related physiological and metabolic traits in peanut seedlings *Peanut Sci.* **43** 24–35

[132] Chakraborty K, Bishi S K, Singh A L, Zala P V, Mahatma M K and Kalariya K A *et al* 2018 Rapid induction of small heat shock proteins improves physiological adaptation to high temperature stress in peanut *J Agron. Crop Sci.* **204** 285–97

[133] Yu E, Fan C, Yang Q, Li X, Wan B and Dong Y *et al* 2014 Identification of heat responsive genes in *Brassica napus* siliques at the seed-filling stage through transcriptional profiling *PLoS One* **9** e101914

[134] Rahaman M, Mamidi S and Rahman M 2018 Genome-wide association study of heat stress-tolerance traits in spring-type *Brassica napus* L. under controlled conditions *Crop J.* **6** 115–25

[135] Wang Y, Sun F, Cao H, Peng H, Ni Z and Sun Q *et al* 2012 TamiR159 directed wheat TaGAMYB cleavage and its involvement in anther development and heat response *PLoS One* **7** e48445

[136] Chung E, Kim K M and Lee J H 2013 Genome-wide analysis and molecular characterization of heat shock transcription factor family in *Glycine max J. Genet. Genom.* **40** 127–35
[137] Xu W, Cai S Y, Zhang Y, Wang Y, Ahammed G J and Xia X J *et al* 2016 Melatonin enhances thermotolerance by promoting cellular protein protection in tomato plants *J. Pineal. Res.* **61** 457–69
[138] Valdés-López O, Batek J, Gomez-Hernandez N, Nguyen C T, Isidra-Arellano M C and Zhang N *et al* 2016 Soybean roots grown under heat stress show global changes in their transcriptional and proteomic profiles *Front. Plant Sci.* **7** 517
[139] Gillman J D, Biever J J, Ye S, Spollen W G, Givan S A and Lyu Z *et al* 2019 A seed germination transcriptomic study contrasting two soybean genotypes that differ in terms of their tolerance to the deleterious impacts of elevated temperatures during seed fill *BMC Res. Notes* **12** 1–8
[140] Hewezi T, Léger M and Gentzbittel L 2008 A comprehensive analysis of the combined effects of high light and high temperature stresses on gene expression in sunflower *Ann. Bot.* **102** 127–40
[141] Mazzeo M F, Cacace G, Iovieno P, Massarelli I, Grillo S and Siciliano R A 2018 Response mechanisms induced by exposure to high temperature in anthers from thermo-tolerant and thermo-sensitive tomato plants: a proteomic perspective *PLoS One* **13** e0201027
[142] Spicher L, Glauser G and Kessler F 2016 Lipid antioxidant and galactolipid remodeling under temperature stress in tomato plants *Front. Plant Sci.* **7** 167
[143] Chaturvedi P, Doerfler H, Jegadeesan S, Ghatak A, Pressman E and Castillejo M A *et al* 2015 Heat-treatment-responsive proteins in different developmental stages of tomato pollen detected by targeted mass accuracy precursor alignment (tMAPA) *J. Proteo. Res.* **14** 4463–71
[144] Lin H H, Lin K H, Syu J Y, Tang S Y and Lo H F 2016 Physiological and proteomic analysis in two wild tomato lines under waterlogging and high temperature stress *J. Plant Biochem. Biotechnol.* **25** 87–96
[145] Bita C E, Zenoni S, Vriezen W H, Mariani C, Pezzotti M and Gerats T 2011 Temperature stress differentially modulates transcription in meiotic anthers of heat-tolerant and heat-sensitive tomato plants *BMC Genom.* **12** 1–8
[146] Frank G, Pressman E, Ophir R, Althan L, Shaked R and Freedman M *et al* 2009 Transcriptional profiling of maturing tomato (*Solanum lycopersicum* L.) microspores reveals the involvement of heat shock proteins, ROS scavengers, hormones, and sugars in the heat stress response *J. Exp. Bot.* **60** 3891–908
[147] Fragkostefanakis S, Simm S, Paul P, Bublak D, Scharf K D and Schleiff E 2015 Chaperone network composition in *Solanum lycopersicum* explored by transcriptome profiling and microarray meta-analysis *Plant Cell Environ.* **38** 693–709
[148] Paupière M J, Müller F, Li H, Rieu I, Tikunov Y M, Visser R G and Bovy A G 2017 Untargeted metabolomic analysis of tomato pollen development and heat stress response *Plant Reprod.* **30** 81–94
[149] Muhlemann J K, Younts T L and Muday G K 2018 Flavonols control pollen tube growth and integrity by regulating ROS homeostasis during high-temperature stress *Proc. Natl. Acad. Sci. USA* **115** E11188–97
[150] Hancock R D, Morris W L, Ducreux L J, Morris J A, Usman M and Verrall S R *et al* 2014 Physiological, biochemical and molecular responses of the potato (*Solanum tuberosum* L.) plant to moderately elevated temperature *Plant Cell Environ.* **37** 439–50

[151] Tang R, Zhu W, Song X, Lin X, Cai J and Wang M *et al* 2016 Genome-wide identification and function analyses of heat shock transcription factors in potato *Front. Plant Sci.* **7** 490

[152] Ginzberg I, Barel G, Ophir R, Tzin E, Tanami Z and Muddarangappa T *et al* 2009 Transcriptomic profiling of heat-stress response in potato periderm *J. Exp. Bot.* **60** 4411–21

[153] Liu C W, Chang T S, Hsu Y K, Wang A Z, Yen H C and Wu Y P *et al* 2014 Comparative proteomic analysis of early salt stress responsive proteins in roots and leaves of rice *Proteomics* **14** 1759–75

[154] Jiang J, Liu X, Liu C, Liu G, Li S and Wang L 2017 Integrating omics and alternative splicing reveals insights into grape response to high temperature *Plant Physiol.* **173** 1502–18

[155] Liu G T, Wang J F, Cramer G, Dai Z W, Duan W and Xu H G *et al* 2012 Transcriptomic analysis of grape (*Vitis vinifera* L.) leaves during and after recovery from heat stress *BMC Plant Biol.* **12** 1–0

[156] Qi Y, Wang H, Zou Y, Liu C, Liu Y and Wang Y *et al* 2011 Over-expression of mitochondrial heat shock protein 70 suppresses programmed cell death in rice *FEBS Lett.* **585** 231–9

[157] Katiyar-Agarwal S, Agarwal M and Grover A 2003 Heat-tolerant basmati rice engineered by over-expression of hsp101 *Plant Mol. Biol.* **51** 677–86

[158] El-Esawi M A and Alayafi A A 2019 Overexpression of rice Rab7 gene improves drought and heat tolerance and increases grain yield in rice (*Oryza sativa* L.) *Genes* **10** 56

[159] Qin D, Wang F, Geng X, Zhang L, Yao Y and Ni Z *et al* 2015 Overexpression of heat stress-responsive TaMBF1c, a wheat (*Triticum aestivum* L.) Multiprotein Bridging Factor, confers heat tolerance in both yeast and rice *Plant Mol. Biol.* **87** 31–45

[160] Xue G P, Sadat S, Drenth J and McIntyre C L 2014 The heat shock factor family from *Triticum aestivum* in response to heat and other major abiotic stresses and their role in regulation of heat shock protein genes *J. Exp. Bot.* **65** 539–57

[161] Fu J, Momčilović I, Clemente T E, Nersesian N, Trick H N and Ristic Z 2008 Heterologous expression of a plastid EF-Tu reduces protein thermal aggregation and enhances CO_2 fixation in wheat (*Triticum aestivum*) following heat stress *Plant Mol. Biol.* **68** 277–88

[162] Xue G P, Drenth J and McIntyre C L 2015 TaHsfA6f is a transcriptional activator that regulates a suite of heat stress protection genes in wheat (*Triticum aestivum* L.) including previously unknown Hsf targets *J. Exp. Bot.* **66** 1025–39

[163] Zang X, Geng X, Wang F, Liu Z, Zhang L and Zhao Y *et al* 2017 Overexpression of wheat ferritin gene TaFER-5B enhances tolerance to heat stress and other abiotic stresses associated with the ROS scavenging *BMC Plant Biol.* **17** 1–3

[164] Hu X J, Chen D, Lynne McIntyre C, Fernanda Dreccer M, Zhang Z B and Drenth J *et al* 2018 Heat shock factor C2a serves as a proactive mechanism for heat protection in developing grains in wheat via an ABA-mediated regulatory pathway *Plant Cell Environ.* **41** 79–98

[165] Zhang L, Geng X, Zhang H, Zhou C, Zhao A and Wang F *et al* 2017 Isolation and characterization of heat-responsive gene TaGASR1 from wheat (*Triticum aestivum* L.) *J. Plant Biol.* **60** 57–65

[166] Casaretto J A, El-Kereamy A, Zeng B, Stiegelmeyer S M, Chen X and Bi Y M *et al* 2016 Expression of OsMYB55 in maize activates stress-responsive genes and enhances heat and drought tolerance *BMC Genom.* **17** 1–5

[167] Nelson D E, Repetti P P, Adams T R, Creelman R A, Wu J and Warner D C *et al* 2007 Plant nuclear factor Y (NF-Y) B subunits confer drought tolerance and lead to improved corn yields on water-limited acres *Proc. Natl. Acad. Sci. USA* **104** 16450–5

[168] Trapero Mozos A, Morris W L, Ducreux L J, McLean K, Stephens J and Torrance L *et al* 2018 Engineering heat tolerance in potato by temperature dependent expression of a specific allele of HEAT-SHOCK COGNATE 70 *Plant Biotechnol. J.* **16** 197–207

[169] De Ronde J A, Cress W A, Krüger G H, Strasser R J and Van Staden J 2004 Photosynthetic response of transgenic soybean plants, containing an *Arabidopsis* P5CR gene, during heat and drought stress *J. Plant Physiol.* **161** 1211–24

[170] Montero-Barrientos M, Hermosa R, Nicolás C, Cardoza R E, Gutiérrez S and Monte E 2008 Overexpression of a *Trichoderma* HSP70 gene increases fungal resistance to heat and other abiotic stresses *Fungal Genet. Biol.* **45** 1506–13

[171] Shan Q, Wang Y, Li J and Gao C 2014 Genome editing in rice and wheat using the CRISPR/Cas system *Nature Protoc.* **9** 2395–410

[172] Shen L, Hua Y, Fu Y, Li J, Liu Q and Jiao X *et al* 2017 Rapid generation of genetic diversity by multiplex CRISPR/Cas9 genome editing in rice *Sci. China Life Sci.* **60** 506–15

[173] Shan Q, Wang Y, Li J, Zhang Y, Chen K and Liang Z *et al* 2013 Targeted genome modification of crop plants using a CRISPR-Cas system *Nat. Biotechnol.* **31** 686–8

[174] Kim D, Alptekin B and Budak H 2018 CRISPR/Cas9 genome editing in wheat *Funct. Integr. Genom.* **18** 31–41

[175] Shi J, Gao H, Wang H, Lafitte H R, Archibald R L and Yang M *et al* 2017 ARGOS 8 variants generated by CRISPR-Cas9 improve maize grain yield under field drought stress conditions *Plant Biotechnol. J.* **15** 207–16

IOP Publishing

Molecular Response and Genetic Engineering for Stress in Plants, Volume 1

Abiotic stress

Pawan Shukla, Anirudh Kumar, Rakesh Kumar and Manish K Pandey

Chapter 8

Oxidative stress responses in plants to abiotic stress tolerance

Abhaypratap Vishwakarma[1,2,*], Srinivas Suresh[3], Priscilla Kagolla[3], Devade Pandurang Ramrao[3], Naveen Kumar Singh[4], Pawan Shukla[5], Rakesh Kumar[3] and Gaurav Kumar[1]

[1]*Department of Plant Molecular Biology, University of Delhi South Campus, New Delhi, India*

[2]*Deshbandhu College, University of Delhi, New Delhi, India*

[3]*Department of Life Science, Central University of Karnataka, Kalaburagi, India*

[4]*Department of Life Science, Central University of South Bihar, Gaya, India*

[5]*Seribiotech Research Laboratory (SBRL), Kodathi, Carmelram Post, Bangalore, India*

Being immobile in nature, plants are frequently exposed to a variety of abiotic stressors such as drought, heat, cold, salinity, and flooding that cause a reduction of overall plant growth and yield. A perturbance in the physiological metabolism of plants by abiotic stresses can cause generation of excess reactive oxygen species (ROS), for example superoxide (O_2^-), hydrogen peroxide (H_2O_2), and hydroxyl radical ($OH^•$), as byproducts. A low level of ROS is a prerequisite for the various complex signaling networks of a plant. However, excess ROS have deleterious effects on cells' organelles and may eventually lead to cell death. Therefore, an efficient antioxidant defense system and alternate pathways are required to maintain ROS homeostasis and optimal physiological metabolism. In other words, manipulating ROS levels can alter the levels of hormones and transcription factors, leading to upregulation of antioxidant defense systems to combat abiotic stresses. This chapter summarizes the essential biochemical and molecular events required to counter abiotic stresses through ROS regulation.

8.1 Introduction

Stress is defined as an adverse state imposed by the environment that affects the standard growth and functioning of a plants. Stress is mainly divided into biotic and abiotic

* Author to whom any correspondence should be addressed Email: abhayps08@gmail.com.

doi:10.1088/978-0-7503-4921-5ch8

stresses. Biotic stressors include insects, rodents, and microbial pathogens, whereas abiotic stressors include osmotic balance, drought, temperature, flood, hypoxia, and metal toxicity [1, 2]. The frequency of abiotic stresses are steadily increasing because of global climate change [3, 4] and a contemporary increase in man-made pollutants [5]. Salinity stress has posed one of the major problems in agricultural history [4]. Wicke *et al* reported that 1,125 million ha (Mha) of agricultural land is distressed by salinity throughout world [6]. Statistically, one out of five irrigated lands is distressed by salinity, and, alarmingly, this rate is continuously growing by loss of 1.5 ha of agricultural land per year [7]. In agriculture soil salinity can be defined by measuring the soluble salt concentration, when the ECe value of the soil is 4 dS m^{-1} or more [8]. A rise in temperature and CO_2 level due to global climate change ultimately alters rainfall patterns and their distribution; thereby the deficiency of water leads to drought stress. In addition, loss of water from lands through evaporation, mainly driven by high temperatures and dry winds, can further aggravate the existing drought conditions [9]. The Institute for Agro-Environmental Sciences, NARO has presented data on the global loss of crop yield caused by drought stress from 1983 to 2009. Wheat displayed the maximum loss by 161 Mha (75% of global harvested area by each crop), followed by 124 Mha (82%) for maize, 102 Mha (62%) for rice, and 67 Mha (91%) for soy [10]. Flood is another major abiotic stress which affects every country in the world. On a global scale flood affects >1,700 million km^2 land, equal to twice the area of the USA [11]. Climate forecasting suggests that flooding events may increase in frequency and severity [12]. Based on the rate of rising population, it is estimated that the world will require 87% more food by 2050 [13]. Therefore, understanding abiotic stress responses is very important for plant scientists to improve crop survival and yield (figure 8.1).

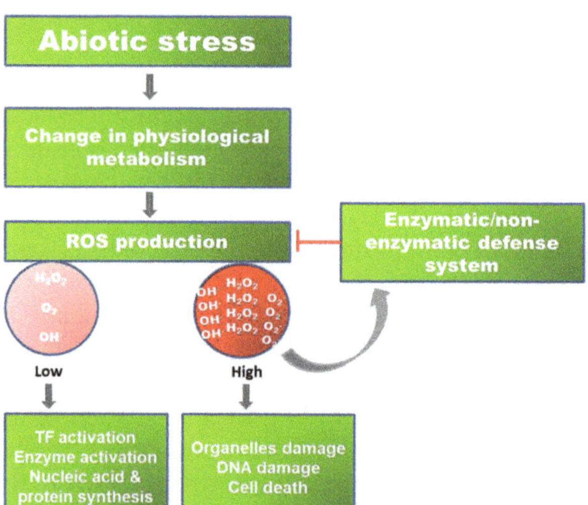

Figure 8.1. Schematic model of ROS production and defense system in response to abiotic stresses. Different abiotic stresses can generate ROS by changing the physiological metabolism. ROS level is controlled by the enzymatic/nonenzymatic defense system. A low accumulation of ROS is often desired by plant cells to initiate various signaling pathways, and activation of transcription factors, nucleic acids, and protein synthesis. In contrast, higher levels of ROS accumulation creates oxidative stress, thereby inducing DNA damage and cell death.

Occurrence of these abiotic stresses causes alteration in a plant's metabolism and thereby production of harmful ROS, for instance O_2^-, H_2O_2, $OH^\bullet$, and $^1O_2$. When these ROS molecules are overproduced and accumulate in the cells, the condition is known as oxidative stress. In plant cells mitochondria, chloroplasts, and peroxisomes are the major sources for ROS production [14–16]. The major causes of oxidative stress include: (1) de novo synthesis of ROS as an integral part of stress signaling and immune response for defense, and (2) imbalance between ROS production and its detoxification [17]. To survive adverse conditions and minimize the excess ROS, plants have evolved diverse alternative pathways and complex signal transduction networks to activate an antioxidant defense system [18]. When ROS production exceeds beyond the limits of detoxification, ROS leak to other parts of the cell, which leads to oxidative damage. These leaky ROS react with all kind of biomolecules and hence incur damage to the DNA, lipids in membrane (lipid peroxidation), and enzymes, and thereby promote programmed cell death [19, 20]. The present chapter discusses the oxidative stress response mechanism in plants to combat abiotic stresses.

8.2 Antioxidant defense system

The excessive ROS produced during abiotic stress not only affects a plant's developmental and physiological processes but can damage the genetic material, as well. It can also cause cell death by inducing lipid oxidation in the cell membrane [21–24]. Therefore, to minimize the causative effects of ROS, the plant activates various biochemical, physiological, and molecular mechanisms in a well-orchestrated manner during adverse conditions. Antioxidants are critical components of the plant's defense system as they counter the detrimental effects of ROS by scavenging the free radicals generated during stress. The whole antioxidant system is comprised of enzymatic and nonenzymatic antioxidants. The enzymatic antioxidant machinery includes catalase (CAT), superoxide dismutase (SOD), ascorbate peroxidase (APX), monodehydroascorbate reductase (MDHAR), dehydroascorbate reductase (DHAR), glutathione reductase (GR), and glutathione peroxidases (GPX). Nonenzymatic antioxidant machinery includes ascorbic acid (AsA), glutathione (GSH), α-tocopherol, carotenoids, flavonoids, alkaloids, and nonprotein amino acids working in a coordinated fashion with enzymatic antioxidants to suppress excessive ROS production [25, 26]. A major player in this antioxidant system is the ascorbate-glutathione pathway (i.e. Halliwell–Asada cycle), which is comprised of nonenzymatic antioxidants ascorbate and glutathione and enzymatic antioxidants APX, DHAR, MDHAR, and GR. This cycle plays a key part in cellular redox maintenance during stress conditions.

8.2.1 Enzymatic antioxidants

Superoxide dismutases are ubiquitous, highly conserved metalloenzymes that play a pivotal role in the removal of highly reactive superoxide radicals ($O_2^{\bullet-}$). On the basis of metal cofactors present in active sites, SODs are divided into three main categories: Cu/Zn-SOD, Mn-SOD, or Fe-SOD [27]. Protein localization studies have shown the presence of Cu/Zn-SOD in the cytosol and chloroplast, whereas Mn-SOD is in the mitochondrial matrix and peroxisome [25]. These SOD members are considered to be part of the first line of defense against different stress conditions [28, 29]. They remove superoxide radicals ($O_2^{\bullet-}$) by catalysing their dismutation into hydrogen peroxide (H_2O_2). The H_2O_2 produced by SODs is further converted into O_2 and H_2O by the heme-containing antioxidant enzyme CAT [30]. It has been reported that catalase is a highly active enzyme that can convert around 26 million H_2O_2 molecules into O_2 in just one minute [31]. They are categorized into three families: class 1 catalases, which are mostly present in green photosynthetic tissues; class 2, which are highly expressed in vascular tissues; and class 3, which are notably abundant in seeds and reproductive organs [32].

APX is a crucial component of the Halliwell–Asada cycle. It performs the same function as catalase. While catalase eliminates ROS predominately in the peroxisome, APX does the same in the chloroplast and cytosol. APX utilizes AsA as a reducing agent and converts H_2O_2 to water molecules and dehydroascorbate (DHA). Based on the amino acid sequence and subcellular localization, five isoforms of APX have been reported [25]. In comparison to catalase, APX shows more affinity towards H_2O_2 and hence is considered a crucial enzyme during stress conditions [33]. The ascorbate, used by APX during H_2O_2 scavenging, is reproduced by a NADPH-dependent flavin adenine dinucleotide enzyme, MDHAR. It reduces short-lived monodehydroascorbate (MDHA) and regenerates ascorbate [33]. In addition, the AsA pool is also replenished by another antioxidant enzyme, DHAR, that uses GSH as an electron donor [34].

From a plethora of research evidence, reduced glutathione (GSH) has been found to be major alleviator of ROS. It is part of the Halliwell–Asada cycle and is used for regenerating AsA from MDHA and DHA. In this process, GSH is converted to its oxidized form (GSSG). The ratio between GSH and GSSG is used for measuring oxidative stress [35]. The higher GSH/GSSG content is maintained by GR, a crucial enzyme for the plant system. GR uses NADPH as a reducing agent and catalyzes the formation of GSH from GSSG [36]. GR is mostly found in chloroplasts, with a little present in cytosol and mitochondria. Glutathione peroxidases (GPXs) are a thiol-containing enzyme family that catalyze the reduction of H_2O_2 to H_2O or corresponding alcohols using GSH [37]. The presence of cysteine residue in their active site makes plant GPXs distinct from their animal counterpart [38]. Their presence has been identified in different subcellular compartments including the chloroplast and mitochondria. Apart from ROS scavenging, GPXs also help in eliminating the toxic products formed in the lipid peroxidation reaction [39].

8.2.2 Nonenzymatic antioxidants

Nonenzymatic antioxidants not only protect plant cells during adverse conditions but also participate in fundamental activities such as growth and development. Ascorbate is one of the most extensively studied nonenzymatic antioxidant compounds. It is a physiologically active form of AsA and is considered a powerful tool for ROS detoxification. Ascorbate is regarded as a highly critical compound for stress management strategies in plants as it is capable of transferring electrons in several reactions [26, 40]. The primary function of ascorbate is to remove H_2O_2. In the AsA-GSH cycle, AsA is utilized by the APX enzyme for the reduction of H_2O_2 into water molecules. In addition, AsA also helps in regenerating α-tocopherol (vitamin E) from tocopheroxyl radicals and thus provides membrane protection [41, 42]. The α-tocopherol is a lipophilic antioxidant, which effectively eliminates ROS and lipid radicals [43]. As discussed earlier, GSH plays a substantial role in AsA-GSH cycle regulation. Carotenoids are another effective group of antioxidant molecules which protect photosynthetic machinery against ROS [31]. Flavonoids are phenolic compounds belonging to secondary metabolites that provide pigmentation in various parts of plants. They attract pollinators and, consequently, help in pollination and seed dispersal. Flavonoids can absorb high-energy ultraviolet radiation and thereby prevent ROS production and quench ROS once they are produced [44–46]. Furthermore, alkaloids and nonprotein amino acids, such as ornithine and citrulline, are also known as effective nonenzymatic antioxidants [47, 48].

8.3 Involvement of NADPH oxidase (NOX) during abiotic stress

NADPH oxidase is situated at the plasma membrane. This enzyme produces apoplastic O_2^-, where NADPH is used as an electron donor. The primary function of NADPH oxidase was discovered in activating phagocytic cells to kill pathogens [49, 50]. Later, occurrence of NADPH oxidase family members were discerned in a broad range of organisms such as vertebrates, urochordates, echinoderms, insects, nematodes, fungi, slime mold amoeba, algae, and plants [51]. In plants, NADPH oxidases constitute a multigene family, commonly called respiratory burst oxidase homologs (Rbohs), due to their capability for excessive ROS production [52]. Around 150 protein members of NADPH oxidase have been reported up to now. In *Triticum sp.* about 46 Rboh family genes were characterized [53]. In the case of fungi, there are four types of NOXs, such as *NOXA*, *NOXB*, *NOXC*, and *FRE*. In the case of animals, seven types of NOX members have been found, such as *NOX1*, *NOX2*, *NOX3*, *NOX4*, *NOX5*, *DUOX1* and *DUOX2* [54–56].

In response to abiotic stresses, NADPH oxidase produces superoxides, which leads to oxidative burst. The ROS produced in this reaction is essential in activating several complex signaling networks for wound healing, stomatal movement, ozone stress, and salinity stress tolerance, etc [57, 58]. In *Arabidopsis thaliana Rbohl* expression was induced by mannitol while decreased by external application of ABA hormone. Overexpression of *Rbohl* under the CaMV35S promoter confers drought tolerance in *Arabidopsis* [59]. AtRbohD and AtRbohF are found in each part of the plant and have been shown to control a number of abiotic stresses such as Na^+/K^+ homeostasis by improving salt tolerance, whereas AtRbohC and AtRbohE regulate drought-induced leaf senescence [60]. Recently, Huang *et al* demonstrated the functional role of NADPH-oxidases-dependent H_2O_2 signaling in response to salt tolerance by inhibiting NADPH oxidase using diphenylene iodonium [61]. The activation of NADPH oxidases involving CDPKs Ser/Thr protein kinases occurs via the Ca^{2+} binding calmodulin domain along with regulating adenosine triphosphate, phospholipase Dα1, and phosphatidic acid [57, 62]. External application of brassinosteroids (BRs) induced Rbohs-mediated ROS production, which in turn was shown to provide oxidative tolerance in cucumber [63]. In response to abiotic stress ROS might act as a transient signal in accumulation of proline [64]. To mitigate salinity stress, NADPH-oxidases-mediated ROS is required to regulate the proline metabolism in *A. thaliana* [65]. Plants undergo adaptations to hypoxia stress by regulating the ratio of K^+/Ca^{2+} ions in root cells. *AtRbohD* knockout exhibited sensitivity; concomitantly, a reduction in NADPH oxidase activity in response to hypoxia eventually suggested that AtRbohD is crucial for activity of NADPH oxidase [66]. In rice during hypoxic conditions, *OsRbohH* activates ethylene-induced aerenchyma formation in roots, which is triggered by *CDPK5* and *CDPK13* [67]. A recent study has demonstrated that involvement of NADPH oxidase is crucial in generating oxidative burst during hyperhydricity in garlic plantlets [68].

8.4 Hormones' interaction with antioxidants

In the course of abiotic stresses various ROS molecules are generated by excitation and partial reduction of molecular oxygen, which in turn damages the DNA, proteins, and membrane lipids. However, lower amounts of ROS molecules act as signaling molecules to induce several defense alternatives to combat stress responses [69, 70]. The hormones synthesized in plants for normal growth include auxins, gibberellins, cytokinins, jasmonic acid (JA), abscisic acid (ABA), ethylene, salicylic acid (SA), strigolactones, and BRs [71, 72]. JA, ethylene, and SA are mainly responsible for plant defense against biotic stresses [71, 73]. ABA is a major hormone that triggers plant defense against heat, cold, salinity, and drought stress [74–77]. Exogenous ABA treatment enhanced production of ROS in response to water stress; the increase in ROS level triggered antioxidative defense genes [78–80]. ABA-induced H_2O_2 formation mediates nitric oxide production in maize leaves and activates the mitogen-activated protein kinase (MAPK) cascade, thereby inducing antioxidative enzymes [81]. In ABA signaling, ROS, calcium, and nitric oxide are major signaling molecules [82]. Increased calcium in cells is recognized by calcium sensors such as calmodulin (CaM), calcineurin B-like protein (CBP), calcium-dependent protein kinase (CDPK), and Ca^{2+}/CaM-dependent protein kinase (CCaMK). These calcium sensors work in regulation of various downstream target gene expressions, thereby modulating different physiological responses [83–87]. It was demonstrated that Ca^{2+} and calmodulin is involved in ABA-induced antioxidative response by an increase in the accumulation of antioxidative enzymes, e.g. SOD, GR, CAT, and APX [78, 88–91]. Shi *et al* identified a novel role of CCaMK in ABA-induced antioxidative defense and demonstrated that under water stress *OsDMI3* is involved in ABA-induced antioxidative response such as enhanced expression of superoxide dismutase and catalase in rice plants [92]. In tobacco plants, bright yellow-2 cell cultures supplied with exogenous polyethylene glycol (PEG) induced drought stress, ABA and NaCl; in the presence of NaCl and PEG, the transcript levels of superoxide dismutase, catalase, and ascorbate peroxidase were elevated and glutathione reductase remained constant. ABA-supplied tobacco cell cultures showed increased transcript levels of *APX* and *SOD*, while *CAT* and *GPX* remained unchanged [93]. In rice, *OsDMI3* and *OsMPK1* function in the ABA signaling pathway. *OsMPK1* functions downstream of *OsDMI3* in regulation of antioxidative defense response [94]. Analysis of *ZFP36* overexpressing and down-regulated *ZFP36* transgenic lines demonstrated that *ZFP36* induces SOD and APX in an ABA-dependent manner. C2H2-type zinc finger protein *ZFP36* gene has been identified in rice. *ZFP36* expression is regulated by MAPK in ABA signaling, and *ZFP36* regulates expression of NADPH oxidase [95]. Water-stress-induced accumulation of ABA increased the production of ROS, which, in turn, elevated antioxidant enzymes in maize leaves [79]. ABA and its role in antioxidant defense systems in plants has been reviewed, with detailed explanation of the different regulatory mechanisms in ABA-induced antioxidant enzyme production upon various environmental stresses [96].

Auxin plays a crucial role in plant growth by controlling cell division, meristem development, and cell elongation. Recent studies have shown ROS have an influence in auxin signaling, biosynthesis, transport, and metabolism [97, 98]. Exogenous application of auxin enhances H_2O_2 level in plants [99]. Contrastingly, exogenous application of auxin decreased H_2O_2 accumulation in the roots of tomato by activating peroxidases, CAT, and Cu-Zn-SOD [100]. A double mutant of the auxin receptor *tir1afb2* showed a decrease in superoxide anion and H_2O_2 accumulation and increased activity of APX and CAT [101]. Further evidence was shown that the *cat2* mutant exhibits 'up curling phenotype', showing decreased catalase activity and enhanced H_2O_2 accumulation with decreased auxin levels in *Arabidopsis* leaves under photorespiratory conditions [102]. Auxin-mediated cell elongation in plants is influenced by formation of hydrogen peroxide, hydroxyl radicals, and superoxide anion [103].

8.5 Contribution of transcription factors

Transcription factors (TFs) play a crucial role in synchronizing abiotic stress response through binding on cis-acting elements in the promoter region to control the expression of stress-related genes (table 8.1). A wide variety of TF families have been identified in plants, such as AP2/ERF, MYB, NAC, WRKY, bHLH, bZIP, zinc finger, and DREB. MYB TFs are initiated when the ABA signaling is activated by a stress response. Most MYB research has been carried out in *A. thaliana*. The first plant MYB TF was identified in maize [104]. It was named after finding the conserved N-terminal domain MYB domains. Each MYB-binding domain has nonredundant imperfect repeats, i.e. R1, R2, R3, and R4. Depending on the number of repeats, the MYB family is further divided into four subfamilies, i.e. 1R-MYB (MYB related), 2R-MYB (R2R3 MYB), 3R-MYB (R1R2R3 MYB), and 4R-MYB (R1R2R3R4 MYB) [105]. Among these four subfamilies, R2R3 MYB is the largest and is involved in phenyl propanoid metabolism [106, 107], abiotic and biotic stresses [108, 109], hormonal responses [110], GAMYB & CpMYB [111, 112], and formation of B-type cyclins [113]. Several studies have demonstrated that MYBs are involved in the synthesis and regulation of secondary cell walls in *Arabidopsis*,

Table 8.1. List of transcription factors involved in abiotic stress tolerance.

Functional category	Genes	Species	Abiotic stress tolerance	References
AP2/ERF	*JERF3*	*N. benthamiana*	Drought, salt, and freezing stress	[128]
	SUB1A	*O. sativa*	Drought, submerge, and oxidative stress	[129]
	SERF1	*O. sativa*	Salt stress	[130]
WRKY	*GhWRKY17*	*N. benthamiana*	Drought and salt stress	[131]
	GmWRKY27	*G. max*	Drought and salt stress	[132]
	AtWRKY25, AtWRKY26	*A. thaliana*	Heat stress	[119]
	TaWRKY33	*T. aestivum*	Heat stress	[133]
	OsWRKY45	*O. sativa*	Drought	[134]
	OsWRKY72	*O. sativa*	Salt	[135]
	GhWRKY68	*N. benthamiana*	Drought and salt stress	[136]
NAC	*GmNAC29*	*G. max*	Drought and salt stress	[132]
	SNAC3	*O. sativa*	Drought, heat, and oxidative stress	[137]
	GmNAC2	*N. benthamiana*	Drought, salinity, and cold stress	[138]
	TaNAC2	*A. thaliana*	Drought and salt stress	[139]
	TaNAC69	*T. aestivum*	PEG-induced dehydration and mild salt tolerance	[140]
	GmNAC11	*G. max*	Salt tolerance	[141]
Zn finger	*DST*	*O. sativa*	Drought and salt stress	[142]
	ZFP36	*O. sativa*	Drought, and oxidative stress	[95]
	OsTZF1	*O. sativa*	Drought, salt, and oxidative stress	[130]
MYB	*JAMYB*	*O. sativaA. thaliana*	High salinity	[143]
	AtMYB015	*A. thaliana*	Cold stress tolerance	[144]
	AtMYB096	*A. thaliana*	Drought tolerance	[145]
	OsMYB55	*Oryza sativa*	Heat stress tolerance	[146]

namely *AtMYB46*, *AtMYB58*, *AtMYB63*, *AtMYB83*, and *AtMYB103* [114]. On the other hand, *AtMYB4*, *AtMYB7*, and *AtMYB32* act as antipathetic regulators of secondary cell wall synthesis [115]. In transgenic rice plant, *OsMYB3R-2* augmented the tolerance levels to freezing, dehydration, and salt stress by decreasing the levels of ABA [116]. Additional roles of MYB TFs in abiotic stress tolerance are described in table 8.1.

WRKY is a major TF family among plants, which synchronizes various physiological and developmental processes. The WRKY TFs are characterized by a ~60 amino acid region possessing four-stranded β-sheet WRKY DNA-binding domain (DBD) and zinc-finger motifs. Based on these characteristics, WRKY is categorized into three groups. Group I is characterized by two WRKY DBDs, group II by single WRKY DBD and zinc-finger motifs, whereas group III contains single WRKY DBD with a C2HC zinc-finger motif [117]. Several WRKY TFs have been recognized in plants, for instance 74 in *Arabidopsis*, 109 in rice [117], 83 in tomato [118], and so on. Group I WRKY TFs (*AtWRKY25*, *AtWRKY26*, and *AtWRKY33*) were found to stimulate a heat-induced signal transduction pathway, which contributes to heat stress tolerance [119]. Upon submergence treatment, *WRKY22* was significantly increased among other tested WRKY members in *Arabidopsis*. Further, overexpression of *WRKY22* causes an induction in the TIR1 domain that triggers innate immunity in response to pathogen infection either during or after flooding stress [120]. Many more contributions of WRKY TFs in response to abiotic stress is presented in table 8.1.

NAC is a term derived from the first letters of three proteins, i.e. no apical meristem (NAM), *Arabidopsis* transcription activator factor (ATAF), and cup-shaped cotyledon (CUC). The NAC TFs are characterized by the presence of conserved N-terminal regions or domains and variable transcriptional regulatory c-terminal domains. These transcriptional regulatory domains can either function as positive regulators or negative regulators [121]. Recent analysis showed that *A. thaliana* has 117 nonredundant genes and sorghum has 145 NAC TFs, as well as there being 151 in rice, and 63 NAC genes from the genome of *C. canephora*. In *Cicer arietinum* phylogenetic analysis revealed that, out of 71 CaNAC genes, 23 were dehydration-related genes, four were downregulated, and four were upregulated under drought stress conditions in leaves [122]. *CaNAC06* and *CaNAC67* are upregulated genes and *CaNAC02* and *CaNAC04* are downregulated genes. The NAC TFs are found in the plasma membrane and instigated by ABA-dependent or ABA-independent pathways. They bind to the cis-acting promoters of dehydration-responsive element-binding (DREB), cold regulated (COR), and NAC recognition sequence (NACRS) proteins, thereby synchronizing ROS, flower locus T, and glutathione-S transferase expression. During plant stress, the calibration of NAC genes makes them tolerant to abiotic stress.

DREBs are part of the ethylene response factor (ERF) TF family. During stress conditions, the TF binds either to DRE or CRT genes in the promoter region. ERFs are a subfamily of APETALA2/ethylene response element-binding proteins (EREBPs). EREBPs are further subdivided into two subclasses, i.e. DREB1/CBF and DREB2. Overexpression of *AtDREB1A* with stress-inducible rd29A promoter showed improved drought and low-temperature stress tolerance by limiting ion leakage and maintaining higher quantum yield [123]. DREB2 shows a limited number of results, but if one modifies the post-translational mechanism then the *DREB2A* function activates by deleting the amino acid position between the 135th and 165th positions [124]. *AtCBF1* exhibits a dual role: under constitutive expression it contributed to freezing tolerance in *Arabidopsis* and canola [125, 126], while in transgenic tomato *AtCBF1* contributed to drought tolerance [127]. Additionally, DREBs' function in abiotic stress tolerance is listed in table 8.1.

8.6 Conclusions

As plants respire aerobically the production of ROS is inevitable. Previously, ROS were considered as highly toxic molecules which can damage to DNA, protein, and membrane lipids, leading to cell death. However, in the last two decades several research studies have demonstrated that a minimal level of ROS is a prerequisite to activate complex signaling networks in order to control physiological and biochemical reactions in the cell by modulating TFs. An alteration in TFs can activate the enzymatic and/or nonenzymatic antioxidant defense system and hormone-induced antioxidant defense, augmenting abiotic stress tolerance.

Acknowledgments

A.V. thanks DBT-BIRAC for financial assistance and Prof. Indranil Dasgupta for extending his lab facilities during this period. N.K.S. and R.K. are recipients of a UGC start-up grant. G.K. is thankful to UGC for a Kothari postdoctoral fellowship.

Multiple choice questions

1. Where is ROS produced in the plant cell?
   (a) Mitochondria;
   (b) Chloroplasts;
   (c) Peroxisomes;
   (d) All of the above.
2. NADPH oxidase produces …, which in turn generates oxidative burst
   (a) Superoxide;
   (b) Hydrogen peroxide;
   (c) Nitric oxide;
   (d) Nitrogen dioxide.
3. Which class of catalase is/are found in seed and reproductive organs?
   (a) Class 1;
   (b) Class 2;
   (c) Class 3;
   (d) Both class 1 and class 3.
4. DREBs are part of the … TFs family.
   (a) ERF;
   (b) NAC;

   (c) MYB;
   (d) WRKY.
5. Excess ROS can damage …
   (a) DNA;
   (b) Protein;
   (c) Lipid;
   (d) All of the above.

Answers: 1 (d); 2 (a); 3 (c); 4 (a); 5 (d).

Descriptive-type questions

1. How does the ascorbate-glutathione cycle contribute to reactive oxygen species (ROS) detoxification and regeneration of ascorbic acid?
2. Describe ABA signaling in activating an antioxidant defense system.
3. What are the characteristic features of the WRKY domain?
4. Which transcription factor would be the most promising candidate to enhance drought tolerance in wheat?
5. What is the significance of ROS in abiotic stress tolerance?

References

[1] Mittler R 2006 Abiotic stress, the field environment and stress combination *Trends Plant Sci.* **11** 15–9

[2] Lu H, Wei T, Lou H, Shu X and Chen Q 2021 A critical review on communication mechanism within plant-endophytic fungi interactions to cope with biotic and abiotic stresses *J. Fungi* **7** 719

[3] Onyekachi O G, Boniface O O, Gemlack N F and Nicholas N 2019 The effect of climate change on abiotic plant stress: a review *Abiotic and Biotic Stress in Plants* (London: IntechOpen)

[4] Ahmed I, Shukla P and Gautam R 2021 *Omics Technologies for Sustainable Agriculture and Global Food Security* vol 1 (Berlin: Springer) pp 169–89

[5] Wallace J S, Acreman M C and Sullivan C A 2003 The sharing of water between society and ecosystems: from conflict to catchment–based co–management *Philos. Trans. R. Soc. London, Ser.* B **358** 2011–26

[6] Wicke B, Smeets E, Dornburg V, Vashev B, Gaiser T, Turkenburg W and Faaij A 2011 The global technical and economic potential of bioenergy from salt-affected soils *Energy Environ. Sci.* **4** 2669–81

[7] Hossain M S 2019 Present scenario of global salt affected soils, its management and importance of salinity research *Int. Res. J. Biol. Sci.* **1** 1–3

[8] Munns R and Tester M 2008 Mechanisms of salinity tolerance *Annu. Rev. Plant Biol.* **59** 651–81

[9] Cohen I, Zandalinas S I, Huck C, Fritschi F B and Mittler R 2021 Meta-analysis of drought and heat stress combination impact on crop yield and yield components *Physiol. Plant.* **171** 66–76

[10] Kim W, Iizumi T and Nishimori M 2019 Global patterns of crop production losses associated with droughts from 1983 to 2009 *J. Appl. Meteorol. Climatol.* **58** 1233–44

[11] Perata P, Armstrong W and Voesenek L A 2011 Plants and flooding stress *New Phytol.* **190** 269–73

[12] Jongman B, Hochrainer-Stigler S, Feyen L, Aerts J C, Mechler R, Botzen W W, Bouwer L M, Pflug G, Rojas R and Ward P J 2014 Increasing stress on disaster-risk finance due to large floods *Nat. Clim. Change* **4** 264–8

[13] Kromdijk J and Long S P 2016 One crop breeding cycle from starvation? How engineering crop photosynthesis for rising CO_2 and temperature could be one important route to alleviation *Proc. Royal Soc. B: Biol. Sci.* **283** 20152578

[14] Blokhina O and Fagerstedt K V 2010 Reactive oxygen species and nitric oxide in plant mitochondria: origin and redundant regulatory systems *Physiol. Plant.* **138** 447–62

[15] Foyer C H and Harbinson J 2019 *Causes of Photooxidative Stress and Amelioration of Defense Systems in Plants* (Boca Raton, FL: CRC Press) pp 1–42

[16] Corpas F J, Gupta D K and Palma J M 2015 *Reactive Oxygen Species and Oxidative Damage in Plants under Stress* (Berlin: Springer) pp 1–22

[17] Demidchik V 2015 Mechanisms of oxidative stress in plants: from classical chemistry to cell biology *Environ. Exp. Bot.* **109** 212–28

[18] Hirayama T and Shinozaki K 2010 Research on plant abiotic stress responses in the post-genome era: past, present and future *Plant J.* **61** 1041–52

[19] Sharma P, Jha A B, Dubey R S and Pessarakli M 2012 Reactive oxygen species, oxidative damage, and antioxidative defense mechanism in plants under stressful conditions *J. Bot.* **2012** 217037

[20] Sabater B and Martín M 2013 Hypothesis: increase of the ratio singlet oxygen plus superoxide radical to hydrogen peroxide changes stress defense response to programmed leaf death *Front. Plant Sci.* **4** 479

[21] Choudhary A, Kumar A and Kaur N 2020 ROS and oxidative burst: roots in plant development *Plant Divers.* **42** 33–43

[22] Vishwakarma A, Kumari A, Mur L A and Gupta K J 2018 A discrete role for alternative oxidase under hypoxia to increase nitric oxide and drive energy production *Free Radical Biol. Med.* **122** 40–51

[23] Vishwakarma A, Tetali S D, Selinski J, Scheibe R and Padmasree K 2015 Importance of the alternative oxidase (AOX) pathway in regulating cellular redox and ROS homeostasis to optimize photosynthesis during restriction of the cytochrome oxidase pathway in *Arabidopsis thaliana Annal. Bot.* **116** 555–69

[24] Vishwakarma A, Bashyam L, Senthilkumaran B, Scheibe R and Padmasree K 2014 Physiological role of AOX1a in photosynthesis and maintenance of cellular redox homeostasis under high light in *Arabidopsis thaliana Plant Physiol. Biochem.* **81** 44–53

[25] Ahmad P, Jaleel C A, Salem M A, Nabi G and Sharma S 2010 Roles of enzymatic and nonenzymatic antioxidants in plants during abiotic stress *Crit. Rev. Biotechnol.* **30** 161–75

[26] Singh N, Mansoori A, Dey D, Kumar R and Kumar A 2021 Potential of metabolomics in plant abiotic stress management *Omics Technologies for Sustainable Agriculture and Global Food Security* vol II ed A Kumar, R Kumar, P Shukla and H K Patel (Singapore: Springer) pp 193–214

[27] Hasanuzzaman M, Bhuyan M, Zulfiqar F, Raza A, Mohsin S M, Mahmud J A, Fujita M and Fotopoulos V 2020 Reactive oxygen species and antioxidant defense in plants under abiotic stress: revisiting the crucial role of a universal defense regulator *Antioxidants* **9** 681

[28] Berwal M and Ram C 2018 Superoxide dismutase: a stable biochemical marker for abiotic stress tolerance in higher plants *Abiotic and Biotic Stress in Plants* (London: IntechOpen)

[29] Ighodaro O and Akinloye O 2018 First line defence antioxidants-superoxide dismutase (SOD), catalase (CAT) and glutathione peroxidase (GPX): their fundamental role in the entire antioxidant defence grid *Alex. J. Med.* **54** 287–93

[30] García-Caparrós P, De Filippis L, Gul A, Hasanuzzaman M, Ozturk M, Altay V and Lao M T 2020 Oxidative stress and antioxidant metabolism under adverse environmental conditions: a review *Bot. Rev.* 1–46

[31] Mehla N, Sindhi V, Josula D, Bisht P and Wani S H 2017 *Reactive Oxygen Species and Antioxidant Systems in Plants: Role and Regulation under Abiotic Stress* (Berlin: Springer) pp 1–23

[32] Mhamdi A, Queval G, Chaouch S, Vanderauwera S, Van Breusegem F and Noctor G 2010 Catalase function in plants: a focus on Arabidopsis mutants as stress-mimic models *J. Exp. Bot.* **61** 4197–220

[33] Das K and Roychoudhury A 2014 Reactive oxygen species (ROS) and response of antioxidants as ROS-scavengers during environmental stress in plants *Front. Environ. Sci.* **2** 53

[34] Chen Z and Gallie D R 2006 Dehydroascorbate reductase affects leaf growth, development, and function *Plant Physiol.* **142** 775–87

[35] Banerjee A and Roychoudhury A 2019 Role of glutathione in plant abiotic stress tolerance *Reactive Oxygen, Nitrogen and Sulfur Species in Plants: Production, Metabolism, Signaling and Defense Mechanisms* (New York: Wiley) pp 159–72

[36] Dar M I, Naikoo M I, Khan F A, Rehman F, Green I D, Naushin F and Ansari A A 2017 *Reactive Oxygen Species and Antioxidant Systems in Plants: Role and Regulation under Abiotic Stress* (Berlin: Springer) pp 25–52

[37] Dvorak P, Krasylenko Y, Zeiner A, Samaj J and Takac T 2020 Signaling toward ROS-scavenging enzymes in plants *Front. Plant Sci.* **11** 2178

[38] Ursini F, Maiorino F M, Brigelius-Flohé R, Aumann K, Roveri A and Schomburg D *et al* 1995 [5] Diversity of glutathione peroxidases *Meth. Enzymol.* **252** 38–53

[39] Bela K, Horváth E, Gallé Á, Szabados L, Tari I and Csiszár J 2015 Plant glutathione peroxidases: emerging role of the antioxidant enzymes in plant development and stress responses *J. Plant Physiol.* **176** 192–201

[40] Hasanuzzaman M, Bhuyan M, Anee T I, Parvin K, Nahar K, Mahmud J A and Fujita M 2019 Regulation of ascorbate-glutathione pathway in mitigating oxidative damage in plants under abiotic stress *Antioxidants* **8** 384

[41] Naz H, Akram N A and Ashraf M 2016 Impact of ascorbic acid on growth and some physiological attributes of cucumber (*Cucumis sativus*) plants under water-deficit conditions *Pak. J. Bot.* **48** 877–83

[42] Seminario A, Song L, Zulet A, Nguyen H T, González E M and Larrainzar E 2017 Drought stress causes a reduction in the biosynthesis of ascorbic acid in soybean plants *Front. Plant Sci.* **8** 1042

[43] Szarka A, Tomasskovics B and Bánhegyi G 2012 The ascorbate-glutathione-α-tocopherol triad in abiotic stress response *Int. J. Mol. Sci.* **13** 4458–83

[44] Kumar A, Gul M Z, Zeeshan A, Bimolata W, Qureshi I A and Ghazi I A 2013 Differential antioxidative responses of three different rice genotypes during bacterial blight infection *Aust. J. Crop Sci.* **7** 1893–900

[45] Brunetti C, Di Ferdinando M, Fini A, Pollastri S and Tattini M 2013 Flavonoids as antioxidants and developmental regulators: relative significance in plants and humans *Int. J. Mol. Sci.* **14** 3540–55

[46] Tyagi K, Shukla P, Rohela G K, Shabnam A A and Gautam R 2020 *Plant Phenolics in Sustainable Agriculture* (Berlin: Springer) pp 431–49

[47] Vranova V, Rejsek K, Skene K R and Formanek P 2011 Non-protein amino acids: plant, soil and ecosystem interactions *Plant Soil* **342** 31–48

[48] Kumar A, John M M, Gul M Z, Bimolata W and Ghazi I A 2011 Differential responses of non-enzymatic antioxidative system underwater deficit condition in rice (*Oryza sativa* L.) *Int. Proc. of Chemical Biological and Environmental Engineering* vol 9 (Singapore: IACSIT Press)

[49] DeLeo F R, Allen L-A H, Apicella M and Nauseef W M 1999 NADPH oxidase activation and assembly during phagocytosis *J. Immunol.* **163** 6732–40

[50] Babior B, Lambeth J and Nauseef W 2002 The neutrophil NADPH oxidase *Arch. Biochem. Biophys.* **397** 342–4

[51] Kawahara T, Quinn M T and Lambeth J D 2007 Molecular evolution of the reactive oxygen-generating NADPH oxidase (Nox/Duox) family of enzymes *BMC Evol. Biol.* **7** 1–21

[52] Sagi M and Fluhr R 2006 Production of reactive oxygen species by plant NADPH oxidases *Plant Physiol.* **141** 336–40

[53] Hu C-H, Wang P-Q, Zhang P-P, Nie X-M, Li B-B, Tai L, Liu W-T, Li W-Q and Chen K-M 2020 NADPH oxidases: the vital performers and center hubs during plant growth and signaling *Cells* **9** 437

[54] Aguirre J, Ríos-Momberg M, Hewitt D and Hansberg W 2005 Reactive oxygen species and development in microbial eukaryotes *Trends Microbiol.* **13** 111–8

[55] Bedard K, Lardy B and Krause K-H 2007 NOX family NADPH oxidases: not just in mammals *Biochimie* **89** 1107–12

[56] Hu C-H, Wei X-Y, Yuan B, Yao L-B, Ma T-T, Zhang P-P, Wang X, Wang P-Q, Liu W-T and Li W-Q 2018 Genome-wide identification and functional analysis of NADPH oxidase family genes in wheat during development and environmental stress responses *Front. Plant Sci.* **9** 906

[57] Kaur G, Sharma A, Guruprasad K and Pati P K 2014 Versatile roles of plant NADPH oxidases and emerging concepts *Biotechnol. Adv.* **32** 551–63

[58] Qu Y, Yan M and Zhang Q 2017 Functional regulation of plant NADPH oxidase and its role in signaling *Plant Signal. Behav.* **12** e1356970

[59] He H, Yan J, Yu X, Liang Y, Fang L, Scheller H V and Zhang A 2017 The NADPH-oxidase AtRbohI plays a positive role in drought-stress response in *Arabidopsis thaliana* *Biochem. Biophys. Res. Commun.* **491** 834–9

[60] Torres M A, Dangl J L and Jones J D 2002 Arabidopsis gp91phox homologues AtrbohD and AtrbohF are required for accumulation of reactive oxygen intermediates in the plant defense response *Proc. Natl Acad. Sci.* **99** 517–22

[61] Huang Y, Cao H, Yang L, Chen C, Shabala L, Xiong M, Niu M, Liu J, Zheng Z and Zhou L 2019 Tissue-specific respiratory burst oxidase homolog-dependent H_2O_2 signaling to the plasma membrane H^+-ATPase confers potassium uptake and salinity tolerance in Cucurbitaceae *J. Exp. Bot.* **70** 5879–93

[62] Kobayashi M, Ohura I, Kawakita K, Yokota N, Fujiwara M, Shimamoto K, Doke N and Yoshioka H 2007 Calcium-dependent protein kinases regulate the production of reactive oxygen species by potato NADPH oxidase *Plant Cell* **19** 1065–80
[63] Xia X-J, Wang Y-J, Zhou Y-H, Tao Y, Mao W-H, Shi K, Asami T, Chen Z and Yu J-Q 2009 Reactive oxygen species are involved in brassinosteroid-induced stress tolerance in cucumber *Plant Physiol.* **150** 801–14
[64] Fabro G, Kovács I, Pavet V, Szabados L and Alvarez M E 2004 Proline accumulation and AtP5CS2 gene activation are induced by plant-pathogen incompatible interactions in Arabidopsis *Mol. Plant-Microbe Interact.* **17** 343–50
[65] Ben Rejeb K, Lefebvre-De Vos D, Le Disquet I, Leprince A S, Bordenave M, Maldiney R, Jdey A, Abdelly C and Savouré A 2015 Hydrogen peroxide produced by NADPH oxidases increases proline accumulation during salt or mannitol stress in A rabidopsis thaliana *New Phytol.* **208** 1138–48
[66] Wang W, Chen D, Zhang X, Liu D, Cheng Y and Shen F 2018 Role of plant respiratory burst oxidase homologs in stress responses *Free Radical Res.* **52** 826–39
[67] Yamauchi T, Yoshioka M, Fukazawa A, Mori H, Nishizawa N K, Tsutsumi N, Yoshioka H and Nakazono M 2017 An NADPH oxidase RBOH functions in rice roots during lysigenous aerenchyma formation under oxygen-deficient conditions *Plant Cell* **29** 775–90
[68] Tian J, Cheng Y, Kong X, Liu M, Jiang F and Wu Z 2017 Induction of reactive oxygen species and the potential role of NADPH oxidase in hyperhydricity of garlic plantlets *in vitro* *Protoplasma* **254** 379–88
[69] Apel K and Hirt H 2004 Reactive oxygen species: metabolism, oxidative stress, and signal transduction *Annu. Rev. Plant Biol.* **55** 373–99
[70] Mittler R, Vanderauwera S, Gollery M and Van Breusegem F 2004 Reactive oxygen gene network of plants *Trends Plant Sci.* **9** 490–8
[71] Bari R and Jones J D 2009 Role of plant hormones in plant defence responses *Plant Mol. Biol.* **69** 473–88
[72] Nakashima K and Yamaguchi-Shinozaki K 2013 ABA signaling in stress-response and seed development *Plant Cell Rep.* **32** 959–70
[73] Navarro L, Bari R, Achard P, Lisón P, Nemri A, Harberd N P and Jones J D 2008 DELLAs control plant immune responses by modulating the balance of jasmonic acid and salicylic acid signaling *Curr. Biol.* **18** 650–5
[74] Lata C and Prasad M 2011 Role of DREBs in regulation of abiotic stress responses in plants *J. Exp. Bot.* **62** 4731–48
[75] Eremina M, Rozhon W and Poppenberger B 2016 Hormonal control of cold stress responses in plants *Cell. Mol. Life Sci.* **73** 797–810
[76] Ahammed G J, Li X, Zhou J, Zhou Y-H and Yu J-Q 2016 *Plant Hormones under Challenging Environmental Factors* (Berlin: Springer) pp 1–21
[77] Gai Z, Wang Y, Ding Y, Qian W, Qiu C, Xie H, Sun L, Jiang Z, Ma Q and Wang L 2020 Exogenous abscisic acid induces the lipid and flavonoid metabolism of tea plants under drought stress *Sci. Rep.* **10** 1–13
[78] Jiang M and Zhang J 2002 Involvement of plasma-membrane NADPH oxidase in abscisic acid- and water stress-induced antioxidant defense in leaves of maize seedlings *Planta* **215** 1022–30
[79] Jiang M and Zhang J 2002 Water stress-induced abscisic acid accumulation triggers the increased generation of reactive oxygen species and up-regulates the activities of antioxidant enzymes in maize leaves *J. Exp. Bot.* **53** 2401–10

[80] Ye N, Zhu G, Liu Y, Li Y and Zhang J 2011 ABA controls H2O2 accumulation through the induction of OsCATB in rice leaves under water stress *Plant Cell Physiol.* **52** 689–98
[81] Zhang Q 2007 Strategies for developing green super rice *Proc. Natl. Acad. Sci.* **104** 16402–9
[82] Jammes F, Song C, Shin D, Munemasa S, Takeda K, Gu D, Cho D, Lee S, Giordo R and Sritubtim S 2009 MAP kinases MPK9 and MPK12 are preferentially expressed in guard cells and positively regulate ROS-mediated ABA signaling *Proc. Natl. Acad. Sci.* **106** 20520–5
[83] DeFalco T A, Bender K W and Snedden W A 2010 Breaking the code: Ca^{2+} sensors in plant signalling *Biochem. J* **425** 27–40
[84] Harper J F, Breton G and Harmon A 2004 Decoding Ca^{2+} signals through plant protein kinases *Annu. Rev. Plant Biol.* **55** 263–88
[85] Kim M C, Chung W S, Yun D-J and Cho M J 2009 Calcium and calmodulin-mediated regulation of gene expression in plants *Mol. Plant* **2** 13–21
[86] Yang T and Poovaiah B 2003 Calcium/calmodulin-mediated signal network in plants *Trends Plant Sci.* **8** 505–12
[87] Singh N K, Shukla P and Kirti P 2020 A CBL-interacting protein kinase AdCIPK5 confers salt and osmotic stress tolerance in transgenic tobacco *Sci. Rep.* **10** 1–14
[88] Sang J, Zhang A, Lin F, Tan M and Jiang M 2008 Cross-talk between calcium-calmodulin and nitric oxide in abscisic acid signaling in leaves of maize plants *Cell Res.* **18** 577–88
[89] Ahmed I, Yadav D, Shukla P, Vineeth T, Sharma P and Kirti P 2017 Constitutive expression of Brassica juncea annexin, AnnBj2 confers salt tolerance and glucose and ABA insensitivity in mustard transgenic plants *Plant Sci.* **265** 12–28
[90] Ahmed I, Yadav D, Shukla P and Kirti P B 2018 Heterologous expression of Brassica juncea annexin, AnnBj2 confers salt tolerance and ABA insensitivity in transgenic tobacco seedlings *Funct. Integr. Genomics* **18** 569–79
[91] Yadav D, Boyidi P, Ahmed I and Kirti P B 2018 Plant annexins and their involvement in stress responses *Environ. Exp. Bot.* **155** 293–306
[92] Shi B, Ni L, Zhang A, Cao J, Zhang H, Qin T, Tan M, Zhang J and Jiang M 2012 OsDMI3 is a novel component of abscisic acid signaling in the induction of antioxidant defense in leaves of rice *Mol. Plant* **5** 1359–74
[93] Bueno P, Piqueras A, Kurepa J, Savouré A, Verbruggen N, Van Montagu M and Inzé D 1998 Expression of antioxidant enzymes in response to abscisic acid and high osmoticum in tobacco BY-2 cell cultures *Plant Sci.* **138** 27–34
[94] Shi B, Ni L, Liu Y, Zhang A, Tan M and Jiang M 2014 OsDMI3-mediated activation of OsMPK1 regulates the activities of antioxidant enzymes in abscisic acid signalling in rice *Plant, Cell Environ.* **37** 341–52
[95] Zhang H, Liu Y, Wen F, Yao D, Wang L, Guo J, Ni L, Zhang A, Tan M and Jiang M 2014 A novel rice C2H2-type zinc finger protein, ZFP36, is a key player involved in abscisic acid-induced antioxidant defence and oxidative stress tolerance in rice *J. Exp. Bot.* **65** 5795–809
[96] Jiang M-Y and Zhang J-H 2004 Abscisic acid and antioxidant defense in plant cells *Acta Bot. Sin.* **46** 1–9
[97] Mangano S, Denita-Juarez S P, Choi H-S, Marzol E, Hwang Y, Ranocha P, Velasquez S M, Borassi C, Barberini M L and Aptekmann A A 2017 Molecular link between auxin and ROS-mediated polar growth *Proc. Natl. Acad. Sci.* **114** 5289–94
[98] Zhang W, Lu L-Y, Hu L-Y, Cao W, Sun K, Sun Q-B, Siddikee A, Shi R-H and Dai C-C 2018 Evidence for the involvement of auxin, ethylene and ROS signaling during primary

root inhibition of arabidopsis by the allelochemical benzoic acid *Plant Cell Physiol.* **59** 1889–904

[99] Peer W A, Cheng Y and Murphy A S 2013 Evidence of oxidative attenuation of auxin signalling *J. Exp. Bot.* **64** 2629–39

[100] Tyburski J, Dunajska K, Mazurek P, Piotrowska B and Tretyn A 2009 Exogenous auxin regulates H_2O_2 metabolism in roots of tomato (*Lycopersicon esculentum* Mill.) seedlings affecting the expression and activity of CuZn-superoxide dismutase, catalase, and peroxidase *Acta Physiol. Plant.* **31** 249

[101] Iglesias M J, Terrile M C, Bartoli C G, D'Ippólito S and Casalongué C A 2010 Auxin signaling participates in the adaptative response against oxidative stress and salinity by interacting with redox metabolism in *Arabidopsis Plant Mol. Biol.* **74** 215–22

[102] Gao X, Yuan H M, Hu Y Q, Li J and Lu Y T 2014 Mutation of A rabidopsis CATALASE2 results in hyponastic leaves by changes of auxin levels *Plant, Cell Environ.* **37** 175–88

[103] Schopfer P, Liszkay A, Bechtold M, Frahry G and Wagner A 2002 Evidence that hydroxyl radicals mediate auxin-induced extension growth *Planta* **214** 821–8

[104] Paz-Ares J, Ghosal D, Wienand U, Peterson P and Saedler H 1987 The regulatory c1 locus of *Zea mays* encodes a protein with homology to myb proto-oncogene products and with structural similarities to transcriptional activators *EMBO J.* **6** 3553–8

[105] Dubos C, Stracke R, Grotewold E, Weisshaar B, Martin C and Lepiniec L 2010 MYB transcription factors in *Arabidopsis Trends Plant Sci.* **15** 573–81

[106] Grotewold E, Drummond B J, Bowen B and Peterson T 1994 The myb-homologous P gene controls phlobaphene pigmentation in maize floral organs by directly activating a flavonoid biosynthetic gene subset *Cell* **76** 543–53

[107] Hichri I, Barrieu F, Bogs J, Kappel C, Delrot S and Lauvergeat V 2011 Recent advances in the transcriptional regulation of the flavonoid biosynthetic pathway *J. Exp. Bot.* **62** 2465–83

[108] Segarra G, Van der Ent S, Trillas I and Pieterse C 2009 MYB72, a node of convergence in induced systemic resistance triggered by a fungal and a bacterial beneficial microbe *Plant Biol.* **11** 90–6

[109] Lippold F, Sanchez D H, Musialak M, Schlereth A, Scheible W-R, Hincha D K and Udvardi M K 2009 AtMyb41 regulates transcriptional and metabolic responses to osmotic stress in Arabidopsis *Plant Physiol.* **149** 1761–72

[110] Urao T, Yamaguchi-Shinozaki K, Urao S and Shinozaki K 1993 An *Arabidopsis* myb homolog is induced by dehydration stress and its gene product binds to the conserved MYB recognition sequence *Plant Cell* **5** 1529–39

[111] Gubler F, Kalla R, Roberts J K and Jacobsen J V 1995 Gibberellin-regulated expression of a myb gene in barley aleurone cells: evidence for Myb transactivation of a high-pI alpha-amylase gene promoter *Plant Cell* **7** 1879–91

[112] Iturriaga G, Leyns L, Villegas A, Gharaibeh R, Salamini F and Bartels D 1996 A family of novel myb-related genes from the resurrection plant *Craterostigma plantagineum* are specifically expressed in callus and roots in response to ABA or desiccation *Plant Mol. Biol.* **32** 707–16

[113] Ito M, Araki S, Matsunaga S, Itoh T, Nishihama R, Machida Y, Doonan J H and Watanabe A 2001 G2/M-phase–specific transcription during the plant cell cycle is mediated by c-Myb–like transcription factors *Plant Cell* **13** 1891–905

[114] Zhong R, Richardson E A and Ye Z-H 2007 The MYB46 transcription factor is a direct target of SND1 and regulates secondary wall biosynthesis in *Arabidopsis Plant Cell* **19** 2776–92

[115] *Zhong R, Lee C, Zhou J, McCarthy R L* and *Ye Z-H* 2008 A battery of transcription factors involved in the regulation of secondary cell wall biosynthesis in *Arabidopsis Plant Cel*l **20** 2763–82

[116] Dai X, Xu Y, Ma Q, Xu W, Wang T, Xue Y and Chong K 2007 Overexpression of an R1R2R3 MYB gene, OsMYB3R-2, increases tolerance to freezing, drought, and salt stress in transgenic *Arabidopsis Plant Physiol.* **143** 1739–51

[117] Phukan U J, Jeena G S and Shukla R K 2016 WRKY transcription factors: molecular regulation and stress responses in plants *Front. Plant Sci.* **7** 760

[118] Bai Y, Sunarti S, Kissoudis C, Visser R G and Van Der Linden C 2018 The role of tomato WRKY genes in plant responses to combined abiotic and biotic stresses *Front. Plant Sci.* **9** 801

[119] Li S, Fu Q, Chen L, Huang W and Yu D 2011 *Arabidopsis thaliana* WRKY25, WRKY26, and WRKY33 coordinate induction of plant thermotolerance *Planta* **233** 1237–52

[120] Hsu F-C, Chou M-Y, Chou S-J, Li Y-R, Peng H-P and Shih M-C 2013 Submergence confers immunity mediated by the WRKY22 transcription factor in Arabidopsis *Plant Cell* **25** 2699–713

[121] Puranik S, Sahu P P, Srivastava P S and Prasad M 2012 NAC proteins: regulation and role in stress tolerance *Trends Plant Sci.* **17** 369–81

[122] Ha C V, Nasr Esfahani M, Watanabe Y, Tran U T, Sulieman S, Mochida K, Van Nguyen D and Tran L-S P 2014 Genome-wide identification and expression analysis of the CaNAC family members in chickpea during development, dehydration and ABA treatments *PLoS One* **9** e114107

[123] Kasuga M, Miura S, Shinozaki K and Yamaguchi-Shinozaki K 2004 A combination of the *Arabidopsis* DREB1A gene and stress-inducible rd29A promoter improved drought-and low-temperature stress tolerance in tobacco by gene transfer *Plant Cell Physiol.* **45** 346–50

[124] Sakuma Y, Maruyama K, Qin F, Osakabe Y, Shinozaki K and Yamaguchi-Shinozaki K 2006 Dual function of an *Arabidopsis* transcription factor DREB2A in water-stress-responsive and heat-stress-responsive gene expression *Proc. Natl. Acad. Sci.* **103** 18822–7

[125] Jaglo-Ottosen K R, Gilmour S J, Zarka D G, Schabenberger O and Thomashow M F 1998 *Arabidopsis* CBF1 overexpression induces COR genes and enhances freezing tolerance *Science* **280** 104–6

[126] Jaglo K R, Kleff S, Amundsen K L, Zhang X, Haake V, Zhang J Z, Deits T and Thomashow M F 2001 Components of the *Arabidopsis* C-repeat/dehydration-responsive element binding factor cold-response pathway are conserved in *Brassica napus* and other plant species *Plant Physiol.* **127** 910–7

[127] Hsieh T-H, Lee J-T, Charng Y-Y and Chan M-T 2002 Tomato plants ectopically expressing *Arabidopsis* CBF1 show enhanced resistance to water deficit stress *Plant Physiol.* **130** 618–26

[128] Wu L, Zhang Z, Zhang H, Wang X-C and Huang R 2008 Transcriptional modulation of ethylene response factor protein JERF3 in the oxidative stress response enhances tolerance of tobacco seedlings to salt, drought, and freezing *Plant Physiol.* **148** 1953–63

[129] Fukao T, Yeung E and Bailey-Serres J 2011 The submergence tolerance regulator SUB1A mediates crosstalk between submergence and drought tolerance in rice *Plant Cell* **23** 412–27

[130] Jan A, Maruyama K, Todaka D, Kidokoro S, Abo M, Yoshimura E, Shinozaki K, Nakashima K and Yamaguchi-Shinozaki K 2013 OsTZF1, a CCCH-tandem zinc finger protein, confers delayed senescence and stress tolerance in rice by regulating stress-related genes *Plant Physiol.* **161** 1202–16

[131] Yan H, Jia H, Chen X, Hao L, An H and Guo X 2014 The cotton WRKY transcription factor GhWRKY17 functions in drought and salt stress in transgenic *Nicotiana benthamiana* through ABA signaling and the modulation of reactive oxygen species production *Plant Cell Physiol.* **55** 2060–76

[132] Wang F, Chen H W, Li Q T, Wei W, Li W, Zhang W K, Ma B, Bi Y D, Lai Y C and Liu X L 2015 Gm WRKY 27 interacts with Gm MYB 174 to reduce expression of Gm NAC 29 for stress tolerance in soybean plants *Plant J.* **83** 224–36

[133] He G-H, Xu J-Y, Wang Y-X, Liu J-M, Li P-S, Chen M, Ma Y-Z and Xu Z-S 2016 Drought-responsive WRKY transcription factor genes TaWRKY1 and TaWRKY33 from wheat confer drought and/or heat resistance in *Arabidopsis BMC Plant Biol.* **16** 1–16

[134] Qiu Y and Yu D 2009 Over-expression of the stress-induced OsWRKY45 enhances disease resistance and drought tolerance in *Arabidopsis Environ. Exp. Bot.* **65** 35–47

[135] Song Y, Chen L, Zhang L and Yu D 2010 Overexpression of OsWRKY72 gene interferes in the abscisic acid signal and auxin transport pathway of *Arabidopsis J. Biosci.* **35** 459–71

[136] Jia H, Wang C, Wang F, Liu S, Li G and Guo X 2015 GhWRKY68 reduces resistance to salt and drought in transgenic *Nicotiana benthamiana PLoS One* **10** e0120646

[137] Fang Y, Liao K, Du H, Xu Y, Song H, Li X and Xiong L 2015 A stress-responsive NAC transcription factor SNAC3 confers heat and drought tolerance through modulation of reactive oxygen species in rice *J. Exp. Bot.* **66** 6803–17

[138] Jin H, Huang F, Cheng H, Song H and Yu D 2013 Overexpression of the GmNAC2 gene, an NAC transcription factor, reduces abiotic stress tolerance in tobacco *Plant Mol. Biol. Rep.* **31** 435–42

[139] Mao X, Zhang H, Qian X, Li A, Zhao G and Jing R 2012 TaNAC2, a NAC-type wheat transcription factor conferring enhanced multiple abiotic stress tolerances in *Arabidopsis J. Exp. Bot.* **63** 2933–46

[140] Xue G-P, Way H M, Richardson T, Drenth J, Joyce P A and McIntyre C L 2011 Overexpression of TaNAC69 leads to enhanced transcript levels of stress up-regulated genes and dehydration tolerance in bread wheat *Mol. Plant* **4** 697–712

[141] Hao Y J, Wei W, Song Q X, Chen H W, Zhang Y Q, Wang F, Zou H F, Lei G, Tian A G and Zhang W K 2011 Soybean NAC transcription factors promote abiotic stress tolerance and lateral root formation in transgenic plants *Plant J.* **68** 302–13

[142] Huang J, Sun S, Xu D, Lan H, Sun H, Wang Z, Bao Y, Wang J, Tang H and Zhang H 2012 A TFIIIA-type zinc finger protein confers multiple abiotic stress tolerances in transgenic rice (*Oryza sativa* L.) *Plant Mol. Biol.* **80** 337–50

[143] Yokotani N, Ichikawa T, Kondou Y, Iwabuchi M, Matsui M, Hirochika H and Oda K 2013 Role of the rice transcription factor JAmyb in abiotic stress response *J. Plant Res.* **126** 131–9

[144] Agarwal M, Hao Y, Kapoor A, Dong C-H, Fujii H, Zheng X and Zhu J-K 2006 A R2R3 type MYB transcription factor is involved in the cold regulation of CBF genes and in acquired freezing tolerance *J. Biol. Chem.* **281** 37636–45

[145] Seo P J, Xiang F, Qiao M, Park J-Y, Lee Y N, Kim S-G, Lee Y-H, Park W J and Park C-M 2009 The MYB96 transcription factor mediates abscisic acid signaling during drought stress response in Arabidopsis *Plant Physiol.* **151** 275–89

[146] El-Kereamy A, Bi Y-M, Ranathunge K, Beatty P H, Good A G and Rothstein S J 2012 The rice R2R3-MYB transcription factor OsMYB55 is involved in the tolerance to high temperature and modulates amino acid metabolism *PLoS One* **7** e52030

Chapter 9

Potential impacts of ultraviolet-B radiation on crop plants and its consequences

Pooja Singh[1], Krishna Kumar Choudhary[1]*

[1]*Botany Section, Mahila Mahavidyalaya, Banaras Hindu University, Varanasi 221005, India*

Ultraviolet-B (UV-B) radiation is an integral component of the solar spectrum. Depletion of stratospheric ozone contributes to an increase in terrestrial UV-B radiation. Despite the fact that the stabilization and recovery of stratospheric ozone is expected in future, continuous climate change modifies the UV-B exposure of plants and animals. Plants are inevitably exposed to UV-B radiation since they are sessile in nature, and unavoidably dependent on sunlight for photosynthesis. Various studies have been performed to assess the direct and indirect influence of ambient and supplemental UV-B radiation on plant growth and development, as a result of which plants show unusual patterns of growth and symptoms of visual injury like chlorotic patches, which eventually become necrotic, causing early leaf senescence. Leaf modifications such as reduced leaf area, increased epicuticular wax, and leaf thickness, as well as stunted growth are commonly observed morphological changes induced by high UV-B radiation. These changes mediate architectural changes in plants. Solar radiation absorbed by the chlorophyll pigments present in plant photosystems has been regarded as the primary vector for UV-B damage. Damage to the photosynthetic machinery and its components (photosystem I, photosystem II, D1 and D2 proteins, oxygen-evolving complex, ATPase) halts the process of photosynthesis and thereby decreases photosynthetic efficiency. UV-B induces reactive oxygen species generation in plant cells that subsequently alters the pattern of gene expression of antioxidative enzymes like superoxide dismutase, peroxidase, ascorbate peroxidase and glutathione reductase. Acclimation to UV-B involves accumulation of secondary metabolites in the vacuoles of epidermal cells. These include phenolic compounds, flavonoids, alkaloids, terpenes,

* Author to whom any correspondence should be addressed Email: kkc@bhu.ac.in.

doi:10.1088/978-0-7503-4921-5ch9

anthocyanins, glucosinolates, carotenoids, lignin, etc, which protect the underlying tissues from UV-B injury. Yield is always a key factor in determining the impact of stress on plants. Decrease in photosynthesis, particularly due to reduction in chlorophyll content and damage to photosystems, results in lower production of biomass and yield of many crops. Various studies have been conducted on crop plants, including *Pisum sativum* L., *Triticum asestivum* L., *Linum usitatissimum* L., *Zea mays* L., *Oryza sativa* L., etc, all of which reported a significant reduction in grain yield. Alteration in oil quality and protein content as a result of elevated UV-B signifies a serious concern to human health and development. Keeping in view UV-B-induced damage in plants, this chapter will highlight the deleterious effects imposed by UV-B radiation on crop plants.

9.1 Introduction

Climate change impacts the lives of terrestrial organisms. Ultraviolet-B (UV-B) exposure has an increasing effect on these organisms by mediating changes in phenology, aerosols, species distribution, snow cover, wildfires, shifting land-use patterns, and extreme weather events [1–3]. The Environmental Impacts Assessment Panel of the United Nations Environment Programme studies the interactive changes in stratospheric ozone, temperature, and UV-B radiation on the environment on a regular basis. According to their most recent report, thanks to the Montreal Protocol, fluctuations in surface UV-B radiation have been fairly low (approximately 4% each decade) over the last 20 years [4]. However, recent reports suggest that water vapors present in the stratosphere could damage the ozone layer. Apart from this, four additional ozone-depleting compounds have been detected, namely CFC-112a (CF2ClCCl3), CFC-113a (CF3CCl3), CFC-112 (CFCl2CFCl2), and HCFC-133a (CF3CH2Cl), which were previously not included in the Montreal Protocol. This might increase the penetration of UV-B radiation, as a result posing a threat to the physiology and productivity of many crops in the upcoming future.

The impact of increased UV-B radiation on plants (figure 9.1) has been widely studied and is considered to be major threat for crop growth and development [5]. Despite the fact that UV-B is only a minor part of the Sun's spectrum, its high intensity greatly influences life forms [6]. Regardless of fluctuations in stratospheric ozone, climate change is continually shifting the timing, intensity, and duration of UV-B radiation on plants [2]. Plants absorb UV-B light directly, which has a negative impact on their metabolic activities and development. However, various reports have revealed that plants exhibit a wide range of absorption, physiological reactions, and sensitivity to UV-B radiation [7–13]. Differential responses in plant development and physiology are primarily caused by differences in plant species and length of UV-B exposure. The most common morphological changes induced by enhanced UV-B radiation include decreased leaf area, reduced leaf thickness, increased epicuticular wax content of leaves, and significant decrease in crop yield and its quality [11, 14–17]. A regulatory response mediated by the UVR8 photoreceptor mediates architectural change in plants. In fact, the role of UVR8 in the perception and sensing of UV-B radiation has been well studied and reviewed [18–21]. Exposure of plants for an extended period of time under UV-B decreases chlorophyll pigments

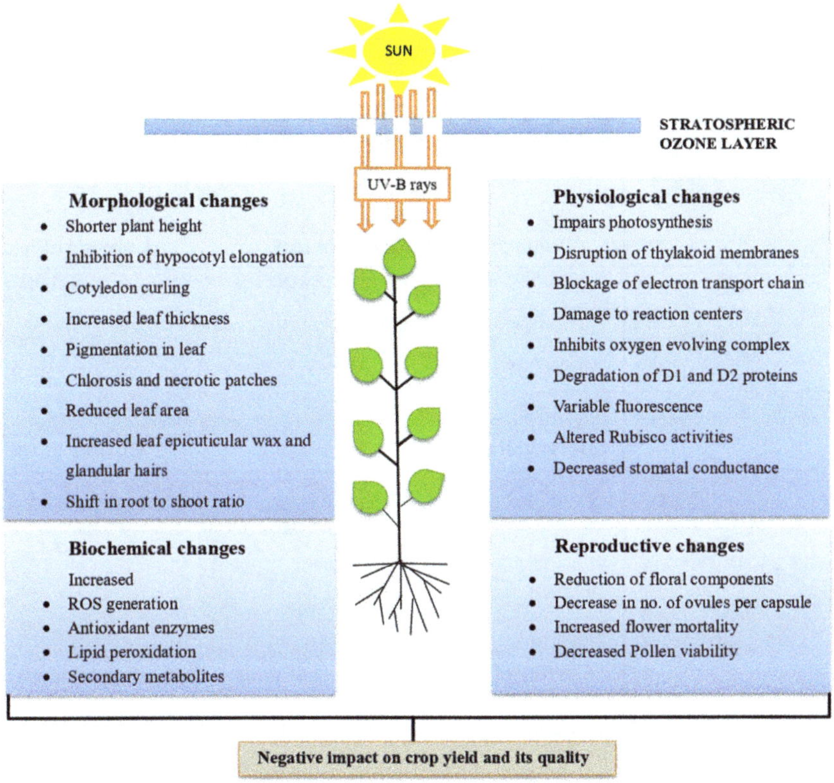

Figure 9.1. Detrimental effects of UV-B radiation on plants.

owing to structural damage, which results in a major reduction in photosynthetic rate and an increase in reactive oxygen species (ROS) generation, hindering growth and development of plants, and eventually plant mortality [9, 11, 17].

UV-B damage has elicited a wide spectrum of responses in plants including the buildup of secondary metabolites such as phenolic acids, flavonoids, terpenes, glucosinolates, alkaloids, anthocyanin, ascorbic acid, and beta-carotene, which affect a variety of physiological processes in plants [22]. Another defense includes alterations in metabolic and molecular processes in plants, such as the activation of antioxidant systems that counteract ROS generation, which can potentially cause DNA damage [23]. This chapter will focus on the current understanding of UV-B-induced damage with special emphasis on plant physiology.

9.2 Visual injury symptoms under UV-B exposure

UV-B stress initially results in bronzing, glazing, and cupping of leaves, followed by the emergence of irregular patches, i.e. chlorotic patches. After continuous exposure, these chlorotic patches turn into brown necrotic spots and eventually die [24, 25]. The formation of these spots occurs due to decrease in chlorophyll content [24]. For

Figure 9.2. Visible leaf injury symptoms in (a) mung bean, (b) soybean, and (c) pea plants under elevated UV-B exposure.

instance, in cotton leaves these chlorotic patches in the interveinal zone eventually turn necrotic, causing early leaf senescence [26, 27]. With prolonged exposure to UV-B, the leaves become cup-shaped or involuted and eventually dry up. Additionally, the conversion of indole-3-acetic-acid to 3-methyleneoxindole takes place, which promotes the upper side of leaves to grow slower while normal growth proceeds on the lower side, resulting in leaf cupping [26]. The downward curling (epinasty) of leaves is another UV-B-mediated leaf response. This photomorphogenetic response is modulated by the expression of auxin-responsive genes as leaf blades accumulate with auxin levels [28]. Similarly, another mechanism includes the glossy and glabrous appearance of leaves due to increased deposition of waxy material. Through this mediating mechanism, they are protected from the harmful effects of UV-B radiation. Excess UV-B exposure causes leaf bronzing and reddening as a result of increased production of polyphenolic chemicals such as flavonoids. In figure 9.2 symptoms of visual leaf injury of three important crop plants, namely mung bean, soybean, and pea, have been depicted, in addition to illustrating different strategies adapted by the plant as defense mechanisms.

9.3 Growth and morphological characteristics

UV-B radiation induces morphological changes in plants (table 9.1) including stem and leaf elongation, changes in leaf thickness, reduced leaf area, burning and

Table 9.1. Effects of UV-B radiation on various plants.

Plant species	UV-B dose	Negative impacts	References
Glycine max L.	0–10 kJ m^{-2} d^{-1}	Decrease in flower lengths by 28%; pollen production by 30%–50%; staminal length by 28%–81%.	[184]
	8.64–9.5 kJ m^{-2} d^{-1}	Decrease in root biomass and root oxidative capacity by 25.2% and 39.9%; flavonoid and ABA contents were significantly increased.	[11]
	5.8–7.2 kJ m^{-2} d^{-1}	Maximum reduction in Pn by 62%; Ci increased by 60%; gs and WUE in all cultivars increased significantly; total biomass reduced by maximum 36%; yield parameters (NSPPo, NSPP, WSPPo, WSPP, HI, TW) were significantly reduced; maximum reduction in soluble protein by 20%; QRI and CSRI exhibited sensitivity toward elevated UV-B.	[87]
	5.8–7.2 kJ m^{-2} d^{-1}	Reduction in chlorophyll by 46% and 36%; increased kaempferol content by 67% and 83%; quercetin levels by 77% and 95%; lignin content increased by 50% and 60%; wax content increased by 71% and 88%; oil content reduced significantly by 30% and 23%; increased acid value (by 47% and 38%); iodine value (by 15% and 1%); saponification value decreased by 25%; activities of CAT, APX, GR, POX, and SOD increased significantly in both cultivars.	[17]
Oryza sativa L.	1.12 Wm^{-2}	Suppression and degradation of Rubisco while LHCII changes greatly with leaf age.	[112]
	8–16 kJ m^{-2} d^{-1}	Decreased plant height by 12%; number of viable leaves and total leaf area increased by 58% and 82%; reduction in yield by 79%; decreased Pn by 60%; highest leaf phenolic content 65%; decreased pollen germination by 67%; spikelet fertility reduced by 51%.	[39]
	1.0 Wm^{-2}	Total protein content increased by 12.5% and 13.2%; total nitrogen content increased by 12.5% and 14.7%.	[189]
Triticum aestivum L.	8.85 kJ m^{-2} d^{-1}	Decrease in plant height by 10%; reduction in total leaf area up to 6%; increase in leaf thickness by 15%; flavonoids and major UV-B absorbing nonphotosynthetic pigments markedly increased; biomass and yield were significantly inhibited.	[40]
	15 μWcm^{-2}	Reduction in maize height; decreased yield due to reduction in kernels.	[188]

(*Continued*)

Table 9.1. (*Continued*)

Plant species	UV-B dose	Negative impacts	References
	20 μWcm^{-2}	Total phenolics, DPPH, and ABTS were significantly increased by 26.3, 25.1 and 12%; activities of SOD, POD, and CAT were decreased while APX activity increased.	[146]
Zea mays L.	5.4 kJ cm^{-2} d^{-1}	Significant increase in POD and SOD activity; reduction in chl content; anthocyanin levels were more pronounced.	[127]
Ocimum basilicum L.	18.7 kJ m^{-2} h^{-1}	Shorter plant height, width and leaf area; increase in phenolic contents (anthocyanin by 9%–23%); phenolics (28%–126%); flavonoids (80%–169%).	[41]
Nicotiana tabacum L.	9.75–20.76 μWcm^{-2}	Increase in lutein content by 10.05%; ROS increase by 50% in leaves; increase in carotenoid content that enhances TAC of antioxidant system.	[124]
Vicia faba	4.6 kJ m^{-2}	Dehydroascorbate radical increased four-fold.	[120]
Vigna radiata L.	7.2 kJ m^{-2} d^{-1}	Leaf injury increased up to 27%; leaf area reduced by 20%; total biomass reduced significantly by 29%; RSR reduced by 22%; elevated response of total flavonoids by 36%; kaempferol by 70%; quercetin by 164%; PAL up to 105%; decrement in Pn up to 69%, gs; significant increase in Ci up to 64%; JA and SA increase by 94% and 123%; antioxidant activities of SOD, POD, CAT, and APX increased by 55.8%, 33.8%, 88.9%, and 46.2%; decline in protein content by 49.2%; increment in phenol content by 37%; reduction in seed yield by 18.5%.	[12]
Withania coagulans L.	2.8 Wm^{-2}	Decreased total chlorophyll content by 48.3%; carotenoid content significantly improved by 1.36 fold; decrease in photosynthetic rate by 7.6 fold; lipid peroxidation and increase in MDA content by 46.8%; protein and proline content was significantly high.	[88]
Prunella vulgaris L.	35 μWcm^{-2} nm^{-1}	Spica width, stem weight, spica weight, and whole-plant significantly increased; reduced total chl and carotenoids; H_2O_2, MDA, and proline are markedly increased by 64.81%, 485.92%, and 42.67%.	[102]
Alternanthera sessilis	0–40 J cm^{-2}	Increase in betanidin pigment by 39%; betanin level by 33%; highest flavonoid content estimated (51%); DPPH inhibition by 68.78%.	[153]

Phaeolus vulagris L.	10.2 kJ m^{-2}	Shorter seedling; increase in cuticular wax; detection of fatty acids and steroids; enhanced phenolic compounds.	[51]
Lonicera japonica L.	8.4–22.4 μWcm^{-2}	Significant increase in chlorophyll content at lower dose while it is upregulated by 10.05% at higher dose; CAT activity increased by 13.18% and SOD by 26.46%.	[125]
Brassica oleracea L.	1.2 Wm^{-2}	Fresh and dry weights were slightly decreased and leaf area by 10%–25% lower than control; TFC levels slightly increased for upper leaves; TPC levels for upper leaves decreased by 25%.	[154]
Pisum sativum L.	7.2 kJ m^{-2} d^{-1}	Reduced leaf area by 20%; number of pods by 30%; total biomass decreases by 29%; marked increase in kaempferol by 72%; quercetin by 114%; H_2O_2 content by 56%; significant increase in CAT (11%), APX (12%), and SOD (55%); jasmonic and salicylic content were increased by 91% and 59%; yield parameters decreased significantly (NSPPo by 35%, NSPP by 34%, WSPPo by 26%, WSPP by 45%, TW by 21%, and HI by 24%); seed quality parameters deteriorated significantly (TFAA by 26%, TSS by 17%, and SP by 18%).	[13]
	7.1–8.6 kJ m^{-2}	Reduction in yield by 18.9%; LPO increases by 75.3%; ascorbic acid content decreased by 25.3%; maximum increment in thiol content 45.8%; SOD and POD activities were significantly increased by 51.3% and CAT decreased by 22.4%.	[126]
Linum usitatissimum L.	7.2 kJ m^{-2} d^{-1}	Yield parameters were significantly affected (no. of seeds decreased by 34 and 45%; number of ovules per capsule decreased by 39 and 35%); protein content declined by 51.4%; total sugar content decreased by 47.9%; nitrogen content decreased by 26.3%; maximum decrease in phosphorus content by 36.9%; oil content decreased by 47%; acid value increased by 34.5%; saponification value reduced by 9.6%; enhancement in fatty acid prolife contents.	[187]
Hordeum vulgare L.	38–56 μWcm^{-2}	FPC, BPC, and TPC were increased significantly at both doses; ferulic acid content increased by 16.6%; p-coumaric acid decreased significantly by 39.33% and 30.2% in free form.	[147]
Ginkgo biloba	7.14–21.42 kJ m^{-2} d^{-1}	Marked increase in quercetin and kaempferol levels; MDA content was significantly higher; CAT activity reaches maximum by 105.79 nmol g^{-1} min^{-1}; flavonoid synthetic genes (*FLS*, *F3'H*) were upregulated.	[128]

(*Continued*)

Table 9.1. (*Continued*)

Plant species	UV-B dose	Negative impacts	References
Olea europaea L.	6.5–12.4 kJ m^{-2} d^{-1}	GR and APX activities were significantly decreased by 75% and 36% at low dose while promoted a significant increase (59%) in GR at higher dose and no difference in APX activity; levels of 4'methoxy luteolin and 4' or 3'-methoxy luteolin (flavonoids) decreased significantly.	[129]
Cucumis sativus L.	0.912 ± 0.074 kJ m^{-2}	Reduction in biomass, chlorophyll, and flavonol concentration; no effect on gas exchange parameters.	[103]
Helianthus annuus L.	7.2 kJ m^{-2} d^{-1}	Reduction of pollen viability by 15.1% and 14.5% in both varieties; number of whorls decreased by 14.3% and 15.6%; significant reduction in disc florets by 20.9% and ray florets by 11.5%.	[179]

Abbreviations: ABA: abscisic acid; APX: ascorbate peroxidase; ABTS: 2,2′-azinobis(3-ethylbenzothiazoline 6-sulfonic acid); BPC: bound phenolic content; Ci: internal CO_2; CSRI: cumulative stress response index; CAT: catalase; DPPH: 2,2-diphenyl-1-picryl-hydrazyl-hydrate; FPC: free phenolic content; F'3H: flavanone 3'-hydroxylase; FLS: flavonol synthase; gs: stomatal conductance; GR: glutathione reductase; HI: harvest index; LHCII: light-harvesting complex II; LPO: lipid peroxidation; MDA: malondialdehyde; NSPPo: number of seeds per pod; NSSP: number of seeds per plant; Pn: Photosynthetic rate; POD: peroxidase; QRI: quality response index; RSR: root-shoot ratio; SC: starch content; SP: soluble protein; SOD: superoxide dismutase; TAC: total antioxidant activity; TPC: total phenolic compounds; TFC: total flavonoid compound; TW: test weight; TSS: total soluble sugars; TFAA: total free amino acids; WUE: water use efficiency; WSPPo: weight of seeds per pod; WSPP: weight of seeds per plant.

discoloration, axillary branching, shorter plant height, reduced number of pods, and shifts in root-to-shoot ratio [29–33]. UV-B exposure slows the elongation of the main branch, resulting in greater compactness and shortness of plants. The primary cause for plant height reduction is due to shorter internodes, rather than fewer nodes [24, 34–38]. In pea plants, shorter internodes were observed to be related to fewer cells rather than shorter cell length [38]. Similarly, various plants have been reported with shorter plant height upon exposure to enhanced UV-B radiation [39–41]. A possible mechanism for decreased stem elongation by UV-B, according to [42], could be due to changes in phytohormone levels, particularly IAA, involved in stem elongation.

Likewise, plant height and the amount of leaf area is also a delicate growth characteristic that responds to elevated UV-B radiation. Reduction in leaf area is one of the most consistent responses observed with regard to UV-B exposure. This could be attributed to reduced epidermal cell number, changes in leaf anatomy, and both cell division and cell expansion [38, 43–45]. Similarly, specific leaf area (SLA) shows increased leaf thickness upon UV-B exposure. For example, when exposed to enhanced UV-B radiation the SLA of basil plants was considerably reduced and increased leaf thickness was noticed [41]. This increment in leaf thickness was due to inclusion of spongy mesophyll cells.

The epicuticular wax layer is a key leaf surface characteristic that serves as a first line of defense in response to stress [46, 47]. Increased wax acts as a protective mechanism as it reflects incident UV-B radiation up to 10%–30% [48]. For instance, in cotton leaves, exposure to UV-B radiation results in a 200% increase in epicuticular wax [26]. Similarly, another study conducted on two cultivars of *Glycine max* L. resulted in a 71% and 88% increase in wax content [17]. Enhanced UV-B radiation does not alter the chemical composition of wax; instead, it modifies the reflectance of UV-B on the leaf [49, 50]. However, a recent study, on *Phaseolus vulgaris* L., suggests that UV-B radiation modifies the wax content by enhancing the proportion of alkanes [51]. These acclimation responses not only protect plants from excessive UV-B energy, but also improve plant tolerance to other adverse conditions such as excessive heat and mechanical treatment during post-harvest [26, 52, 53].

UV-B-induced photomorphogenic responses include suppression in hypocotyl development, cotyledon curling, and transcription of genes involved with the phenylpropanoid and flavonoid production pathways [18, 52, 54–57]. COP1, hypocotyl 5 (HY5), and HY-5 homolog (HYH) are response regulators in photomorphogenesis [58–60]. Whereas the repressors of UV-B photomorphogenesis (RUP-1 and RUP-2) are both negative regulators of the UVR8 signaling pathway [61]. UV-B reduces hypocotyl growth not only through the UVR8 photoreceptor but also through DNA damage responses at higher doses. UV-B, for example, inhibits hypocotyl development in *Arabidopsis thaliana* via mediating DNA damage [62].

9.4 Physiological and biochemical characteristics

9.4.1 Photosynthesis

Decline in photosynthesis rate is linked to damage of photosynthetic machinery and decline in gs (stomatal conductance). Damage to photosynthetic machinery includes

D1/D2 reaction center proteins, oxygen-evolving complex (OEC), and the components on the donor and acceptor sites of photosystem II (PS II), as well as alterations of genes involved in PS II reaction center synthesis [63]. In photosynthetic organisms, the PS II complex is a primary target of UV-B radiation [64]. UV-B inactivates quinone electron acceptors (QA and QB) on the acceptor side and the oxygen evolving complex (OEC) on the donor side, causing the electron transport chain to become blocked [63–65].

9.4.2 Photosystem II

Photosystem II is a multifaceted protein complex, embedded in the grana of the thylakoid membrane of a chloroplast. UV-B has been linked to a variety of deleterious consequences at PS II sites, including the following:

- *Oxygen-evolving complex (OEC).* UV-B interferes with the functioning of the OEC complex by disrupting the links between manganese atoms via Mn cluster absorption or by damaging the intermediates involved in the water development process. Increased susceptibility to UV-B radiation ranges from S0 to S3, resulting in the production of an OH radical rather than a Mn-OH bond, preventing oxygen release [63, 66–70].
- *D1 protein and D2 protein.* These are key structural proteins, having a molecular weight of 38.021 KDa. Upon exposure to UV-B irradiation, they reduce to 20 KDa fragments, which are thereafter completely destroyed by protease enzymes in a light-dependent manner [71]. On the other hand, the D2 protein is responsible for binding important electron transporters including plastoquinone (PQ); thus, under UV-B radiation, there is a considerable drop in D2 content [72–74]. These two proteins are particularly sensitive to UV-B, and exposure to UV-B promotes D1 and D2 protein degradation, resulting in PS II impairment, which can be evaluated by reduced evolution of oxygen or variable chlorophyll florescence. This fluctuating chlorophyll florescence is frequently used to assess PS II radiation damage.
- *Tyrosine.* Tyrosine can be attenuated or photooxidized to 3, 4-dihydroxy-phenylalanine (DOPA) and di-tyrosine as a result of UV-B exposure [74, 75].
- *Cytochrome b6f.* The Cytochrome b6f complex is made up of four main subunits: cytochrome f, subunit IV, and Rieske iron-sulfur protein, as well as four smaller hydrophobic subunits (PetM, PetN, PetL, and Pet G) [76]. The electron transport chain connection between photosystems I and II is mediated by this complex, thus oxidizing PQH2 generated by PS II and decreasing plastocyanin levels [75]. According to previous studies, Cytochrome b6f is the least impacted by UV-B as it has two quinine-binding sites (one for quinol oxidation and other for quinone reduction).

9.4.3 Photosystem I

The effect of UV-B on photosystem I (PS I) is comparatively less as compared to PS II [63, 74, 77]. This lack of effect can be confirmed: when an artificial donor couple is added to UV-B-treated chloroplasts, it practically retains all functions [78]. PS I-mediated cyclic photophosphorylation may be inhibited by UV-B radiation [79].

9.4.4 ATPase complex

ATPase is an enzyme complex that comprises two parts: a coupling factor C0 (CF_0), a proton channel, and coupling factor 1 (CF_1), which mediates the binding of inorganic phosphate (P) to adenosine diphosphate (ADP) to form adenosine triphosphate (ATP) [80]. UV-B has a substantial impact on ATPase because it reduces the amount of coupling factor CF1, which may be related to changes in thylakoid permeability that disturb the proton gradient across the stroma and lumen [74, 81, 82]. The amount and activity of ATPase in thylakoids were reduced under supplemental UV-B exposure; the amount and function of ATP-synthase was also reduced significantly [82].

The most variable characteristic that regulates plant responses to UV-B is stomatal conductance (gs). As a result, intraspecific variations in UV-B effects are frequently related to the inherent capacity of stomatal conductance; plants with higher stomatal uptake rates exhibit larger effects [83]. Reduced stomatal conductance (gs) is observed under UV-B exposure [84–86]. A greenhouse experiment conducted with pea plants under supplementary UV-B radiation [44] revealed a 65% drop in adaxial stomatal conductance, which is responsible for increased stomatal CO_2 uptake limitation, and a decrease in water utilization efficiency (WUE). Similarly, in soybean plants, a significant reduction in Photosynthetic rate (Pn), stomatal conductance (gs), and WUE were observed [12, 87]. In a medicinal plant, *Withania coagulans*, decrease in photosynthetic rate (Pn) by 7.6 fold on exposure to supplemental UV-B radiation has been reported [88] (table 9.1).

Chorophyll fluorescence has been generally applied to detect stress-induced alterations in photosynthesis. Under UV-B, the Fv/Fm ratio decreases, indicating photoinhibition. In UV-B-exposed plants, inactivation of the PS II reaction center is related to a decrease in Fm and Fv/Fm [89, 90]. A reduction in flow of electrons from the water-splitting complex could result in donor-side damage, reducing Fm under UV-B-irradiated plants [91]. Fv, which measures a leaf's or chloroplast's photosynthetic competence, decreases to some extent in both ambient and UV-B conditions. The impact of UV-B radiation on PS II activity is generally measured as a reduction in evolution of oxygen or a change in fluorescence [92]. For instance, soybean cultivars exhibit significant reduction in Fm, Fv, and Fv/Fm [87] (table 9.1). Delayed quenching and damage to PS II reaction centers may cause reductions in Fv/Fm, decreasing PS II photochemistry and quantum efficiency [93]. Under UV-B stress, the drop in Fv/Fm ratio was reported to be accompanied by an increase in superoxide radical generation [94], which is linked with greater malondialdehyde (MDA) concentration.

9.4.5 Photosynthetic pigments

In higher plants, chlorophylls a and b are the primary photosynthetic pigments, whereas carotenoids are accessory pigments that absorb light in the 400–500nm range [95]. They defend against photooxidative agents by quenching the single-excited chlorophyll (1Chl*) and triplet-excited chlorophyll (3Chl*) within photosystem reaction centers, inhibiting the formation of ROS and subsequently scavenging them through xanthophylls and carotenes [96–100]. In addition to carotenoids, other pigments found in plants include xanthophylls, anthocyanin, and others, which protect plants against stress. Ultrastructural damage to chloroplasts and alterations in chlorophyll pigments result in a decline of photosynthesis [101]. Chlorophyll reduction after exposure to UV-B radiation varies significantly among different crop species [17, 87, 88, 102, 103] (table 9.1). Reduction in chlorophyll is due to lipid peroxidation (LPO) of chloroplast membranes, which involves the photo-conversion of protochlorophyllide to chlorophyllide by protochlorophyllideoxidoreductase [104]. Damage to this enzyme under UV-B exposure results in a slower rate of chlorophyll accumulation.

9.4.6 Calvin cycle enzyme: Rubisco

Rubisco is the key enzyme involved in the C3 cycle, reflected by the fact that its quantity and activity affect the photosynthetic rate and plant biomass [105]. Increased UV-B exposure may decrease Rubisco activities [106, 107]. The concentration of carboxylating enzymes is significantly reduced when high UV-B irradiation is combined with low photosynthetically active radiation (PAR) levels [108]. However, high PAR (>1000 $mol^2 s^1$) in conjunction with low UV-B levels does not affect the activity of Rubisco. UV-B modulates the carboxylation efficiency as well as concentration of Rubisco proteins [109]. The reason behind this is the reduced electron transport rate beyond PS II, which mediates a reduction in regeneration of RUBP that further reduces the Rubisco carboxylation rate [110]. However, UV-B-induced damage to the Rubisco protein and activity has also been linked to increased ROS generation in cells [63, 111]. Supplementary UV-B radiation lowers Rubisco levels while light-harvesting complex II (LHCII) content increases gradually at different leaf stages of two rice cultivars, since Rubisco is a soluble protein whereas LHCII binds strongly to the thylakoid membrane in the chloroplast, it is possible that the difference in the localization of protein corresponds to the difference in the sensitivity to supplementary UV-B radiation. [112] (table 9.1). In mung beans, a decrease in the amount of Rubisco under UV-B treatment implies an increase in H_2O_2 activity and proteolytic activity [113]. Similarly, another study conducted on *Canavalia ensiformis* L. revealed that a decrement in Rubisco levels is associated with decreased Rubisco activity and that UV-B further inhibits its activation and induction by Rubisco activase [114].

9.4.7 Generation of reactive oxygen species

ROS are oxygen-containing chemical species that are chemically reactive. The superoxide ion (O_2), hydrogen peroxide (H_2O_2), and hydroxyl radical (OH) are the most studied forms of ROS. The chloroplasts, peroxisomes, and mitochondria are the key biosynthetic regions for ROS generation. However, chloroplasts and peroxisomes are the predominant sources of ROS generation in the presence of light, whereas the mitochondrion is the primary ROS producer in the absence of light [115]. At low UV-B irradiation, ROS have been recognized to have a crucial role as chemical messengers in plant cells. However, at high concentrations, they damage organelle membranes, such as chloroplasts, which in turn generates secondary stress, specifically oxidative stress [116, 117]. The authors of [118] postulated that UV-B stimulates NADPH oxidase, localized in the membrane, resulting in the formation of ROS.

Intrinsic target molecules that have been oxidized can also be used as reporter molecules for ROS. The presence of ROS is quantified by means of LPO, which is accessed by an increase in MDA concentration [119], results from oxidation of DNA and membrane lipids, respectively. The quantity of oxidized dehydroascorbate radical in *Vicia faba* leaves increased four-fold after a brief exposure to 0.46 kJ m^{-2} UV-B, indicating oxidative pressure [120] (table 9.1). Moreover, any molecule with the ability to scavenge radicals can provide information pertaining to ROS; it is not solely the Asada–Halliwell antioxidant system that needs to be investigated [121].

Plants metabolize ROS by activating defense systems, including low-molecular-weight enzymes such as glutathione, ascorbate, phenols, carotenoids, xanthophylls, and other high-molecular-weight enzymes like ascorbate peroxidase, superoxide (SOD), catalase (CAT), and peroxidase (POD) [122, 123]. SOD is the major enzyme that converts superoxide anions to H_2O_2, while CAT and POX catalyze the breakdown of H_2O_2. Induction of SOD activity under elevated UV-B radiation in soybean cultivars depicted its better capability to quench superoxide radicals [12, 17]. Table 9.1 gives several studies reporting an increase in antioxidant enzymes in different plants such as *Nicotiana tabacum* L. [124], *Lonicera japonica* L. [125], *Pisum sativum* L. [126], *Zea mays* L. [127], *Gingko biloba* [128], *Olea europaea* L. [129], *Cucumis sativus* L. [103], and *Prunella vulgaris* L. [102].

Further, jasmonic acid (JA) plays an important role in the activation of antioxidative defense systems in plants against UV-B radiation. Exogenous application of 1 mM JA was able to restore SOD activity in bluegrass under UV-B radiation stress [130]. In wheat seedlings, JA induced antioxidative defense mechanisms against UV-B to some extent, as [131], detected a considerable drop in peroxidase and catalase activities.

9.4.8 Secondary metabolites

Synthesis and accumulation of secondary metabolites in leaf tissues as a result of elevated UV-B exposure is another adaptation mechanism evolved in plants. UV-B radiation enters the epidermal layer after being reflected by the epicuticular wax layer. This epidermal layer is known to contain major secondary metabolites such as flavonoids, phenolics, terpenoids, carotenoids, alkaloids, sterols, anthraquinones, and lignin [132–134]. These metabolites protect the underlying tissues from injury after UV-B exposure [135–137].

Phenolic compounds play an essential role in plant nutrition and health. Among abiotic stresses, particularly UV-B radiation, these phenolic compounds are biosynthesized via the phenylpropanoid pathway and often accumulate in the vacuoles of the upper epidermis of leaves, thus acting as UV-filtering pigments and antioxidants [138–141]. Plant phenolic compounds induced under UV-B include phenolic acids, lignans, neolignans, stillbenes, flavonoids, aromatic polyketides, hydrolysable tannins, coumarins, monolignols, and aromatic amino acids, which serve as antioxidants, signaling molecules, pigments, electron transporters as well as connecting the primary and secondary metabolic pathways [142, 143]. According to reports, UV-B radiation of comparatively low intensity may induce a large amount of phenolics [144]. It can produce phenol-rich fruits and vegetables, as well as other secondary metabolites in plants [145]. It has been shown that UV-B irradiation increased the activity and gene expression of important enzymes involved in the phenylpropyl metabolism pathway, resulting in the buildup of phenolics in plants such as barley and wheat [146, 147].

Plants also contain a large amount of flavonoids, which are polyphenols. They are variable in shape and are found in nature as glycosides. As a result of hydroxylation, glycosylation, acylation, or methoxylation, about 6500 such structures have been found. Genes responsible for flavonoids biosynthesis are also regulated by UV-B, and flavonoids accumulate in cellular compartments like cell walls, vacuoles, and chloroplasts [18, 148]. UVR8, a UV-B photoreceptor, interacts with MYB13, a transcription factor, to influence flavonoid production. MYB-13 directly binds to the promoters of chalcone synthase, chalcone flavanone isomerase, and flavonol synthase 1 (FLS), promoting the transcription of flavanol production and accumulation [149].

UV-B-absorbing compounds such as flavonoids and hydroxycinnamic acids are increased in plants in response to elevated UV-B levels [148, 150]. As antioxidants, flavonoids have the ability to filter UV-B radiation [151, 152]. In soybean plants, during flowering and podding stages, the concentrations of flavonoids, quercetein, and rutin were considerably increased under UV-B [11]. The highest estimated flavonoids levels were recorded in aquatic plants such as *Alternanthera sessilis* under UV-B exposure for 8 hr followed by a recovery period of 24 hr [153] (table 9.1). In *G. biloba*, flavonoids synthesis genes involving FLS, F3'H, were upregulated in response to UV-B exposure, indicating their importance in flavonoid synthesis to long UV-B exposure [128]. *O. Europaea* leaves are abundant in luteolin-7-O-glucoside, a flavonoid that accounts for high tolerance to stress [129]. Supplemental UV-B radiation combined with four light treatments led to considerable reduction in biomass, as well as chlorophyll and flavanol concentration in the leaf epidermis of *C. sativus*. On the other hand, gas exchange and chlorophyll parameters do not have significant effects [103]. The ability and sensitivity of plants to absorb UV-B radiation varies, depending on leaf location and growth stage, resulting in differential total flavonoid compound and total phenolic compound biosynthesis [154] (table 9.1). In leguminous plants, UV-B disturbs the composition of root exudates responsible for reduced nodule size and number [57]. Thus, increased flavonoid metabolites in leaf tissues indicates a defensive mechanism adapted by the plant in response to UV-B stress [155].

Glucosinolates (GS) are sulfur-rich metabolites that play an important role in pathogen and disease resistance [156, 157]. They are mostly nonbiologically active and located in the vacuoles; however, when plant tissue is disrupted, UV-B radiation hydrolyzes them into biologically active forms [158]. These are derived from different precursors like aromatic glucosinolate (from phenylalanine), aliphatic glucosinolate (methionine), and indolyl (tryptophan). For instance, in broccoli, both low and high UV-B dosages induce aliphatic glucosinolate, especially methylsulfinylalkylglucosinolate, whereas indolyl GS is influenced by several days of low UV-B radiation or a single strong dose of UV-B irradiation. However, after three days of UV-B exposure (five days), aliphatic GS accumulated, whereas the overall indolyl GS content returned to baseline but the profile of individual indolyl GS changed [158–160].

Jasmonic acid production and signal transduction also plays an important role in plant defensive responses under UV-B stress. In pea and mung bean cultivars, increased level of endogenous JA is associated with a stronger tolerance mechanism to UV-B exposure [12, 13]. JA is primarily involved in activation of the antioxidative defense mechanism (via superoxide anion radicals, NADPH oxidase, peroxidase, and other enzymes), stimulation of secondary metabolite production (anthocyanin, beta-carotene, and stillbene), storage of soluble carbohydrates and amino acids (i.e. methionine and isoleucine), and regulation of stomata [161–164]. JA also promotes the accumulation of phenolics and flavonoids that have been thought to be a defense mechanism against UV-B, entering the mesophyll [165]. Furthermore, anthocyanins exhibited the capacity to protect plants from photooxidative harm by shielding chloroplasts from high-energy quanta and scavenging of ROS. Exogenous application of JAs has increased their anthocyanin content in many plants [166]. Furthermore, UV-B radiation causes damage to the membrane and increased LPO in various crop plants [57]. Pre-treatment with 2.5 mM MeJA can reduce LPO in plants caused by UV-B exposure [167]. The application of JA to the plants alone, on the other hand, results in an increase in MDA content [168]. Larger accumulation of endogenous JA in pea cultivars exhibited higher increase in quercetin (up to 118%) and kaempferol (up to 72%) content [13]. Mung bean cultivars showed similar results, with 70% increases in kaempferol and 164% increases in quercetin flavonols [12]. In general, JA has been shown to be advantageous under UV-B exposure by enhancing plant defenses via inducing antioxidative enzymes, photosynthetic activities, metabolites, and several other physiological parameters, which result in better crop production.

Salicylic acid (SA) also plays an important role in plants under enhanced UV-B exposure. There are two mechanisms to SA synthesis: isochorismate synthase (ICS) and phenylalanine ammonia lyase (PAL), both of which begin with chorismate [169]. The dominance of routes varies depending upon the different species; some may have equal dominance [170]. The levels of SA in shoots are substantially higher than in roots [171]. Since the PAL and ICS routes are independent, and the ICS system accounts for 95% of SA synthesis, it may be assumed that a tiny fraction of SA produced under UV-B stress is via the PAL pathway, as seen in *Arabidopsis* and *Nicotiana benthamiana* [172]. In a recent study, increased SA levels resulted in increased oxidative stress, which plants were unable to withstand, potentially resulting in plant mortality [173]. Since it has inhibitory effects on ascorbate peroxidase activity and catalase activity, increased levels of SA may enhance oxidative stress in plants [174–176]. As a result, plants accumulate more H_2O_2, which is accompanied by an increase in SA levels that further makes the plant more susceptible to UV-B [12].

9.5 Reproductive fitness, crop yield and its quality

The process of reproducing sexually is a complex part in a plant's life cycle, and UV-B radiation affects several stages of reproductive development [177]. The impact of UV-B on the sexual reproduction of plants can be determined by genetic factors including floral tissues and organs. UV-B is known to diminish the size of cowpea flowers and cotton sepals [26, 178]. Similarly, *Helianthus annuus* L. when supplemented with UV-B and ozone, results in reduction of floral components [179]. Since DNA is the potential target of UV-B, the haploid genome of pollen makes it highly vulnerable against UV-B. The pollen wall transmits about 20% UV-B, which has a direct impact on pollen germination [180]. UV-B augments pollen viability via leakage in membrane due to protein and lipid distortion of pollen membrane [181–185]. Seed abortion rates were observed higher in pollinated plants grown under enhanced UV-B environment [181].

Yield is always a key factor in determining the impact of abiotic stress on cash crops. The duration of photoassimilate formation and grain filling is affected by UV-B, which is a critical contributor in yield loss. Carbon assimilation becomes altered in different parts of the plant due to reduction in photosynthesis. Carbohydrate translocation from source to sink is halted under UV-B exposure, thereby causing accumulation of carbohydrates at the source levels and consequently biomass partitioning of photosynthates. The distribution of sucrose and starch differs with the plant's growth stage. The authors of [87] reported a reduction in the yield of test cultivars of *P. sativum* L. subjected to elevated UV-B radiation. Similarly, significant reductions in grain yield of *Triticum aestivum* L., *P. sativum* L., *L. usitatissimum* L., *Z. mays* L., and *Oryza sativa* L. have been reported resulting from UV-B exposure [126, 186, 188, 189] (table 9.1). The authors of [190] observed significant negative impact of UV-B during the transitioning of plants from vegetative to reproductive phase. The reproductive organ demands a high level of photosynthates throughout flowering and fruiting stages, hence why photosynthates are transferred from the leaves and roots to the reproductive organs. During the anthesis stage of cereal crops, the active assimilatory source are the flag leaves. UV-B radiation induces senescence of flag leaves by causing a loss of chlorophyll and soluble protein in the leaves.

Changes in oil quantity and quality due to enhanced UV-B has also been noticed in several studies, as changes in the chemical composition of the oil is a serious concern linked to human health. In crop plants, the vegetative parts are principally responsible for the development of reproductive organs and seeds, and any stress to them can alter yield and seed composition directly or indirectly [191]. Alteration in environmental factors during the synthesis of fatty acids may also have a major impact on oil composition and fatty acid content [192]. For instance, linseed cultivars were reported with significant decrease in oil content after UV-B exposure [191]. Low oil content induced through UV-B stress is mainly due to alterations in the precursor avaliability needed for seed filling, as well as altered embryo growth and the pericarp-to-embryo ratio [193]. In *H. annuus* L. supplemental UV-B radiation induced significant reductions in amino acids, total sugars, and oil content [179]. Similarly, in *L. usitatissimum* L. seed quality parameters including protein content were reduced by 51.4%, total sugar content by 47.9%, nitrogen content by 26.3%, and phosphorus content by 36.9%, along with reductions in the content of Ca, Mg, and Zn (60.6, 44.6 and 38.6%), respectively, under supplemental UV-B [187] (table 9.1). Apart from reductions in oil content, the chemical properties of the oil were also modified. Under enhanced UV-B radiation, an increased acid value indicates that triglycerides are converted to glycerols, resulting in rancidification of the oil, which gives an undesirable odor, taste, and nutritional value. Higher iodine level indicates more unsaturated bonds, rendering it more prone to oxidative rancidity [194]. The tendency of oil to oxidize is reflected by a higher iodine value under elevated UV-B. The saponification value is primarily used to calculate the average molecular weight and chain length of fatty acids in oil. The presence of long fatty chains is revealed by a low saponification value under elevated UV-B. Overall, oil quality measures (iodine, acid value, and saponification value) confirmed that elevated UV-B had a detrimental impact on oil composition, content, and quality [17].

9.6 Conclusions

UV-B radiation has a severe impact on biological entities, particularly plants because of their sessile nature. Elevated UV-B radiation induces diverse morphological effects on plants such as formation of chlorotic and necrotic patches due to decrease in chlorophyll content, reduction in elongation of branches and stems resulting in shorter plant height, reduced leaf area that in turn increases the thickness of the leaf, and modification of epicuticular wax content to avoid the reflectance of UV-B on the leaf. Other photomorphogenic responses include inhibition of hypocotyl elongation, cotyledon curling, coiling of tendrils, etc. Apart from this, elevated UV-B affects crop growth directly through damage to photosynthetic machinery, including damage to reaction centers, particularly PS II, D1 and D2 proteins, OEC, ATPase synthase, and reduction in chlorophyll pigments due to lipid peroxidation of the chloroplast membrane, thereby decreasing photosynthetic efficiency, measured in terms of chlorophyll fluorescence and stomatal conductance. A decrease in chlorophyll fluorescence is accompanied by an increase in ROS generation. Intrinsic target molecules that have been oxidized by ROS can be quantified by an increase in LPO and MDA concentrations. Accumulation of secondary metabolites such as phenolics, flavonoids, terpenes, glucosinolates, anthocyanins, beta-carotene, etc, in the epidermal layer provides protection from higher UV-B doses. Plant vegetative parts are principally responsible for the development of reproductive organs and seeds, and UV-B mediates altered allocation and accumulation patterns of photosynthates, ultimately resulting in significant loss of yield and quality.

Acknowledgments

P.S. is thankful to CSIR, New Delhi, India for a Junior Research Fellowship. K.K. C. is grateful for financial assistance from a UGC-BSR Start-up Grant, New Delhi, India and to a seed grant, Institute of Eminence (IoE), Banaras Hindu University, Varanasi, India.

Multiple choice questions

1. Which of the following leaf injury symptoms are exhibited by plants after UV-B exposure?
    (a) Glazing;
    (b) Cupping;
    (c) Bronzing;
    (d) All of the above.
2. Curling of leaves in plants after UV-B exposure is related to …
    (a) Auxin;
    (b) Gibberellin;
    (c) Ethylene;
    (d) Abscisic acid.
3. The first line of defense in plants against UV-B exposure is …
    (a) Increase in reactive oxygen species;
    (b) Deposition of epicuticular wax and pigments;
    (c) Enhanced activities of antioxidants;
    (d) Chlorosis.
4. Higher malondialdehyde content in plants under UV-B is due to …
    (a) Reduced photosynthesis;
    (b) Increased lipid peroxidation;
    (c) Increased antioxidative enzymes;
    (d) Increased pigmentation.
5. The photosystem II complex, on exposure to UV-B radiation gets impaired due to degradation of D1 and D2 protein which can be accessed through...
    (a) Chlorophyll fluorescence;
    (b) Malondialdehyde content;
    (c) DPPH;
    (d) Stomatal conductance.

Answers: 1 (d); 2 (a); 3 (b); 4 (b); 5 (a).

Descriptive-type questions

1. How does ultraviolet-B radiation reach the Earth's surface and what is its negative impact on plants?
2. What are the natural and anthropogenic factors responsible for the depletion of the stratospheric ozone layer?
3. How is UV-B radiation affecting the photosynthetic efficiency of plants? Discuss in detail.
4. Discuss the various defense strategies adopted by plants against UV-B exposure.
5. Explain the negative impact of UV-B radiation on crop yield and its quality.

References

[1] Bais F, Bernhard G, McKenzie R L, Aucamp P J, Young P J, Ilyas M, Jöckel P and Deushi M 2019 Ozone–climate interactions and effects on solar ultraviolet radiation *Photochem. Photobiol. Sci.* **18** 602–40

[2] Bornman J F, Barnes P W, Robson T M, Robinson S A, Jansen M A K, Ballare C L and Flint S D 2019 Linkages between stratospheric ozone, UV radiation and climate change and their implications for terrestrial ecosystems *Photochem. Photobiol. Sci.* **18** 681–716

[3] Sulzberger B, Austin A T, Cory R M, Zepp R G and Paul N D 2019 Solar UV radiation in a changing world: roles of cryosphere-land-water-atmosphere interfaces in global 642 biogeochemical cycles *Photochem. Photobiol. Sci.* **18** 747–74

[4] Neale R E, Barnes P W and Robson T M *et al* 2021 Environmental effects of stratospheric ozone depletion, UV radiation, and interactions with climate change: UNEP environmental effects assessment panel, update 2020 *Photochem. Photobiol. Sci.* **20** 1–67

[5] Bornman J F, Barnes P W, Robinson S A, Ballare C L, Flint S D and Caldwell M M 2015 Solar ultraviolet radiation and ozone depletion-driven climate change: effects on terrestrial ecosystems *Photochem. Photobiol. Sci.* **14** 88–107

[6] Müller-Xing R, Xing Q and Goodrich J 2014 Footprints of the Sun: memory of UV and light stress in plants *Front. Plant Sci.* **5** 474

[7] Frohnmeyer H and Staiger D 2003 Ultraviolet-B radiation-mediated responses in plants. Balancing damage and protection *Plant Physiol.* **133** 1420–28

[8] Ravindran K C, Indrajith A, Pratheesh P V, Sanjiviraja K and Balakrishnan V 2010 Effect of ultraviolet-B radiation on biochemical and antioxidant defence system in *Indigoferatinctoria* L. seedlings *Int. J. Eng. Sci. Technol.* **2** 226–32

[9] Katerova Z, TodorovaD, Tasheva K and Sergiev I 2012 Influence of ultraviolet radiation on plant secondary metabolite production *Genet. Plant Physiol.* **2** 113–44

[10] Kumari R and Prasad M N 2013 *Medicinal Plant Active Compounds Produced by UV-B Exposure Sustainable Agriculture Reviews* (Dordrecht: Springer) 225–54

[11] Mao B, Wang Y, Zhao T H, Tian R R, Wang W and Ye J S 2017 Combined effects of elevated O_3 concentrations and enhanced UV-B radiation of the biometric and biochemical properties of soybean roots *Front. Plant Sci.* **8** 1568

[12] Choudhary K K and Agrawal S B 2014 Cultivar specificity of tropical mung bean (*Vignaradiata* L.) to elevated ultraviolet-B: changes in antioxidative defense system, nitrogen metabolism and accumulation of jasmonic and salicylic acids *Environ. Exp. Bot.* **99** 122–32

[13] Choudhary K K and Agrawal S B 2014 Ultraviolet-B induced changes in morphological, physiological and biochemical parameters of two cultivars of pea (*Pisumsativum* L.) *Ecotoxicol. Environm. Safety* **100** 178–87

[14] Gao W, Zheng Y, Slusser J R and Heisler G M 2003 Impact of enhanced ultraviolet-B irradiance on cotton growth, development, yield, and qualities under field conditions *Agric. For. Meteorol.* **120** 241–8

[15] Gupta S, Gupta V, Singh V and Varadwaj P K 2018 Extrapolation of significant genes and transcriptional regulatory networks involved in *Zea mays* in response in UV-B stress *Genes Genomics* **40** 973–90

[16] Kakani V G, Reddy K R, Zhao D and Gao W 2004 Senescence and hyperspectral reflectance of cotton leaves exposed to ultraviolet-B radiation and carbon dioxide *Physiologiaplantarum* **121** 250–7

[17] Choudhary K K and Agrawal S B 2016 Assessment of fatty acid profile and seed mineral nutrients of two soybean (*Glycine max* L.) cultivars under elevated ultraviolet-B: role of ROS, pigments and antioxidants *Photochem. Photobiol.* **92** 134–43
[18] Tilbrook K, Arongaus A B, Binkert M, Heijde M, Yin R and Ulm R 2013 The UVR8 UV-B photoreceptor: perception, signaling and response *The Arabidopsis Book* (Rockville, MD: American Society of Plant Biologists) https://doi.org/10.1199/tab.0164
[19] Jenkins G I 2017 Photomorphogenic responses to ultraviolet-B light *Plant Cell Environ.* **40** 2544–57
[20] Yin R and Ulm R 2017 How plants cope with UV-B: from perception to response *Curr. Opin. Plant Biol.* **37** 42–8
[21] Jaiswal D, Pandey-Rai S and Agrawal S B 2021 Untangling the UV-B radiation-induced transcriptional network regulating plant morphogenesis and secondary metabolite production *Environ. Exp. Bot.* **21** 104655
[22] Kasote D M, Katyare S S, Hegde M V and Bae H 2015 Significance of antioxidant potential of plants and its relevance to therapeutic applications *Int. J. Biol. Sci.* **11** 982
[23] Rácz A and Hideg É 2021 Narrow-band 311 nm ultraviolet-B radiation evokes different antioxidant responses from broad-band ultraviolet *Plants* **10** 1570
[24] Zhao D, Reddy K R, Kakani V G, Read J and Sullivan J 2003 Growth and physiological responses of cotton (*Gossypiumhirsutum* L.) to elevated carbon dioxide and ultraviolet-B radiation under control *Plant Cell Environ.* **26** 771–82
[25] Smith J L, Burritt D J and Bannister P 2000 Shoot dry weight, chlorophyll and UV-B-absorbing compounds as indicators of a plant's sensitivity to UV-B radiation *Ann. Bot.* **86** 1057–063
[26] Kakani V G, Reddy K R, Zhao D and Sailaja K 2003 Field crop responses to ultraviolet-B radiation: a review *Agric. For. Meteorol.* **120** 191–218
[27] Reddy K R, Kakani V G, Zhao D, Mohammed A R and Gao W 2003 Cotton responses to ultraviolet-B radiation: experimentation and algorithm development *Agric. For. Meteorol.* **120** 249–65
[28] Fierro A C, Leroux O, De Coninck B, Cammue B P, Marchal K, Prinsen E, Van Der Straeten D and Vandenbussche F 2015 Ultraviolet-B radiation stimulates downward leaf curling in *Arabidopsis thaliana Plant Physiol. Biochem.* **93** 9–17
[29] Hectors K, Prinsen E, De Coen W, Jansen M A and Guisez Y 2007 *Arabidopsis thaliana* plants acclimated to low dose rates of ultraviolet B radiation show specific changes in morphology and gene expression in the absence of stress symptoms *New Phytol.* **175** 255–70
[30] Hayes S, Sharma A, Fraser D P, Trevisan M, Cragg-Barber C K, Tavridou E, Fankhauser C, Jenkins G I and Franklin K A 2017 UV-B perceived by the UVR8 photoreceptor inhibits plant thermomorphogenesis *Curr. Biol.* **27** 120–7
[31] Robson T M, Klem K, Urban O and Jansen M A 2015 Re-interpreting plant morphological responses to UV-B radiation *Plant, Cell Environ.* **38** 856–66
[32] Kanto T, Matsuura K, Yamada M, Usami T and Amemiya Y 2008 UV-B radiation for control of strawberry powdery mildew *VI Int. Strawberry Symp.* 359–62
[33] Jansen M 2002 Ultraviolet-B radiation effects on plants: induction of morphogenic responses *Physiol. Plant.* **116** 423–9
[34] Tevini M and Teramura A H 1989 UV-B effects on terrestrial plants *Photochem. Photobiol.* **50** 479–87

[35] Barnes P W, Maggard S, Holman S R and Vergara B S 1993 Intraspecific variation in sensitivity to UV-B radiation in rice *Crop Sci.* **33** 1041–6

[36] Santos A, Almeida J M, Santos I and Salema R 1998 Biochemical and ultrastructural changes in pollen of *Zea mays* L. grown under enhanced UV-B radiation *Ann. Bot.* **82** 641–5

[37] Searles P S, Caldwell M M and Winter K 1995 The response of five tropical dicotyledon species to solar ultraviolet-B radiation *Am. J. Bot.* **82** 445–53

[38] Gonzalez R, Mepsted R, Wellburn A R and Paul N D 1998 Non-photosynthetic mechanisms of growth reduction in pea (*Pisumsativum* L.) exposed to UV-B radiation *Plant Cell Environ.* **21** 23–32

[39] Mohammed A R and Tarpley L 2011 Morphological and physiological responses of nine southern US rice cultivars differing in their tolerance to enhanced ultraviolet-B radiation *Environ. Exp. Bot.* **70** 174–84

[40] Feng H, Li S, Xue L, An L and Wang X 2007 The interactive effects of enhanced UV-B radiation and soil drought on spring wheat *S. Afr. J. Bot.* **73** 429–34

[41] Dou H, Niu G and Gu M 2019 Pre-harvest UV-B radiation and photosynthetic photon flux density interactively affect plant photosynthesis, growth, and secondary metabolites accumulation in basil (*Ocimumbasilicum* L.) plants *Agronomy* **9** 434

[42] Mark U and Tevini M 1996 Combination effects of UV-B radiation and temperature on sunflower (*Helianthus annuus* L., cv. polstar) and maize (*Zea mays* L, cv. Zenit 2000) seedlings *J. Plant Physiol.* **148** 49–56

[43] Tevini M, Iwanzik W and Teramura A H 1983 Effects of UV-B radiation on plants during mild water stress II. Effects on growth, protein and flavonoid content *Zeitschriftfürpflanzenphysiologie* **110** 459–67

[44] Nogués S, Allen D J, Morison J I and Baker N R 1998 Ultraviolet-B radiation effects on water relations, leaf development, and photosynthesis in droughted pea plants *Plant Physiol.* **117** 173–81

[45] Hofmann R W, Campbell B D, Fountain D W, Jordan B R, Greer D H, Hunt D Y and Hunt C L 2001 Multivariate analysis of intraspecific responses to UV-B radiation in white clover (*Trifoliumrepens* L.) *Plant Cell Environ.* **24** 917–27

[46] Bondada B R, Oosterhuis D M, Murohy J B and Kim M S 1996 Effect of water stress on the epicuticular wax composition and ultrastructure of cotton (*Gossypiumhirsutum* L.) leaf, bract, and boll *Environ. Exp. Bot.* **36** 61–9

[47] Baker E A 1982 Chemistry and morphology of plant epicuticular waxes *The Plant Cuticle* ed D F Cutler, K L Alvin and C E Price (London: Academic) pp 139–66

[48] Caldwell M M, Robberecht R and Flint S D 1983 Internal filters: prospects for UV-acclimation in higher plants *Physiol. Plant.* **58** 445–50

[49] Tevini M and Steinmüller D 1987 Influence of light, UV-B radiation, and herbicides on wax biosynthesis of cucumber seedling *J. Plant Physiol.* **131** 111–21

[50] Barnes P W, Ballaré C L and Caldwell M M 1996 Photomorphogenic effects of UV-B radiation on plants: consequences for light competition *J. Plant Physiol.* **148** 15–20

[51] Santos D Y *et al* 2021 UV-B and drought stress influenced growth and cellular compounds of two cultivars of *phaseolus vulgaris* L. (Fabaceae) *Photochem. Photobiol.* **97** 166–79

[52] Wargent J J, Gegas V C, Jenkins G I, Doonan J H and Paul N D 2009 UVR8 in *Arabidopsis thaliana* regulates multiple aspects of cellular differentiation during leaf development in response to ultraviolet B radiation *New Phytol.* **183** 315–26

[53] Teklemariam T and Blake T J 2003 Effects of UVB preconditioning on heat tolerance of cucumber (*Cucumissativus* L.) *Environ. Exp. Bot.* **50** 169–82
[54] Favory J J *et al* 2009 Interaction of COP1 and UVR8 regulates UV-B-induced photomorphogenesis and stress acclimation in Arabidopsis *EMBO J.* **28** 591–601
[55] Kliebenstein D J, Lim J E, Landry L G and Last R L 2002 *Arabidopsis* UVR8 regulates ultraviolet-B signal transduction and tolerance and contains sequence similarity to human regulator of chromatin condensation1 *Plant Physiol.* **130** 234–43
[56] Fehér B, Kozma-Bognár L, Kevei É, Hajdu A, Binkert M, Davis S J, Schäfer E, Ulm R and Nagy F 2011 Functional interaction of the circadian clock and UV RESISTANCE LOCUS 8-controlled UV-B signaling pathways in *Arabidopsis thaliana Plant J.* **67** 37–48
[57] Choudhary K K and Agrawal S B 2017 Effect of UV-B radiation on leguminous plants *Sustainable Agriculture Reviews* (Cham: Springer) pp 115–62
[58] Rizzini L *et al* 2011 Perception of UV-B by the *Arabidopsis* UVR8 protein *Science* **332** 103–6
[59] Ulm R, Baumann A, Oravecz A, Máté Z, Ádám É, Oakeley E J, Schäfer E and Nagy F 2004 Genome-wide analysis of gene expression reveals function of the bZIP transcription factor HY5 in the UV-B response of *Arabidopsis Proc. Natl. Acad. Sci. USA* **101** 1397–402
[60] Brown B A, Headland L R and Jenkins G I 2009 UV-B action spectrum for UVR8-mediated HY5 transcript accumulation in Arabidopsis *Photochem. Photobiol.* **85** 1147–55
[61] Gruber H, Heijde M, Heller W, Albert A, Seidlitz H K and Ulm R 2010 Negative feedback regulation of UV-B–induced photomorphogenesis and stress acclimation in Arabidopsis *Proc. Natl. Acad. Sci.* **107** 20132–7
[62] Biever J J, Brinkman D and Gardner G 2014 UV-B inhibition of hypocotyl growth in etiolated *Arabidopsis thaliana* seedlings is a consequence of cell cycle arrest initiated by photodimer accumulation *J. Exp. Bot.* **65** 2949–61
[63] Kataria S, Jajoo A and Guruprasad K N 2014 Impact of increasing ultraviolet-B (UV-B) radiation on photosynthetic processes *J. Photochem. Photobiol.* B **137** 55–66
[64] Melis A 1999 Photosystem-II damage and repair cycle in chloroplasts: what modulates the rate of photodamage *in vivo*? *Trends Plant Sci.* **4** 130–5
[65] Zavafer A, Koinuma W, Chow W S, Cheah M H and Mino H 2017 Mechanism of photodamage of the oxygen evolving Mn cluster of photosystem II by excessive light energy *Sci. Rep.* **7** 1–8
[66] Sicora C, Szilárd A, Sass L, Turcsányi E, Máté Z and Vass I 2006 UV-B and UV-A radiation effects on photosynthesis at the molecular level *Environmental UV Radiation: Impact on Ecosystems and Human Health and Predictive Models* (Dordrecht: Springer) pp 121–35
[67] Szilárd A, Sass L, Deák Z and Vass I 2007 The sensitivity of Photosystem II to damage by UV-B radiation depends on the oxidation state of the water-splitting complex *Biochim. Biophys. Acta (BBA)-Bioenerg.* **1767** 876–82
[68] Vass I 2012 Molecular mechanisms of photodamage in the Photosystem II complex *Biochim. Biophys. Acta (BBA)-Bioenerg.* **1817** 209–17
[69] Ayash A, Mohammad I and Nassour R 2018 The effect of ultraviolet radiation on photosynthetic pigments in *Euglena gracilis Tishreen Univ. J. Basic Sci. Ser.* **40** (http://journal.tishreen.edu.sy/index.php/bassnc/article/view/4351)
[70] Mosadegh H, Trivellini A, Lucchesini M, Ferrante A, Maggini R, Vernieri P and MensualiSodi A 2019 UV-B physiological changes under conditions of distress and eustress in sweet basil *Plants* **8** 396

[71] Bergo E, Segalla A, Giacometti G M, Tarantino D, Soave C, Andreucci F and Barbato R 2003 Role of visible light in the recovery of photosystem II structure and function from ultraviolet-B stress in higher plants *J. Exp. Bot.* **54** 1665–73
[72] Booij-James I S, Dube S K, Jansen M A, Edelman M and Mattoo A K 2000 Ultraviolet-B radiation impacts light-mediated turnover of the photosystem II reaction center heterodimer in *Arabidopsis* mutants altered in phenolic metabolism *Plant Physiol.* **124** 1275–84
[73] Tilbrook K, Dubois M, Crocco C D, Yin R, Chappuis R, Allorent G, Schmid-Siegert E, Goldschmidt-Clermont M and Ulm R 2016 UV-B perception and acclimation in *Chlamydomonasreinhardtii Plant Cell* **28** 966–83
[74] Parihar P, Singh R, Singh S, Singh M P, Singh V P and Prasad S M 2017 Ultraviolet radiation targets in the cellular system: current status and future directions *UV-B Radiation: From Environmental Stressor to Regulator of Plant Growth* (New York: Wiley) pp 155–74
[75] Vass I, Szilárd A and Sicora C 2005 Adverse effects of UV-B Light on the structure and function of the photosynthetic apparatus *Handbook of Photosynthesis* (London: Taylor & Francis) pp 43–63
[76] Kurisu G, Zhang H, Smith J L and Cramer W A 2003 Structure of the cytochrome b6f complex of oxygenic photosynthesis: tuning the cavity *Science* **302** 1009–14
[77] Hollósy F 2002 Effects of ultraviolet radiation on plant cells *Micron* **33** 179–97
[78] Kulandaivelu G and Noorudeen A M 1983 Comparative study of the action of ultraviolet-C and ultraviolet-B radiation on photosynthetic electron transport *Physiol. Plant.* **58** 389–94
[79] Pang Q and Hays J B 1991 UV-B-inducible and temperature-sensitive photoreactivation of cyclobutane pyrimidine dimers in *Arabidopsis thaliana Plant Physiol.* **95** 536–43
[80] Taiz L and Zeiger E 2010 Secondary metabolites and plant defense *Plant Physiology* (Sunderland, MA: Sinauer Associates, Inc) pp 283–308
[81] Yu G H, Li W, Yuan Z Y, Cui H Y, Lv C G, Gao Z P, Han B, Gong Y Z and Chen G X 2013 The effects of enhanced UV-B radiation on photosynthetic and biochemical activities in super-high-yield hybrid rice Liangyoupeijiu at the reproductive stage *Photosynthetica* **51** 33–44
[82] Zhang J, Hu X, Henkow L, Jordan B R and Strid Å 1994 The effects of ultraviolet-B radiation on the CF0F1-ATPase *Biochim. Biophys. Acta (BBA)-Bioenerg.* **1185** 295–302
[83] Mansfield T A and Freer-Smith P H 1984 The role of stomata in resistance mechanisms *Gaseous Air Pollutants and Plant Metabolism* (London: Butterworth-Heinemann) pp 131–46
[84] Dai Q, Coronel V P, Vergara B S, Barnes P W and Quintos A T 1992 Ultraviolet-B radiation effects on growth and physiology of four rice cultivars *Crop Sci.* **32** 1269–74
[85] Middleton E M and Teramura A H 1993 The role of flavonol glycosides and carotenoids in protecting soybean from ultraviolet-B damage *Plant Physiol.* **103** 741–52
[86] Pal M, Sengupta U K, Srivastava A C, Jain V and Meena R C 1999 Changes in growth and photosynthesis of mungbean induced by UV-B radiation *Indian J. Plant Physiol.* **4** 79–84
[87] Choudhary K K and Agrawal S B 2015 Effect of elevated ultraviolet-B on four tropical soybean cultivars: quantitative and qualitative aspects with special emphasis on gas exchange, chlorophyll fluorescence, biomass and yield *Acta Physiol. Plant.* **37** 31
[88] Tripathi D, Meena R P and Pandey-Rai S 2021 Short term UV-B radiation mediated modulation of physiological traits and withanolides production in *Withaniacoagulans* (L.) Dunal under in-vitro condition *Physiol. Mol. Biol. Plants* **27** 1823–35
[89] Long S P, Humphries S and Falkowski P G 1994 Photoinhibition of photosynthesis in nature *Annu. Rev. Plant Biol.* **45** 633–62

[90] Singh S, Agrawal M and Agrawal S B 2013 Differential sensitivity of spinach and amaranthus to enhanced UV-B at varying soil nutrient levels: association with gas exchange, UV-B-absorbing compounds and membrane damage *Photosynth. Res.* **115** 123–38
[91] Demmig-Adams B, Adams W W, Winter K, Meyer A, Schreiber U, Pereira J S, Krüger A, Czygan F C and Lange O L 1989 Photochemical efficiency of photosystem II, photon yield of O_2 evolution, photosynthetic capacity, and carotenoid composition during the midday depression of net CO_2 uptake in *Arbutus unedo* growing in Portugal *Planta* **177** 377–87
[92] Vass I, Sass L, Spetea C, Bakou A, Ghanotakis D F and Petrouleas V 1996 UV-B-induced inhibition of photosystem II electron transport studied by EPR and chlorophyll fluorescence. Impairment of donor and acceptor side components *Biochemistry* **35** 8964–73
[93] Baker N R and Rosenqvist E 2004 Applications of chlorophyll fluorescence can improve crop production strategies: an examination of future possibilities *J. Exp. Bot.* **55** 1607–21
[94] Jin Y H and Tao D L 2000 PS II photoinhibition and O_2 production *Acta. Bot. Sin* **42** 10
[95] Frank H A and Cogdell R J 1996 Carotenoids in photosynthesis *Photochem. Photobiol.* **63** 257–64
[96] Müller P, Li X P and Niyogi K K 2001 Non-photochemical quenching. A response to excess light energy *Plant Physiol.* **125** 1558–66
[97] Mozzo M, Dall'Osto L, Hienerwadel R, Bassi R and Croce R 2008 Photoprotection in the antenna complexes of photosystem II: role of individual xanthophylls in chlorophyll triplet quenching *J. Biol. Chem.* **283** 6184–92
[98] Latowski D, Kuczyńska P and Strzałka K 2011 Xanthophyll cycle–a mechanism protecting plants against oxidative stress *Redox Rep.* **16** 78–90
[99] EichhornBilodeau S, Wu B S, Rufyikiri A S, MacPherson S and Lefsrud M 2019 An update on plant photobiology and implications for cannabis production *Front. Plant Sci.* **10** 296
[100] Bhatt T and Patel K 2020 Carotenoids: potent to prevent diseases review *Nat. Prod. Bioprospecting* **10** 109–17
[101] Sullivan J and Rozema J 1999 UV-B effects on terrestrial plant growth and photosynthesis *Stratospheric Ozone Fepletion: The Effects of Enhanced UV-B Radiation on Terrestrial Ecosystems* (Leiden: Backhuys Publishers) pp 39–57
[102] Zhang X R, Chen Y H, Guo Q S, Wang W M, Liu L, Fan J, Cao L P and Li C 2017 Short-term UV-B radiation effects on morphology, physiological traits and accumulation of bioactive compounds in *Prunella vulgaris* L. *J. Plant Interact.* **12** 348–54
[103] Palma C F, Castro-Alves V, Rosenqvist E, Ottosen C O, Strid Å and Morales L O 2021 Effects of UV radiation on transcript and metabolite accumulation are dependent on monochromatic light background in cucumber *Physiol. Plant.* **173** 750–61
[104] Marwood C A and Greenberg B M 1996 Effect of supplementary UV-B radiation on chlorophyll systems during chloroplast development in *Sspirodelaoligarrhiza J. Photochem. Photobiol.* **64** 664–70
[105] Lorimer G H 1981 The carboxylation and oxygenation of ribulose 1,5-bisphosphate: the primary events in photosynthesis and photorespiration *Annu. Rev. Plant Physiol.* **32** 349–82
[106] Vu C V, Allen L H Jr and Garrard L A 1984 Effects of enhanced UV-B radiation (280–320 nm) on ribulose-1,5-bisphosphate carboxylase in pea and soybean *Environ. Exp. Bot.* **24** 131–43
[107] Strid Å, Chow W S and Anderson J M 1990 Effects of supplementary ultraviolet-B radiation on photosynthesis in *Pisumsativum Biochim. Biophys. Acta (BBA)-Bioenerg.* **1020** 260–8

[108] Barbato R, Frizzo A, Friso G, Rigoni F and Giacometti G M 1995 Degradation of the D1 protein of photosystem-II reaction centre by ultraviolet-B radiation requires the presence of functional manganese on the donor side *Eur. J. Biochem.* **227** 723–9

[109] Nogués S and Baker N R 1995 Evaluation of the role of damage to photosystem II in the inhibition of CO_2 assimilation in pea leaves on exposure to UV-B radiation *Plant Cell Environ.* **18** 781–7

[110] Onoda Y, Hikosaka K and Hirose T 2005 The balance between RuBP carboxylation and RuBP regeneration: a mechanism underlying the interspecific variation in acclimation of photosynthesis to seasonal change in temperature *Funct. Plant Biol.* **32** 903–10

[111] Allen D J, Nogués S and Baker N R 1998 Ozone depletion and increased UV-B radiation: is there a real threat to photosynthesis? *J. Exp. Bot.* **49** 1775–88

[112] Takeuchi A, Yamaguchi T, Hidema J, Strid A and Kumagai T 2002 Changes in synthesis and degradation of Rubisco and LHCII with leaf age in rice (*Oryza sativa* L.) growing under supplementary UV-B radiation *Plant Cell Environ.* **25** 695–706

[113] He J M, She X P, Meng Z N and Zhao W M 2004 Reduction of rubisco amount by UV-B radiation is related to increased H~2O~2 content in leaves of *Mung Bean* seedlings *J. Plant Physiol. Mol. Biol.* **30** 291–6

[114] Choi B Y and Roh K S 2003 UV-B radiation affects chlorophyll and activation of rubisco by rubiscoactivase in *Canavaliaensiformis* L. leaves *J. Plant Biol.* **46** 117

[115] Choudhary K M, Chaudhary N and Agrawal S B 2017 Reactive oxygen species: generation, damage, and quenching in plants during stress *Reactive Oxygen Species in Plants: Boon Or Bane-Revisiting the Role of ROS* (Hoboken, NJ: Wiley) p 89

[116] Yokawa K, Kagenishi T and Baluška F 2016 UV-B induced generation of reactive oxygen species promotes formation of BFA-induced compartments in cells of Arabidopsis root apices *Front. Plant Sci.* **6** 1162

[117] Correia C M, Torres-Pereira M S and Torres-Pereira J M 1999 Growth, photosynthesis and UV-B absorbing compounds of Portuguese Barbela wheat exposed to ultraviolet-B radiation *Environ. Pollut.* **104** 383–8

[118] Rao M V, Paliyath G and Ormrod D P 1996 Ultraviolet-B-and ozone-induced biochemical changes in antioxidant enzymes of *Arabidopsis thaliana Plant Physiol.* **110** 125–36

[119] Schmitz-Hoerner R and Weissenböck G 2003 Contribution of phenolic compounds to the UV-B screening capacity of developing barley primary leaves in relation to DNA damage and repair under elevated UV-B levels *Phytochemistry* **64** 243–55

[120] Hideg É, Mano J, Ohno C and Asada K 1997 Increased levels of monodehydroascorbate radical in UV-B-irradiated broad bean leaves *Plant Cell Physiol.* **38** 684–90

[121] Jayaraj J and Punja Z K 2008 Transgenic carrot plants accumulating ketocarotenoids show tolerance to UV and oxidative stresses *Plant Physiol. Biochem.* **46** 875–83

[122] Mittler R 2002 Oxidative stress, antioxidants and stress tolerance *Trends Plant Sci.* **7** 405–10

[123] Cervilla L M, Blasco B, Rìos J J, Romero L and Ruiz J M 2007 Oxidative stress and antioxidants in tomato (*Solanumlycopersicum* L.) plants subjected to boron toxicity *Ann. Bot.* **100** 747–56

[124] Shen J, Jiang C, Yan Y, Liu B and Zu C 2017 Effect of increased UV-B radiation on carotenoid accumulation and total antioxidant capacity in tobacco (*Nicotianatabacum* L.) leaves *Genet. Mol. Res.* **16** gmr16018438

[125] Wu X, Zhang S, Li X, Zhang F, Fan Y, Liu Q, Wan X and Lin T 2021 Postharvest UV-B radiation increases enzyme activity, polysaccharide and secondary metabolites in honeysuckle (*Lonicera japonica* Thunb.) *Ind. Crops Prod.* **171** 113907
[126] Agrawal S B and Mishra S 2009 Effects of supplemental ultraviolet-B and cadmium on growth, antioxidants and yield of *Pisumsativum* L *Ecotoxicol. Environ. Safety* **72** 610–8
[127] Shen X, Dong Z and Chen Y 2015 Drought and UV-B radiation effect on photosynthesis and antioxidant parameters in soybean and maize *Acta Physiol. Plant.* **37** 25
[128] Zhao B, Wang L, Pang S, Jia Z, Wang L, Li W and Jin B 2020 UV-B promotes flavonoid synthesis in *Ginkgo biloba* leaves *Ind. Crops Prod.* **151** 112483
[129] Dias M C, Pinto D C, Freitas H, Santos C and Silva A M 2020 The antioxidant system in *Olea europaea* to enhanced UV-B radiation also depends on flavonoids and secoiridoids *Phytochemistry* **170** 112199
[130] Zhang X Z and Ervin E H 2005 Effects of methyl jasmonate and salicylic acid on UV-B tolerance associated with free radical scavenging capacity in poapratensis *Int. Turfgrass Soc. Res. J.* **10** 910–15
[131] Liu X, Chi H, Yue M, Zhang X, Li W and Jia E 2012 The regulation of exogenous jasmonic acid on UV-B stress tolerance in wheat *J. Plant Growth Regul.* **31** 436–47
[132] Gu X D, Sun M Y, Zhang L, Fu H W, Cui L, Chen R Z, Zhang D W and Tian J K 2010 UV-B induced changes in the secondary metabolites of *Morus alba* L. leaves *Molecules* **15** 2980–93
[133] Takshak S and Agrawal S B 2019 Defense potential of secondary metabolites in medicinal plants under UV-B stress *J. Photochem. Photobiol.,* B **193** 51–88
[134] Yavaş İ, Ünay A, Ali S and Abbas Z 2020 UV-B radiations and secondary metabolites *Turkish J. Agri.-Food Sci. Technol.* **8** 147–57
[135] Cen Y-P and Bornman J F 1993 The effect of exposure to enhanced UV-B radiation on the penetration of monochromatic and polychromatic UV-B radiation in leaves of *Brassica napus Physiol. Plant.* **87** 249–55
[136] Liu L, Dennis C, Gitz I I I and McClure J W 1995 Effects of UV-B on flavonoids, ferulic acid, growth and photosynthesis in barley primary leaves *Physiol. Plant.* **93** 734–8
[137] Olsson L C, Veit M, Weissenbock G and Bornman J F 1998 Differential flavonoid response to enhanced UV-B radiation in *Brassica napus Phytochemistry* **49** 1021–028
[138] Takshak S and Agrawal S B 2016 Ultraviolet-B radiation: A potent elicitor of phenylpropanoid pathway compounds *J. Sci. Res.* **60** 79–96
[139] Surjadinata B B, Jacobo-Velázquez D A and Cisneros-Zevallos L 2017 UVA, UVB and UVC light enhances the biosynthesis of phenolic antioxidants in fresh-cut carrot through a synergistic effect with wounding *Molecules* **22** 668
[140] Nascimento L B D S, Leal-Costa M V, Menezes E A, Lopes V R, Muzitano M F, Costa S S and Tavares E S 2015 Ultraviolet-B radiation effects on phenolic profile and flavonoid content of *Kalanchoepinnata J. Photochem. Photobiol. B Biol.* **148** 73–81
[141] Wang G, Deng S, Liu J, Ye C, Zhou X and Chen L 2017 Cell damage caused by ultraviolet B radiation in the desert cyanobacterium *Phormidiumtenue* and its recovery process *Ecotoxicol. Environ. Safety* **144** 315–20
[142] Dubos C, Stracke R, Grotewold E, Weisshaar B, Martin C and Lepiniec L 2010 MYB transcription factors in *Arabidopsis Trends Plant Sci.* **15** 573–81
[143] Maeda H and Dudareva N 2012 The shikimate pathway and aromatic amino acid biosynthesis in plants *Annu. Rev. Plant Biol.* **63** 73–105

[144] Yang R, Hui Q, Jiang Q, Liu S, Zhang H and Wu J *et al* 2019 Effect of manitoba-grown red-osier dogwood extracts on recovering Caco-2 cells from H_2O_2-induced oxidative damage *Antioxidants* **8** 250–64

[145] Esua O J, Chin N L, Yusof Y A and Sukor R 2019 Effects of simultaneous UV-radiation and ultrasonic energy postharvest treatment on bioactive compounds and antioxidant activity of tomatoes during storage *Food Chem.* **270** 113–22

[146] Chen Z, Ma Y, Yang R, Gu Z and Wang P 2019 Effects of exogenous Ca^{2+} on phenolic accumulation and physiological changes in germinated wheat (*Triticumaestivum* L.) under UV-B radiation *Food Chem.* **288** 368–76

[147] Wang M, Leng C, Zhu Y, Wang P, Gu Z and Yang R 2022 UV-B treatment enhances phenolic acids accumulation and antioxidant capacity of barley seedlings *LWT* **153** 112445

[148] Agati G and Tattini M 2010 Multiple functional roles of flavonoids in photoprotection *New Phytol.* **186** 786–93

[149] Qian C, Chen Z, Liu Q, Mao W, Chen Y, Tian W, Liu Y, Han J, Ouyang X and Huang X 2020 Coordinated transcriptional regulation by the UV-B photoreceptor and multiple transcription factors for plant UV-B responses *Mol. Plant* **13** 777–92

[150] Jansen M A, Hectors K, O'Brien N M, Guisez Y and Potters G 2008 Plant stress and human health: Do human consumers benefit from UV-B acclimated crops? *Plant Sci.* **175** 449–58

[151] Agati G, Azzarello E, Pollastri S and Tattini M 2012 Flavonoids as antioxidants in plants: location and functional significance *Plant Sci.* **196** 67–76

[152] Hideg É and Strid Å 2017 The effects of UV-B on the biochemistry and metabolism in plants. UV-B radiation and plant life *Mol. Biol. Ecol.* **11** 90–110

[153] Klein F R, Reis A, Kleinowski A M, Telles R T, Amarante L D, Peters J A and Braga E J 2018 UV-B radiation as an elicitor of secondary metabolite production in plants of the genus *Alternanthera Acta Bot. Brasilica.* **32** 615–23

[154] Yoon H I, Kim H Y, Kim J and Son J E 2021 Quantitative analysis of UV-B radiation interception and bioactive compound contents in kale by leaf position according to growth progress *Front. Plant Sci.* **12** 667456

[155] Nazari M and Zarinkamar F 2020 Ultraviolet-B induced changes in *Menthaaquatica* (a medicinal plant) at early and late vegetative growth stages: investigations at molecular and genetic levels *Ind. Crops Prod.* **154** 112618

[156] Stotz U H, Sawada Y, Shimada Y, Hirai M Y, Sasaki E, Krischke M, Brown P D, Saito K and Kamiya Y 2011 Role of camalexin, indoleglucosinolates, and side chain modification of glucosinolate-derived isothiocyanates in defense of *Arabidopsis* against *Sclerotiniasclerotiorum Plant J.* **67** 81–93

[157] Buxdorf K, Yaffe H, Barda O and Levy M 2013 The effects of glucosinolates and their breakdown products on necrotrophic fungi *PLoS One* **8** e70771

[158] Mewis I, Schreiner M, Nguyen C N, Krumbein A, Ulrichs C, Lohse M and Zrenner R 2012 UV-B irradiation changes specifically the secondary metabolite profile in broccoli sprouts: induced signaling overlaps with defense response to biotic stressors *Plant Cell Physiol.* **153** 1546–60

[159] Topcu Y, Dogan A, Sahin-Nadeem H, Polat E, Kasimoglu Z and Erkan M 2018 Morphological and biochemical responses of broccoli florets to supplemental ultraviolet-B illumination *Agri., Ecosyst. Environ.* **259** 1–0

[160] Moreira-Rodríguez M, Nair V, Benavides J, Cisneros-Zevallos L and Jacobo-Velázquez D A 2017 UVA, UVB light, and methyl jasmonate, alone or combined, redirect the biosynthesis of glucosinolates, phenolics, carotenoids, and chlorophylls in broccoli sprouts *Int. J. Mol. Sci.* **18** 2330

[161] Karpets Y V, Kolupaev Y E, Lugovaya A A and Oboznyi A I 2014 Effect of jasmonic acid on the pro-/antioxidant system of wheat coleoptiles as related to hyperthermia tolerance *Russ J. Plant Physiol.* **61** 339–46

[162] Wang S Y, Bowman L and Ding M 2008 Methyl jasmonate enhances antioxidant activity and flavonoid content in blackberries (*Rubus sp.*) and promotes antiproliferation of human cancer cells *Food Chem.* **107** 1261–269

[163] Wasternack C 2014 Action of jasmonates in plant stress responses and development—applied aspects *Biotechnol. Adv.* **32** 31–9

[164] Acharya B R and Assmann S M 2009 Hormone interactions in stomatal function *Plant Mol. Biol.* **69** 451–62

[165] Indrajith A and Ravindran K C 2009 Antioxidant potential of Indian medicinal plant *Phyllanthusamarus* L. under supplementary UV-B radiation *Recent Res. Sci. Technol.* **1**

[166] Shan X, Zhang Y, Peng W, Wang Z and Xie D 2009 Molecular mechanism for jasmonate-induction of anthocyanin accumulation in Arabidopsis *J. Exp. Bot.* **60** 3849–60

[167] Popova L, Ananieva E, Hristova V, Christov K, Georgieva K, Alexieva V and Stoinova Z H 2003 Salicylic acid-and methyl jasmonate-induced protection on photosynthesis to paraquat oxidative stress *Bulg. J. Plant Physiol.* **133** 152

[168] Fedina I S and Benderliev K M 2000 Response of *Scenedesmus incrassatulus* to salt stress as affected by methyl jasmonate *Biol. Plant.* **43** 625–7

[169] Vlot A C *et al* 2008 Identification of likely orthologs of tobacco salicylic acid-binding protein 2 and their role in systemic acquired resistance in *Arabidopsis thaliana Plant J.* **56** 445–56

[170] Lefevere H, Bauters L and Gheysen G 2020 Salicylic acid biosynthesis in plants *Front. Plant Sci.* **11** 338

[171] Duan L, Liu H, Li X, Xiao J and Wang S 2014 Multiple phytohormones and phytoalexins are involved in disease resistance to *Magnaportheoryzae* invaded from roots in rice *Physiol. Plant.* **152** 486–500

[172] Seguel A *et al* 2018 Prohibitin3 forms complexes with ISOCHORISMATE SYNTHASE 1 to regulate stress-induced salicylic acid biosynthesis in *Arabidopsis Plant Physiol.* **176** 2515–31

[173] Horváth E, PálM, Szalai G, Páldi E and Janda T 2007 Exogenous 4-hydroxybenzoic acid and salicylic acid modulate the effect of short-term drought and freezing stress on wheat plants *Biol. Plant.* **51** 480–7

[174] Durner J and Klessig D F 1995 Inhibition of ascorbate peroxidase by salicylic acid and 2,6-dichloroisonicotinic acid, two inducers of plant defense responses *Proc. Natl. Acad Sci USA* **92** 11312–1316

[175] Rao M V, Paliyath G, Ormrod D P, Murr D P and Watkins C B 1997 Influence of salicylic acid on H_2O_2 production, oxidative stress, and H_2O_2-metabolizing enzymes *Plant Physiol.* **115** 137–49

[176] Vicente M R and Plasencia J 2011 Salicylic acid beyond defence: its role in plant growth and development *J. Exp. Bot.* **62** 3321–338

[177] Conner J K and Neumeier R 2002 The effects of ultraviolet-B radiation and intraspecific competition on growth, pollination success, and lifetime female fitness in *Phaceliacampanularia* and *P. purshii* (Hydrophyllaceae) *Am. J. Bot.* **89** 103–10

[178] Singh S K, Surabhi G K, Gao W and Reddy K R 2008 Assessing genotypic variability of cowpea (*Vignaunguiculata* (L.)Walp.) to current and projected ultraviolet-B radiation *J. Photochem. Photobiol.* B **93** 71–81

[179] Tripathi R, Rai K, Singh S, Agrawal M and Agrawal S B 2019 Role of supplemental UV-B in changing the level of ozone toxicity in two cultivars of sunflower: growth, seed yield and oil quality *Ecotoxicology* **28** 277–93

[180] Stadler L J and Uber F M 1942 Genetic effects of ultraviolet radiation in maize. IV. Comparison of monochromatic radiations *Genetics* **27** 84

[181] Demchik S M and Day T A 1996 Effect of enhanced UV-B radiation on pollen quantity, quality, and seed yield in *Brassica rapa* (*Brassicaceae*) *Am. J. Bot.* **83** 573–9

[182] Feder W A and Shrier R 1990 Combination of UV-B and ozone reduces pollen tube growth more than either stress alone *Environ. Exp. Bot.* **30** 451–4

[183] Torabinejad J, Caldwell M M, Flint S D and Durham S 1998 Susceptibility of pollen to UV-B radiation: an assay of 34 taxa *Am. J. Bot.* **85** 360–9

[184] Koti S, Reddy K R, Reddy V R, Kakani V G and Zhao D 2005 Interactive effects of carbon dioxide, temperature, and ultraviolet-B radiation on soybean (*Glycine max* L.) flower and pollen morphology, pollen production, germination, and tube lengths *J. Exp. Bot.* **56** 725–36

[185] Wang S, Xie B, Yin L, Duan L, Li Z, EgrinyaEneji A, Tsuji W and Tsunekawa A 2010 Increased UV-B radiation affects the viability, reactive oxygen species accumulation and antioxidant enzyme activities in maize (*Zea mays* L.) pollen *Photochem. Photobiol.* **86** 110–6

[186] Rathore D, Agrawal S B and Singh A 2003 Influence of supplemental UV-B radiation and minerals on biomass, pigments and yield of two cultivars of wheat (*Triticumaestivum* L.) *Biotronics* **32** 1–5

[187] Tripathi R and Agrawal S B 2013 Interactive effect of supplemental ultraviolet B and elevated ozone on seed yield and oil quality of two cultivars of linseed (*Linum usitatissimum* L.) carried out in open top chambers *J. Sci. Food Agric.* **93** 1016–25

[188] Yin L N and Wang S W 2012 Modulated increased UV-B radiation affects crop growth and grain yield and quality of maize in the field *Photosynthetica* **50** 595–601

[189] Hidema J, Zhang W, Yamamoto M, Sato T and Kumagai T 2005 Changes in grain size and grain storage protein of rice (*Oryza sativa* L.) in response to elevated UV-B radiation under outdoor conditions *J. Radiat. Res.* **46** 143–9

[190] Teramura A H and Sullivan J H 1987 Soybean growth responses to enhanced levels of ultraviolet-B radiation under greenhouse conditions *Am. J. Bot.* **74** 975–9

[191] Bouchereau A, Clossais-Besnard N, Bensaoud A, Leport L and Renard A M 1996 Water stress effects on rapeseed quality *Eur. J. Agron.* **5** 19–30

[192] O'neill C M, Morgan C, Hattori C, Brennan M, Rosas U, Tschoep H, Deng P X, Baker D, Wells R and Bancroft I 2012 Towards the genetic architecture of seed lipid biosynthesis and accumulation in *Arabidopsis thaliana Heredity* **108** 115–23

[193] Omidi H, Tahmasebi Z, Badi H A, Torabi H and Miransari M 2010 Fatty acid composition of canola (*Brassica napus* L.), as affected by agronomical, genotypic and environmental parameters *C. R. Biol.* **333** 248–54

[194] Fox N J and Stachowiak G W 2007 Vegetable oil-based lubricants—a review of oxidation *Tribol. Int.* **40** 1035–46

IOP Publishing

Molecular Response and Genetic Engineering for Stress in Plants, Volume 1
Abiotic stress
Pawan Shukla, Anirudh Kumar, Rakesh Kumar and Manish K Pandey

Chapter 10

Physiological and molecular mechanisms of submergence and waterlogging stress tolerance in crops

Jinu Jacob[1*], K Bini[2], K V Sumiya[3]

[1]*ICAR-Indian Institute of Millets Research (ICAR-IIMR), Hyderabad 500030, India*
[2]*Regional Agricultural Research Station (RARS), Kerala Agricultural University, Kavanattinkara, Kumarakam 686563, Kerala, India*
[3]*ICAR-Krishi Vigyan Kendra (ICAR-KVK), Palakkad 679306, Kerala, India*

Global warming has raised notable concerns over the efficiency of crop production worldwide. Various biotic and abiotic stress factors have been increasing in frequency and severity; an important one among them is submergence and waterlogging stress. Changing climatic patterns have exacerbated the occurrence of floods even in places which had never previously experienced this stress before. Flooding drastically reduces the availability of molecular oxygen, which is vital for plant growth and survival. Still, some species have developed strategies to overcome the adverse effects of flooding through unique morphological, anatomical, physiological, and molecular adaptations. Modern molecular biology tools have been effective in understanding some of the mechanisms of anoxia and flooding tolerance at the molecular level. This chapter attempts to summarize the changes brought about by flooding in soil structure and soil chemistry, plants' adaptations to submergence and waterlogging stress at the morphological and biochemical levels, and the molecular strategies of plants to overcome flooding stress.

10.1 Introduction

Climate change is the major challenge facing agriculture around the world. Extreme events pose a serious threat to food security and global economy. Out of these,

* Author to whom any correspondence should be addressed Email: jinu@millets.res.in.

doi:10.1088/978-0-7503-4921-5ch10

floods are the second gravest disaster for agriculture sector after drought, which is responsible for about of 19% of the total loss [1]. Data from the government of India shows that the country suffered tremendous crop loss over 18.176 million hectares of land due to floods during 2017–19, which was about 9% of the total crop area, of which 10.68 mha was affected in 2019 alone. Flood intensity has been increasing in the recent past, and new areas which were not flood-prone previously were found to be affected. In addition to crop loss, floods negatively impact various socioeconomic aspects like livestock and seed stocks, and cause significant damage to infrastructure and machineries, leading to food insecurity, epidemics, and famine. The final implications are that vulnerable sections of society are left in poverty and susceptible to forced migrations [2].

Water plays an important role in plant–environment interaction, and the amount and intensity of precipitation influences the distribution of plant species in an area [3]. Climate-change-induced increases in the frequency and duration of rainfall has caused serious damage to plant growth and development, leading to the death of plants if it persists for a few days. Based on the water level and extent of inundation, floods are either classified as waterlogging and/or partial/complete submergence [4]. Waterlogging results in full saturation of the soil solution, resulting in submergence of roots while the above-ground organs are exposed in air, while submergence results in complete inundation of the plant's parts. Wetland crops like rice commonly undergo submergence while dryland crops face waterlogging situations.

The flooding tolerance of crops is determined by many morphological and physiological traits that influenced by the environment [5]. The environmental factors which influence flooding are characteristics like temperature, light intensity, concentrations of gases such as carbon dioxide and oxygen, turbidity, and water depth; the extent of harm may depend on these factors. Important characteristics which contribute to survival of the plant after submergence, like chlorophyll retention, photosynthesis under water, carbohydrate accumulation, etc, are also influenced by the quality of the flood water. Soil changes such as hypoxia and anoxia, and the morphological, biochemical, molecular, and physiological aspects of submergence tolerance have been extensively investigated, mainly on model crops like *Arabidopsis* and rice, with recent advances made in other crops also. Much progress has been made in the identification, characterization, and functional analysis of many genes contributing to the traits for adaptation to water stress, and successful attempts have been made to put these to practical use in crop improvement.

Globally, waterlogging or submergence looms large as a major threat for crop production in areas that are low lying or are rain fed. The situation is grave in the wake of intensified global warming effects as well as the changing climate patterns resulting from it. Plants, which are essentally aerobic organisms, become deficient in adequate oxygen supply under waterlogging or submergence. They are forced to derive energy from anaerobic metabolic pathways that are highly nonefficient. Many morphological and anatomical changes are exhibited by plants suffering under waterlogging stress. The external changes are manifestations of the physiological and molecular changes that are happening at the cellular level. Submergence being a recurring event in rain-fed, low-lying areas, knowing the key molecular mechanisms underlying submergence tolerance and developing crop varieties that are tolerant to this stress is essential to meet the food demands of the increasing human population.

10.2 Changes in the physical and chemical properties of soil under waterlogged conditions

Excess water in agricultural fields occurs from intensive or extensive rainfall over a period or overflowing from a nearby body of water. This may result technically in two types of stress situations: flooding and waterlogging. They can be differentiated as flooding is superficial and covers the root portion only, while submergence refers to complete covering of the aerial tissues [6]. When soil is saturated with water for a long time, it provides the soil a certain distinctive gley horizons that are the result of oxidation-reduction processes. The outcome is a partially oxidized A horizon that is rich in organic matter, a mottled zone where oxidation and reduction happen alternately, and a bluish-green zone that is permanently reduced [7, 8]. Saturation with water is influenced by many factors such as the height of the water table, permeability of the soil material, or an impervious layer, etc, and it occur in almost any climatic zone [9]. The characteristics of submerged land vary, from dryland under temporary waterlogging to permanent submerged marshes [9], and the changes in soil chemistry vary accordingly.

10.2.1 Anoxia in waterlogged soils

Oxygen is the first soil component to be reduced after submergence. The gaseous exchange between soil and air is drastically curtailed during submergence and the rate of diffusion is also reduced suddenly by many folds when soil is saturated [10–12]. The optimal soil composition for plant growth is considered to be 50% solid and 50% pore space, and for 25% to be occupied by air. During submergence, the soil pores are saturated with water resulting in a variety of electrochemical changes [13–15].

10.2.2 Change in electrochemical properties of the soil in lowlands

Flooding results in a variety of electrochemical changes including a decrease in redox potential (Eh), an increase in pH of acid soils and a decrease in pH of alkaline soils, changes in specific conductance and ionic strength, cation and anion exchange reactions, sorption and desorption of ions, etc [16]. There is a drastic transformation of the chemical properties of submerged soils, characterized by the presence of reduced Fe^{2+}, Mn^{2+}, NH_4^+, and S^{2-} from their counterparts, decomposition of organic compounds resulting in a group of unstable organic substances, and subsequently the evolution of carbon dioxide and methane.

10.2.3 Changes in pH of the soil

Upon submergence of an aerobic soil, the change and rate of change in the pH of the soil depends on the soil's properties, organic matter content, temperature, etc. In acid soils, there is an increase in the pH; reduction for sodic and calcareous soils converges to almost 7 except for those acid soils lower in iron [17]. Soils rich in organic matter attain this value faster than acidic soils lower in such content, while acid sulfate soils with low iron content may not gain a pH above 5 even after being submerged for many months [18].

10.2.4 Changes in redox potential (Eh) of the soil

Depending on the kind and amount of organic matter, the content of electron acceptors, temperature, duration of submergence, and characteristics of the soil, it undergoes a series of changes under submergence such as an initial reduction, then increases to a maximum, and again a reduction in value [19, 20]. This is because soon after oxygen depletion, reducing substances are released before oxides of Mn^{4+} and Fe^{3+} hydrates can mobilize their buffer capacity.

10.2.5 Changes in specific conductance

This property of soil, which is also influenced by soil characteristics, increases after submergence usually during the first 4 weeks of flooding, reaches a maximum, and declines to a stable value. The initial increase is due to the release of Fe^{2+} and Mn^{2+} from the insoluble Fe^{3+} and Mn^{4+} oxide hydrates, then accumulation of NH_4^+, HCO_3^-, and RCOO– (in calcareous soils) and the dissolution of $CaCO_3$, by CO_2 and organic acids, and displacement of cations from soil colloids.

10.2.6 Changes in mineral composition

The most significant chemical change that is caused by flooding or submergence is the reduction of the soil. The organic nitrogen is converted to ammonia [21] by amino acid deamination, purine degradation, and hydrolysis of urea in submerged soils. Other important changes are the reduction of iron [22], manganese [23, 24], sulfur, the availability of water-soluble phosphorus, etc. The mobility and availability of boron, cobalt, copper, molybdenum, and zinc present in soils are affected by soil submergence. The reduction of elements takes place in the following sequence as O_2, NO_3^-, Mn (IV), Fe (III), SO_4^{2-}, CO_2, N_2, and H^+.

10.2.7 Organic matter decomposition

Apart from rapid degradation of organic matter in aerobic conditions owing to the high energy of aerobic respiration, the action of low-energy-release anerobic respiration in submerged conditions makes the decomposition slow. Methane is the typical product of the anaerobic decomposition of organic matter along with CO, hydrogen, ammonia, amines, mercaptans, hydrogen sulfide, and partially humified residues; while CO, nitrate, sulfate, and humus are produced in aerobic respiration. Carbohydrate metabolism also differs in anerobic soil conditions after the formation of pyruvic acid due to anaerobic conditions [25].

10.3 Morphological and anatomical responses to flooding/waterlogging

Plants under flooded conditions exhibit various kinds of anatomical and morphological response. The most common anatomical change seen in flooded plants is the development of aerenchyma in tissues [26]. Aerenchyma help in increased transport of oxygen and other gases from shoots to roots. The major morphological responses noticed in flooded plants are the formation of adventitious roots, increase in the proportion of plant biomass above water level, and formation of barriers to radial oxygen losses [27]. These modifications facilitate better oxygenation of submerged tissue [28, 29].

10.3.1 Generation of aerenchyma

Aerenchyma formation is the most prominent nonreversible change occurring during waterlogging/flooding [26]. The aerenchyma lacunae found in aerenchyma tissues provide a continuous channel for increased transport of oxygen from aerial shoots to submerged roots and other tissues [30]. Thick casparian strips and aerenchyma are formed in the roots of wheat, barley, and rice [31–33].

Two types of aerenchyma tissue are found in the roots of waterlogging-tolerant plants: primary aerenchyma, present in primary tissues, and secondary aerenchyma, formed in secondary tissues [34, 35]. The arrangement of aerenchyma cells in the root cortex during flooding varies from species to species. Four basic structural root types are found based on the spatial arrangement of aerenchyma tissues and packing of cells in the cortex: graminaceous, cyperaceous, *Apium*-type, and *Rumex*-type. In graminaceous type, the root cortex resembles a bicycle wheel, whereas it resembles a spiderweb in cyperaceous type, a honeycomb-like cortex is observed in *Rumex*, and irregular aerenchyma lacunae are observed in *Apium*.

The aerenchyma tissues are generated in three different ways [26], e.g. lysigeny, schizogeny, and expansigeny. The most common one is lysigeny, which involves the death and collapse of the cells in the cortex zone. Within lysigeny, two different patterns of aerenchyma lacunae formation are identified, radial lysigeny and tangential lysigeny. In radial lysigeny, found in graminaceous type, the aerenchyma lacunae are formed by the breaking down of cells that are aligned radially in the cortex; these cells are separated by the remnant cell walls or radial rows of cells and thus resemble a bicycle wheel. In tangential lysigeny, cell separation and collapse occur in tangential sectors of the root cortex, with intact files of cells in between. The spatial arrangement of aerenchyma lacunae in this type resembles a spiderweb and is found in cyperaceous plants. Sometimes, the aerenchyma tissues formed by cell lysis (lysigeny) does not have a regular identifiable pattern, as is shown in the nonorganized structure of *Apium*-type.

If aerenchyma cells are generated by the expansion of intercellular spaces along the radial sectors, it is known as shizogeny. The process is often followed by lysigeny. In a few cases, shizogeny is the only process generating aerenchyma. In the expansigeny type of origin, aerenchyma lacunae are formed by cell division and cell enlargement and not by cell death (lysigeny) and cell separation (schizogeny). Expansigeny is characterized by a honeycomb structure in the root cortex, as found in the *Rumex*-type.

Formation of secondary aerenchyma is also noticed in certain species, mainly woody plants [36]. Sou *et al* [37] studied the functions of primary and secondary aerenchyma in *Syzygium kunstri* and found that both are involved in longitudinal movement of oxygen from the root base to root tip. Aerenchyma provide an aeration channel from shoots to roots for the transport of oxygen as well as other gases such as carbon dioxide, methane, nitrogen, and ethylene. This helps the plant to grow normally even in waterlogged soil [33]. In many lowland and aquatic plants, aerenchyma cells are constantly present and help to survive submerged, emerged, and floating conditions, whereas in many other species they are inducible. Cell death and lysis preceding aerenchyma lacunae generation occurs due to low oxygen pressure (hypoxia) and ethylene accumulation [38, 39], but also by soil impedance [40] and phosphorus deficiency [41].

Under partial submergence, at least a segment of the shoots are above water level. The atmospheric oxygen captured by the leaves above the water level must be transported to the roots to prevent root anoxia. Oxygenation in the root apex is an essential factor for elongation and soil exploration by the roots under flooded conditions. The oxygen transport from shoots to roots through aerenchyma is mostly by diffusion whereas the long-distance transport of oxygen is by convective mechanisms [28, 42]. The respiration along the roots and radial loss of oxygen toward the rhizosphere (referred to as ROL), generates a reduced oxygen level in the root apex, and consequently a longitudinal gradient is established toward the root apex. The gradient that is generated helps in diffusion of oxygen to the roots.

10.3.2 Barriers to radial oxygen loss to the rhizosphere

When the respiratory demands of the root tissue are fulfilled, the transport of oxygen depends on the loss of oxygen toward the rhizosphere [28]. When ROL is higher, the amount of oxygen reaching the root apex is lower. ROL depends upon the presence of barriers preventing oxygen leakage to soil. Some species possess a high proportion of aerenchyma tissues but no barriers against ROL and hence the apex is poorly oxygenated, restricting root elongation. Some species have developed the capacity to induce ROLs in the outer cortex layers of the roots, which prevents oxygen loss from the roots, inflating the longitudinal diffusion of oxygen and increasing aeration of the root apex and phytotoxin invasion [43]. In rice and maize, the ROL is highest in the root tip and lowest in the root base due to a barrier present at the base that suppresses ROL [44–47]. This promotes root elongation and soil exploration under flooded conditions.

The ROL barrier is constitutive in *Juncus effuses* [48] and *Echinocloa* [49], whereas it is inducible under stagnant conditions in *Caltha palustris* and in some rice cultivars [29]. These barriers are mostly located in the basal region of the root, whereas there are no barriers at the apex [29, 48]. Thus, an aerobic zone is always created around the root apex due to the release of oxygen through the root tips. This aerobic region wards off the accumulation of potentially toxic compounds such as reduced ions of iron (Fe^{2+}) and manganese (Mn^{2+}) in the sensitive root zone. Suberin and lignin are reported to be involved in barrier formation in most species [29, 46, 49, 50].

10.3.3 Formation of adventitious roots

The most conspicuous morphological change that occurs in flooded conditions is the formation of adventitious roots (ARs). ARs replace the older roots and, due to their high porosity, help in better water and nutrient uptake in flooded conditions [51]. These ARs originate from the base of the stem or the waterlogged part of the hypostyle [32]. ARs are connected to the stem through aerenchyma [52]. AR formation has been observed under waterlogged conditions in maize [53], sesame [54], bittersweet [55], etc.

There are three mechanisms for the generation of ARs: (i) by stimulation of the pre-existing root primordia at the shoot base; (ii) by initiation of new root primordia and then subsequent outgrowth, inducing a new root system; and (iii) by reorientation of the root extension and root placement at the soil surface, as seen in woody plants and herbaceous plants. The first two mechanisms are triggered by ethylene, which increases the sensitivity of plant tissues to auxin [56]. Ethylene is the major inducer of responses to waterlogging in tomato [57]. In *Rumex palustris* exogenous application of ethylene was found to revive the production of ARs. In tobacco, the application of naphthaleneacetic acid (NAA) which is an inhibitor of the transport of auxin, reduced A formation in the wild-type [58]. Hence, auxin transport through tissues seems to be essential for AR formation in tobacco. Nitric oxide (NO) induces AR formation in many species [59].

The adaptations to withstand waterlogging observed in tolerant legume species are via generation of aerenchymatous tissue and a barrier to radial oxygen loss in the root cortex and nodules. An increase in the permeability of the oxygen diffusion barrier was also observed in tolerant species [60]. Development of additional nodules like ARs is also observed in a tolerant species, *Melilotus siculus* [61].

10.3.4 Changes in morphology of shoots

Flooding stress on roots also changes shoot morphology. The plants of tolerant species under flooding are mostly taller than the plants under normal conditions. This is achieved by elongation of aerial organs and increase in insertion angles [62, 63]. The most immediate response to flooding is an increase in the petiole insertion angle, called hyponastic growth. The maximum insertion angle (70–80°), i.e. almost a vertical leaf position, is attained within 4 hr after flooding [62]. After the increase in insertion angle, there is an increase in petiole length to maximize the leaf area above water level [63]. Petiole elongation resulting from cell wall loosening is achieved through an increase in the expression of expansins [64]. As in the case of AR formation, an increase in petiole insertion angle and length are also triggered by ethylene application [64].

Similar responses are observed in graminaceous species also. In a grass species, *Paspalaum dilatatatum*, the first morphological response after waterlogging is an enhanced tiller insertion angle [65] and elongation of leaf sheaths, as well as sometimes elongation of leaf blades [66, 67]. The increase in leaf sheath length is attained by an increase in the number of long parenchymatous cells [65]. Stem elongation is induced in rice under waterlogged conditions [68].

Another flooding-induced change noticed in shoots of tolerant species is stem hypertrophy. This is characterized by the formation of white spongy tissues with a lot of intercellular gas spaces within them [69]. They act as secondary aerenchyma, with the increased air space allowing increased exchange of gases between water and plant tissues. Stem hypertrophy is displayed by species like *Lythrum salicaria* [70], *Lotus tenuis* [71], *Glycine max* [72], and *M. siculus* [73].

Increase in nitrogen concentration in plant leaves is a strategy adopted by willow plants to withstand hypoxic or anoxic conditions [74]. Sorghum plants under water-logging maintain their metabolic balance by increasing aerial root formation [75].

An important morphological trait displayed by wood species is hypertrophied lenticels, found at the base of the stem [76]. These special structures allow oxygen entry to shallow roots through aerenchyma and intercellular spaces [72].

Reduced root biomass, increased stem diameter, and delayed flowering have been observed in various genotypes of soybean under waterlogging [77]. Prolonged waterlogging during vegetative phase was shown to reduce grain yield in wheat [78] and to reduce root length and biomass in rice and wheat [79].

10.4 Biochemical adaptations and stress signaling during submergence

Submergence or waterlogging becomes inevitable for many plant species due to their geographical locations, climate change, and various other factors. To tackle complete or partial submergence, plants undergo several metabolic, physiological, and biochemical modifications at the cellular level, which are reflected as morphological, anatomical, and behavioral changes. There are changes in photosynthetic rate, energy catabolism, phytohormone synthesis and regulation, reactive oxygen species production, and signaling at the plant level.

10.4.1 Submergence adaptation strategies: LOQS and LOES

Plants on exposure to flooding choose one of two modes of responses, escape (i.e. low oxygen ecape strategy; LOES) and quiescence (i.e. low oxygen quiescence strategy; LCQS) [80, 81]. LCQS is displayed by plants that can endure long-term underwater submergence such as lowland rice varieties. Plants that adopt this strategy have restricted growth to preserve their energy and have reduced carbon consumption. Slow growth and saving of metabolic resources are features of the same. On the contrary, plants that choose the escape strategy do so mostly in the case of partial submergence. Such plants try to enhanced the growth of their aerial parts, in an attempt to keep their upper leaves above water for facilitating better gas exchange and light absorption. They show enhanced formation of aerenchyma and ARs to achieve the same. Wetland rice, *Rumex*, *Nymphoides peltata,* etc, fall under this category. Rice shows internodal elongation underwater, while in *R. palustris* it is the petiole that elongates. Such plants display reduced leaf thickness and reposition chlorophyll to maximize light absorption. Both strategies differ in the type of phytohormones that are regulated and in the biochemical adjustments involved. The strategy adopted by the plant determines how far the plant tolerates hypoxia. Both mechanisms involve the roles of the triad phytohormones, ethylene, gibberellin (GA), and abscisic acid (ABA).

10.4.2 Energy metabolism under hypoxia

Waterlogged soils experience anoxia or hypoxia, which are conditions of no oxygen or reduced oxygen, respectively. The major impact of hypoxia in plants are adverse effects on the processes that involve oxygen as a major substrate, such as aerobic energy metabolism. Plants in general show reduced stomatal conductance, lesser photosynthetic rate, and early senescence and yield reduction because of waterlogging [82–84]. Prolonged waterlogging affects photosynthesis negatively and the excess excitation energy that is accumulated induces oxidative damage to the photosynthetic apparatus. Waterlogging impairs both water and nutrient absorption by the plant, leading to wilting. Under such conditions, plants also modify their metabolism from aerobic to anaerobic mode.

One of the most characteristic events during flooding is the reduction in concentration of the gases oxygen, CO_2 and ethylene as the rates of diffusion of these gases in water in comparison to those in air are drastically lowered [80, 85]. Oxygen is relatively less soluble in water and impaired gas exchange is a general problem during submergence. Shortage of oxygen causes reduced respiration, which leads to lack of energy (energy deficit). Semi-aquatic species like rice are not severely affected by gas nonavailability during flooding as they have developed gas conduits (i.e. aerenchyma) between submerged and aerial organs [86]. In their absence, plants fail to receive oxygen, and hence the oxygen concentration within the cells is reduced leading to hypoxia. Plants lack an active oxygen transport system and oxygen is mainly distributed through diffusion or convection, which are passive transport systems, and steep oxygen gradients are often found within plant systems [87]. Molecular oxygen makes up an important reagent in several biochemical reactions. Plants, being sessile organisms, experience oxygen deficiency as a major crisis as it is the terminal acceptor of the electron acceptor of the electron transport chain in the mitochondria. A lack of oxygen hence correlates directly to lower energy production. It is also a substrate for various biochemical reactions.

Hypoxia tolerance in many crops is through root development [88]. The previous study indicated that root responses to hypoxia may be an indication of hypoxia tolerance. ARs and aerenchyma are developed in roots, which in turn facilitate exchange of oxygen under flooding [89]. When there is a severe deficiency of oxygen, mitochondrial respiration is negatively affected leading to a much-reduced production of the energy currency of the cell, which is adenosine triphosphate (ATP). Some plants succeed in overcoming the deficiency of ATP by adopting substrate-level ATP production, which is via catabolism of sugars and starch through glycolysis. Lack of oxygen prevents oxidative phosphorylation of energy production, where, as aforementioned, oxygen is the final electron acceptor. In such situations, carbohydrates are broken down via anaerobic methods of fermentative pathways to produce at least a few ATP molecules. In comparison to the net production of 36 ATPs per glucose molecule that is produced during oxidative phosphorylation, the net production through glycolysis is just 2 ATPs per glucose molecule. Therefore, to meet the energy requirement of the cells, glycolysis need to be upregulated multiple fold [87]. But this scenario raises the risk of nicotinamide adenine dinucleotide

(NAD+) depletion, which is an essential substrate for glycolysis. Generally, plants overcome this predicament by enhancing the ethanol and lactate fermentation pathways, both of which uses NADH as one of their substrates and releases NAD^+ [90–93]. The resultant product, pyruvate, needs to be removed as ethanol for NADH to be recycled into NAD+ so that the cycle continues. Ethanol and CO_2 will be leaked into the atmosphere, thus reducing the toxicity. Pyruvate can also be metabolized into alanine and lactate but conversion to ethanol is the major anaerobic respiratory pathway adopted by plants. The cell opts for survival making use of these hypoxic ATP production methods and compromises with the lower ATP output in these methods compared to oxidative phosphorylation.

There are three main fermentation pathways that are displayed by plants under flooding, namely ethanol fermentation, lactate fermentation, and a pathway that generates alanine from glutamate and pyruvate through the alanine aminotransferase (AAT) enzyme. The last one is a plant-specific pathway. In plants the enzymes of these fermentation pathways are not constitutively present and are activated during nonavailability of oxygen as an adaptation mechanism. Cytosolic pH is a deciding factor for the plants to choose between alcoholic and lactic acid fermentation. The lactic acid that is produced reduces the pH of the cytoplasm, thereby inhibiting the lactate dehydrogenase (LDH) but activating pyruvate decarboxylase and alcohol dehydrogenase (ADH) of alcoholic fermentation [93]; plants that are more flooding tolerant were shown to have active alcoholic fermentation [94]. When nitrate reductase generates NO under hypoxic conditions, for the oxidation of NO further to nitrate, NADH is used as a substrate. Plants that exhibited enhanced NO production under flooding were found to have lesser lactate accumulation [95]. ADH activity offered a proportionate increase in flooding tolerance in plants; thus, ADH is the key enzyme controlling the ethanol fermentation pathway and thereby the survival strategies of submerged plants.

Transcriptome studies in grapes revealed that waterlogging results in substantial rearrangement of the primary metabolic pathways in grapes, with enhanced accumulation of glycolytic enzymes and fermentative enzymes [96]. Glycolytic pathway genes were found to be upregulated under flooding in multiple global gene expression studies [97–99]. Most of the metabolic rearrangements that plants undergo under hypoxia are reversible, while the morphological rearrangements, such as formation of AR, stem elongation, etc, become permanent. The anaerobic pathway that the plants adopt to overcome waterlogging leads to severe growth limitations or inhibitions along with accumulation of undesirable, toxic byproducts [100].

10.4.3 Effect of submergence in photosynthesis

Waterlogging leads to closure of stomata, and thereby reduced intake of carbon dioxide. Prolonged waterlogging was found to cause degradation of chlorophyll and carotenoid pigments, leading to yellowing and senescence of leaves [101, 102]. Expression levels of Rubisco, which is the first enzyme in carbon fixation, as well as Rubisco activase were decreased under flooding stress, which ultimately reduced the rate of photosynthesis [103]. Environmental conditions such as heavy and continuous rains reduce the oxygen availability from the environment, and conditions such as flooding start to become detrimental to plant survival when combined with poor light and carbon dioxide availability as required for photosynthesis [104, 105].

Underwater photosynthesis is adversely affected by low CO_2 concentration and nonavailability of light. To add to these, submergence causes chlorophyll degradation and stomatal closure, too [106]. Ethylene, whose concentration increases during submergence, stimulates chlorophyll degradation via stimulating the chlorophyllase enzyme. ROS produced during flooding also were observed to cause damage to the photosynthetic apparatus [107]. The net photosynthetic rate of submerged terrestrial leaves is highly diminished. However, some tolerant species overcome this by developing beneficial leaf traits like enhanced leaf area, reorientation of chloroplasts, etc. Tolerant lines in rice were found to successfully maintain photosynthetic apparatus integrity along with higher chlorophyll content and better stomatal conductance [108]. The content of chlorophyll and chlorophyll fluorescence levels were lower in blackgrass (*Alopecurus myosuroides*) under flooding, and damage to the light-harvesting complex was observed in blackgrass and tomato during waterlogging [109].

10.4.4 Reactive oxygen species metabolism during flooding stress

Prolonged waterlogging significantly reduces the photosynthetic efficiency of plants, and the excess excitation energy that is generated induces production of ROS. Flooding contributes to the generation of harmful ROS species such as superoxide, hydrogen peroxide, and compounds such as malondialdehyde (MDA) [110]. Enhanced ROS production is an important phenomenon that is characteristic of hypoxia and anoxia as well as during reoxygenation following these [111]. ROS can cause irreversible damage to all sorts of biological molecules such as proteins, lipids, nucleic acids, etc. Plants maintain temporary energy production under hypoxia stress by adopting glycolysis and ethanol fermentation. When the stress extends for longer durations, there is an accumulation of toxic molecules generated from the anaerobic metabolism along with an increase in ROS [112, 113].

Lipid peroxidation is increased significantly by enhanced ROS production and the MDA that is produced as a result is often used as a marker of oxidative lipid damage, which is the product of lipoperoxidation of cell membranes [114]. Submergence caused an increase in MDA in some plant species under study, such as the roots of *Triarrhena sacchariflora* [115], Bermuda grass [116], etc. The extent of MDA indicates the ineffectivity of the antioxidant system of plant. The formation of MDA was significantly negatively correlated with ascorbate content and survival rate in rice [117]. Hydrogen peroxide and superoxide levels are significantly increased in many species under flooding. Waterlogging stress was found to enhance the concentration of H_2O_2 and superoxide radicals in many plant species [118, 119].

In order to scavenge ROS from the system, plants activate the production of antioxidant molecules such as glutathione, proline, and ascorbic acid, and antioxidant enzymes such as superoxide dismutase (SOD), catalase, guaiacol peroxidase

(GPX), glutathione reductase (GR), ascorbate peroxidase (APX), dehydroascorbate reductase (DHAR), and monodehydroascorbate reductase (MDHAR). The activities of many antioxidant enzymes such as SOD, catalase, and APX was high during waterlogging in *Sedum* genotypes [120]. Following receding of water after flooding stress, plant survival depends to a large extent on the effective antioxidant mechanisms of the species/genotype [121]. Higher levels of ascorbic acid, which is a natural antioxidant, were observed in plants under hypoxia [122].

Though ROS are toxic to most biomolecules, they act as signaling molecules, too. They mediate ethylene-induced AR formation and programmed cell death (PCD) leading to aerenchyma formation in rice [80, 123].

10.4.5 Role of phytohormones in waterlogging stress tolerance

Phytohormones play significant roles in the growth and development of plants by being part of several signaling networks and in deciding plant responses to various biotic and abiotic stress conditions. Flooding/waterlogging is also characterized by the involvement of various phytohormones which play diverse roles in helping the plant to survive through stress. Some of them are described in detail below.

10.4.5.1 Ethylene

Restricted gas diffusion of oxygen and ethylene are considered to be primary signals of plant responses to flooding. The rate of diffusion of ethylene in water is very low, forcing its accumulation in tissues during flooding. Waterlogging is found to result in significant increase in ethylene concentration in plants. As a result of the high ethylene production, there is reduced shoot and root growth and enhanced/premature senescence and abscission [124]. Enhanced and rapid accumulation of ethylene was observed in waterlogged plants by the authors of [125]. Further studies confirmed the role of ethylene in AR emergence and development. Ethylene accumulation and ethylene-induced AR formation was further found in several plant species, such as sunflower [126], tomato [57], rice [127], *Rumex* [128], etc.

There are several escape and quiescence survival strategies adopted by plants in general against flooding or waterlogging. The 'escape' strategy is mainly through the development of ARs, which originate at the nodes of the stem. They have the advantage of floating over water in a better-oxidized environment than the hypoxic roots that are submerged underneath. Ethylene plays an important role in AR root generation via triggering ROS formation, which indirectly causes PCD at the AR primordia. Hypoxia-induced accumulation of ethylene in the tissues within hours induces hydrogen peroxide produced via NADPH oxidase activity [129, 130].

Ethylene is a gaseous phytohormone and it fails to escape from plants during flooding situations. The accumulated ethylene promotes AR and aerenchyma formation in plants [131, 132]. Aerenchyma are not a constitutive tissue in many plants and most species develop them in response to flooding stress. These tissues are made up of thin-walled cells with enormous gas spaces for easy gas exchange and are formed in roots as well as shoots. Soil waterlogging has been shown to trigger enhanced aerenchyma formation in many plant species; their formation is controlled by ROS and ethylene in most plant species [133]. The formation of ARs is another strategy to survive through flooding. ARs rich in aerenchyma also help in enhanced gas, water, and nutrient exchange. In a study of deepwater rice, elongation of ARs started 8 to 10 hr after waterlogging, and ethylene facilitated production of ARs by inducing PCD of epidermal cells at the root tip [134].

Ethylene plays major regulatory roles during submergence, which is facilitated by the transcription factor family, Group VII ethylene response factor (ERFVII). During submergence, when plants shift their metabolism from aerobic to anaerobic, ethylene has been suggested to play a role by inducing genes that govern aerenchyma development and glycolysis and fermentative pathways. ERFs are ethylene-responsive transcription factors with a conserved APETALA-2 (AP2) domain and are crucial in plant tolerance to waterlogging and submergence stress.

All plant cells are capable of ethylene synthesis, and ethylene concentration within a system is basically decided by the rate of synthesis and diffusion to the external environment. The rates of ethylene biosynthetic enzymes such as ACC synthase and ACC oxidase were found to be increased under flooding in many crop species [135, 136]. Altogether, ethylene plays direct roles in the flood acclimation process in plants by interfering in ARs and aerenchyma formation, shoot elongation, and carbohydrate utilization in the form of ethanolic fermentation [80, 137]. Similarly, stalled AR formation was observed when ethylene or auxin production was inhibited, which hints toward the auxin–ethylene interaction that occurs during AR formation during flooding [57]. GA, ABA, and ethylene are the major phytohormones that are involved in deepwater rice elongation, while ethylene plays the major role in internode elongation. Ethylene, ABA, and jasmonic acid contents showed changes in response to submergence in many crops as revealed by metabolomics [138].

10.4.5.2 Abscisic acid

The main role of ABA in flooding tolerance is in guiding stomatal regulation [139]. Waterlogging significantly reduces ABA concentration in plants [140, 141] and ABA was shown to have a negative role in AR formation, unlike that of ethylene. ABA might be negatively associated with waterlogging tolerance in plants. Yet, an increase in ABA content has been noticed in the upper parts of some plant species such as cotton, wheat, etc [142, 143].

10.4.5.3 Auxin

Auxin and ethylene promote production of each other. Auxin was found to have a role in AR formation as it is the hormone that causes cell division. Auxin functions through polar transport proteins (PINs), and inhibition of PINs was found to inhibit AR production in plants [141, 144].

10.4.5.4 Gibberellin

GA is a phytohormone that plays several significant roles in the growth and development of plants, especially in deciding the size and number of cells. GA content is enhanced under waterlogging, and resistant genotypes in several crops exhibited higher GA content than in their susceptible lines [145]. Exogenous GA application in rice could enhance internode elongation, a survival mechanism used by the crop under flooding [146]. GA also plays a role in waterlogging stress resistance by being in the pathway mediated by SNORKEL genes in rice.

A schematic representation of the various biochemical stress adaptation mechanisms, as well as signaling pathways, is shown in the figure 10.1.

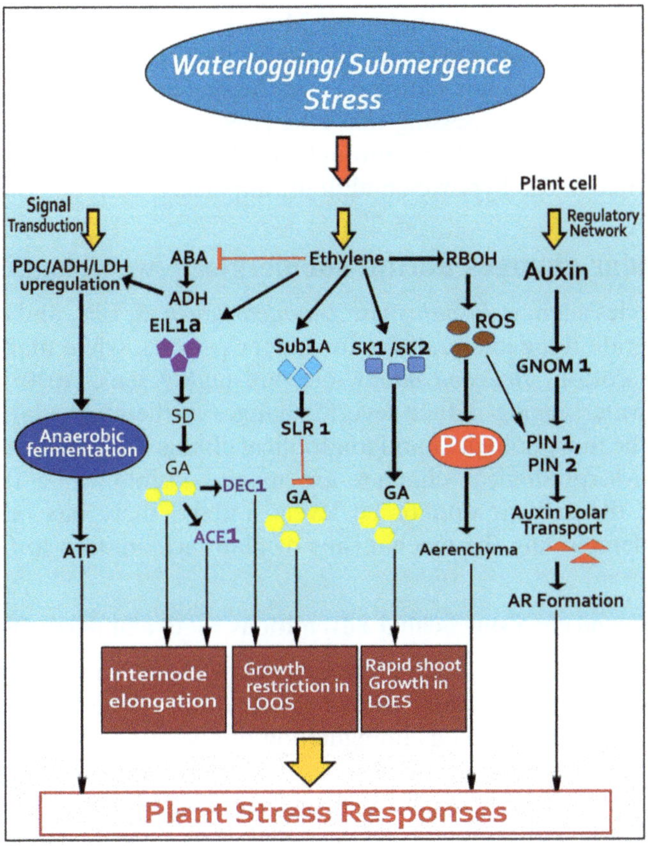

Figure 10.1. Biochemical adaptations and stress signaling during waterlogging. Flooding increases the expression of rboh genes, ethylene biosynthesis genes, such as *ACC synthase* and *ACC oxidase*, and also induces the expression of ethylene response transcription factors (ERFs) such as *SK1/2*, *Sub1A*, *EIL*, etc. The development of adventitious roots (ARs) is mediated through auxin-associated genes like *GNOM1*, *PIN1*, and *PIN2*. rboh-induced ROS production contributes to programmed cell death (PCD) leading to aerenchyma formation. Submerged plants depend upon anaerobic respiration for energy production via the upregulation of key anaerobic fermentation genes such as *pdc*, *adh*, and *ldh*. An ERF gene, *EIL1*, promotes *SK1/2* transcription and GA biosynthesis gene *SD1*, thereby increasing GA production. GA exerts antagonistic effects on internode elongation through ACE1 or DEC1. *Sub1A* negatively regulates GA production through *SLR1* and thereby inhibits shoot elongation.

10.4.6 N-end rule pathway of targeted proteolysis

The N-end rule pathway is a route of protein degradation in which the stability of the protein is decided by the nature of the amino acid at its N-terminal. Though it is commonly found in animals, its role in plants came to light quite recently in hypoxia-related gene expression [147, 148]. This pathway acts as a sensor that detects low oxygen availability and regulates some of the main hypoxia-sensitive transcription factors. ERFVII factors, which are major players in hypoxia sensing, are regulated by this pathway. Plants which do not have this pathway express hypoxia-response genes constitutively and hence prove to be more tolerant to stress. The amino terminus of the ERFVII transcription factors has a characteristic conserved amino acid motif, starting with Met-Cys. One of these factors, *HRE2*, is highly stable under hypoxia and improves stress tolerance via this enhanced stability. *SUB1A* is uncoupled from this pathway, and thereby more stable, and imparts high tolerance to rice varieties possessing this gene [149]. Another ERF factor, *RAP2.12*, degrades under aerobic conditions through the N-end pathway of proteolysis and accumulates in rice under hypoxic conditions, imparting tolerance [148].

10.5 Molecular changes during submergence/waterlogging

Some plant species such as *R. palustris*, *Zea nicaraguensis*, rice, and some *Colocasia* species can tolerate flooded situations for longer periods, while many other species such as maize, cotton, *Brassica napus*, etc, are highly sensitive to flooding [150]. Flooding and waterlogging induce several changes at the molecular level in plants, which decides the morphological and anatomical changes which lead to tolerance or susceptibility. Morphological changes are mostly visible while the biochemical changes are at the cellular and tissue levels. Lately, there has been tremendous progress in understanding the mechanisms of plant acclimation to flooding stress.

10.5.1 Molecular mechanisms behind adventitious roots and aerenchyma formation

The molecular mechanisms behind aerenchyma formation have been primarily understood from submerged maize roots. Upregulation of the SOD enzyme and cell wall loosening enzymes, as well as upregulation of several transcription factors could be observed in the root cortex from whence aerenchyma emerged. Downregulation of a metallothionein mRNA in the cortex was a characteristic feature of submergence in maize and rice along with enhanced ROS content [130, 151]. The induced production of ROS, which are used in the signaling, was due to the action of calcium-dependent plasma membrane respiratory burst oxidase homologues (RBOHs) such as *NADPH oxidase* in plants. Ethylene also contributed to ROS production through *rboh* [132]. Along with this, the genes *LSD1*, *EDS1*, and *PAD4*, which control the generation of ethylene and ROS in the upstream, are also involved in the formation of aerenchyma [152]. When wheat roots were treated with an ethylene precursor, there was enhanced expression of three *rboh* genes, which caused enhanced ROS generation and stimulated lysigenous aerenchyma formation under stagnant conditions, a strategy of wheat seedlings' adaptation to hypoxia stress [132].

Auxin and ethylene are involved in the development of ARs during waterlogging. AR formation also involves ROS production just as in the case of aerenchyma. The epidermal cells undergo PCD, which is mediated by ethylene-promoted ROS molecules from whence AR primordia emerge [153]. A downregulation of metallothionein is also noticed. The molecular mechanisms of AR formation have been well studied in plants and several genes have been identified that contribute to the same, such as auxin-associated genes like *GNOM1* in rice, *PIN1,* and *PIN2* which control polar auxin transport and hence influence AR formation [154]. *rboh*, ERF genes, and the QTL, *ARN* affects AR formation either directly or via ROS regulation [155].

10.5.2 Role of transcription factors in submergence tolerance

Transcription factors are protein molecules that can bind to genomic DNA. They have been identified in regulating the gene expressions that govern plants' response to various environmental and physiological stressors. Hypoxia caused by flooding was shown to dramatically remodel the transcriptome and translatome (ribosome-associated mRNAs) [150], which included a number of transcription factor genes. There was an increase in the expression of members from all transcription families, but the most pronounced were ones from the WRKY family and ERFs (AP2-EREBP/ERF). The aforementioned study also described a set of 49 core hypoxia-responsive genes (HRGs) by analyzing the translatome.

10.5.2.1 Ethylene response factor transcription factor family

ERF is the most prominent transcription factor with proven roles during submergence in plants. Group VII ERFs (ERFVIIs) are stabilized by the N-degron pathway [147, 148]. Out of the various ERFs that have been reported, ERFVIIs is the one that displays significant roles in flooding tolerance. ERFVIIs trigger the expression of a cascade of genes related to hypoxia and they are key players of hypoxia-regulated gene expression. One of the earliest responses to submergence in plants is the induction of transcription factors belonging to the said ERF gene family. The most significant, as well as the most studied, ones among those are *SUB1* and *SK1*/*SK2* in rice, which are discussed in detail below.

10.5.2.2 Submergence1 (Sub1)

A breakthrough in research into the submergence tolerance of crop plants was the identification of the *Sub1* locus in rice. *Sub1* is a major contributor to the submergence tolerance phenotype in rice and it was identified to be a quantitative trait locus (QTL). *Sub1* was identified for the first time from the FR13A (for flood resistant) landrace of rice, grown in Odisha. The genomic region of *Sub1* was later mapped to chromosome 9 using a molecular mapping approach. The *Sub1* gene contributes to the LOQS mechanism of flooding tolerance. It is an ERF-type transcription factor that is transcriptionally induced under flooding stress [156]. It is known that rice accessions lacking the *Sub1A* gene succumb to waterlogging in a week's time [137, 157].

Sub1 consists of three ERFs at the molecular level, *Sub1A, B,* and *C* [156, 158], out of which B and C are found in all rice genotypes while A is found only in transient flood-tolerant lines such as FR13A [159]. *Sub1A* has two alleles, *Sub1A-1*, which is found in tolerant lines, and *Sub1A-2,* which is more common in sensitive lines [156]. Both differ only at the amino acid at the 186th position, which is serine in *Sub1A-1* and proline in *Sub1A-2*. *Sub1A*'s locus spans approximately 200kb and explains the large difference of flooding tolerance exhibited by indica, which are flood-tolerant, and japonica, which are flood-intolerant rice cultivars [160].

Sub1 is an ethylene-induced gene which is upregulated under flood conditions and downregulated when the flood recedes. It imparts submergence tolerance in seedlings for over two weeks of complete submergence. This gene functions by regulating shoot elongation, minimizing carbohydrate consumption, and helping in minimum photosynthesis by retaining chlorophyll content. All these mechanisms help the plant to resume growth when water recedes [161]. These physiological changes are in accordance with the quiescence strategy of survival, wherein carbohydrate reserves are preserved, growth kept at check, and fermentation capacity enhanced. Further, *Sub1A* activates the slender-rice-like 1 (*SLRL1*) gene, which subdues gibberellic-acid-induced internodal elongation in submerged plants and thereby avoiding unnecessary energy wastage and ensuring efficient energy use [162]. Though submergence tolerance is regarded as a single-gene-controlled trait, a few minor QTLs with less significant effects have also been identified subsequently [160]. FR13A is widely recognized as a submergence-tolerant line and is used as a parent in multiple breeding programmes and QTL-association-mapping studies. The submergence tolerance trait from FR13A has been transferred to many elite lines using this QTL introgression via marker-assisted back crossing, and several mega varieties which are good yielders as well as tolerant to submergence were developed and have become widely popular among farmers [163–167].

10.5.2.3 Snorkel 1 and 2 (Sk1 and Sk2)

Ethylene plays a significant regulatory role in submergence, which is facilitated by the ERFVII transcription factor family. Deepwater rice survives flooding stress by its striking ability to elongate its internodes in the form of structures known as 'snorkels', which are hollow structures that facilitate gas exchange with the atmosphere. Though it has been known that the phytohormones auxin, ethylene, and ABA are involved in this process, the genes behind this phenomenon were identified as snorkel 1 *(SK1)* and snorkel 2 *(SK2)* in 2009 by Hattori *et al* [168]. These genes encode ethylene response factors that detect accumulated ethylene under stress and the products of *SK1* and *SK2* cause significant elongation of the internodes (a growth rate of 25 cm per day) via gibberellins. SK1 and SK2 are ERFVII transcription factors. The LOES mechanism of flooding tolerance in rice plants is mainly dictated by the SK locus. SK1 and SK2 are two members of the ERFVII family of transcription factors and were found to be involved in submergence stress tolerance and elongation of internodes during flooding in deepwater rice [168, 169].

Sub1A and *SK1/2* are opposite in their modes of action. *Sub1A* inhibits GA signal transduction [168]. In contrast, *SK1* and *2* enhance the production of GA during flooding, thereby promoting petiole growth, which helps rice to survive in flooding.

SK1 and *SK2* act in a cascade with the OsEIL1a protein in rice, which can bind to *SK1*/*2* promoters or to the semidwarf1 gene (*SD1*), which promotes GA4 synthesis, ultimately resulting in shoot elongation [170]. Rice *Sub1A* limits shoot elongation by suppressing the gibberellic acid concentration while *SK1* and *SK2* impart submergence tolerance in deepwater rice by enhancing internode elongation through enhanced GA signaling.

10.5.2.4 Related APETALA-2.2 (RAP2.2)

One of the members of ERF is known as related APETALA-2.2 (*RAP2.2*), which activates HRG promoters by binding to hypoxia-responsive promoter cis-elements (HRPEs), which are found in most HRGs [171]. WRKY transcription factors are involved in upregulation of RAP2.2 for adaptation to submergence in *Arabidopsis*; the authors of [172] indicated how the two transcription factors WRKY33 and WRKY12 interacted with each other and upregulated the RAP2.2 transcription factor in *Arabidopsis* during submergence stress, helping the plants to overcome tress. *TaERFVII.1* from wheat had induced expression under flooding, specifically in the tolerant lines alone [173]. This gene is important in that, apart from offering flooding tolerance, it ensures plant growth and yield maintenance.

10.5.2.5 Hypoxia-responsive (HRE) ethylene response factor

HRE refers to hypoxia-responsive ERF. HRE1 and HRE2 are upregulated under hypoxic conditions along with alcohol dehydrogenase (ADH1), which together enhance the anaerobic metabolism [171]. Significant upregulation of ERF factors such as HRE2 and RAP2 has been observed in many plants like *Salix* sp., rice, and *A. thaliana.*

10.5.2.6 Other transcription factors involved in flooding tolerance

WRKY is the biggest family of transcription factors in plants and has roles in various biotic and abiotic stress tolerances. Its role in hypoxia has been described in various plants such as persimmon [174], rice [175], sunflower [176], and *A. thaliana* [172]. They regulate transcription by binding to the W box in the promoters. WRKY33 of *A. thaliana* is a key factor in submergence tolerance. It was induced by hypoxia stress [177] and was found to activate the expression of *RAP2.2* by binding to its upstream W box. WRKY33 also builds up the resistance of *Arabidopsis* to oxidative stress [178]. By overexpressing the *WRKY33* gene, *Arabidopsis* transgenic plants could gain enhanced submergence tolerance; the gene acted by upregulating the *RAP2.2* gene [172].

Basic leucine zipper (bZIP) transcription factors possess regions for leucine zippers for dimerization and DNA-binding regions. Their role has been recognized in coleoptile elongation in rice during submergence [175]. They are also positive regulators of the ABA signaling pathway and ensure inhibition of seed germination mediated by ABA under anaerobic conditions [179]. bZIP72 in rice caused seed germination and coleoptile elongation in submerged conditions in rice by activating *ADH1* [180].

When a comparative analysis of tolerant to extremely tolerant rice genotypes was carried out, many transcription factors belonging to the MYB, E2F, ARR-B, bZIP, NAC, AP2/ERF, ARF, MADS-box, AS2, WRKY, ZnF, DOF, and HSF families were found. In addition, the highly tolerant lines displayed regulation of transcription factors of the families HY5 (bZIP), ABF2, ABI5, GBF3, GBF4 and GBF5 (bZIP), DPBF-3 (bZIP), bHLH, and BES/BZR [180a]. Members from many transcription factor families like WRKY, MYB, bHLH, NAC, and bZIP have been induced under waterlogging in many species [100, 181]. The premature internode elongation 1 (PINE1) transcription factor, which enhances GA production and increases internodal length, was identified in rice [182].

10.5.3 Transcriptome studies on waterlogging stress

Transcript profiling allows examination of thousands of genes simultaneously and understanding of their differential expression patterns. Many functional genomics tools such as microarrays, AFLP-based transcript profiling (cDNA-AFLP), and RNA sequencing (RNAseq) have been used for profiling transcript abundances, and they help in identifying differentially expressed genes at different time points in the progression of any developmental stage of stress. High-throughput transcriptomic studies have revealed many details of waterlogging stress tolerance mechanisms in several plant species. Rice has been a model plant for gene expression studies of submergence tolerance and several microarray and RNAseq studies have been conducted on the crop [183, 184]. Recently, many transcriptome studies have been, and continue to be, reported from other crop species, as well (table 10.1).

Genome-wide expression profiles generated in different plant species under anoxic stress created by submergence involves induction of genes belonging to various pathways such as the carbohydrate metabolism, fermentation, ROS metabolism, phytohormone regulation, fatty acid and amino acid synthesis, etc. Differentially expressed genes (DEGs) governing the synthesis of chlorophyll, carbon fixation, glycolysis, oxidative phosphorylation, antioxidant enzymes, phytohormone metabolism, signaling molecules, and anaerobic respiration were the most prominent differentially regulated ones under waterlogging conditions. A rare allele of GA biosynthesis, semidwarf1 (*SD1*), was identified in rice, which provides an adaptation for the crop to deepwater inundation [170]. This gene enhances GA_4 production, which leads to internodal elongation. A gene called leaf gas film 1 (*LGF1*), which guides the synthesis of a long-chain primary alcohol that offers hydrophobicity and formation of a leaf gas film in submerged rice, was identified [185]. The leaf gas film contributed to an almost eight-fold increase in photosynthesis and hence submergence tolerance. The 'accelerator' accelerator of internode elongation 1 (*ACE1*) and the 'decelerator' decelerator of internode elongation 1 (*DEC1*) genes were identified in rice, which have antagonistic effects in controlling internode elongation [186].

Expression of ROS-producing and detoxifying enzymes undergoes significant changes during submergence. Expression of POX (Peroxidase) and SOD was significantly upregulated in *Salix* species [138]. Genes involved in carbohydrate metabolism, like ADH, PDC, aldolase, glucose phosphate isomerase, and sucrose

Table 10.1. List of quantitative trait locii for flooding/waterlogging tolerance identified in various crops.

Crop	Trait	Population	Chromosome housing the QTL	References
Rice	Anaerobic germination	F2, F3 Backcross	2, 7, 11, 1. 3.7. 92, 5, 6, 7	[192][193][194]
	Lowest elongated internode, rate of internode elongation total internode length number of elongated internode	F2RILBC line	1, 3, 123, 121, 1212, 41, 3, 12	[195][196][197][198][199][200]
	Vegetative stage submergence	Backcross lineF2, F3RILDoubled haploid line, recombinant inbred line, F2RILF2, F3	1, 7, 99996, 7, 9, 11, 121, 2, 9, 12	[201][202][203][160][204][192]
	Survival rate	Recombinant inbred line	1, 4, 8, 9, 10,	[205]
	Germination under low temperature and anoxia	F2	5,11	[206]
	Membrane potential and hypoxia	Doubled haploid line	2	[207]
	Submergence tolerance	Backcross line F2	1, 3, 4, 6, 79	[199][208]
Maize	Leaf chlorosis, Leaf senescence	F2	1, 4, 6	[209, 210][211]
	Adventitious root formation	F2	3, 5, 87	[212][213][214][215]
	Constitutive aerenchyma formation	F2	1, 5, 8	[216–218][219]
	ROL formation		3	[53]
Barley	Survival rate	DH	2, 3, 4, 6	[220, 221]
	Aerenchyma formation		4	[142]
	Root porosity		4	[222]
	Chlorophyll fluorescence (ΦPSII)		6,7	[223]
	Root membrane potential		2	[207]
	Yellow leaf percentage		4	[224]

Soybean	Plant growth, seed yield	RIL	6, 18	[225][226]
	Damages and survival	RIL		[227][228][229][230]
	Root traits	RIL		[88]
Cucumber	Tolerance score (TOL), waterlogged shoot dry weight (SDWw), adventitious root formation (ARF), control shoot dry weight (SDWc), control vine length (VLHc), waterlogged vine length (VLHw)		Not defined	[231]
Wheat	Root fresh biomass Shoot fresh biomassChlorophyll contentRoot/shoot dry weightGermination rate index	RILs	1 B1B1D2B7A	[232][233]

synthase, were induced in several crops during flooding stress [187]. *Salix variegata* showed highly upregulated biosynthesis of raffinose, an oligosaccharide which is related to oxidative stress [138]. Overexpression of ADH transgene significantly improved tolerance to waterlogging in *Populus* [188] and enhanced seed germination in soybean under flooding [189].

Genes involved in long-chain fatty acid biosynthesis had enhanced expression and short-chain fatty acid accumulation was decreased [138]. Hypoxia-induced lipid remodeling has been reported in wheat and rice [190, 191]. Submergence was found to alter the lipid composition of *Sub1* near isogenic lines of rice, Swarna, and Swarna Sub1, which differ in their tolerance to submergence. Swarna Sub1 maintained better membrane integrity and photosynthetic efficiency [191].

10.5.4 Genetic studies and quantitative trait locus identifications

Pyramiding of genes and QTLs offer promising ways to develop lines with desired traits without affecting yield or other favorable characterisitics. Using the recent advances in molecular techniques of QTL analysis, fine-mapping, and marker-based technologies, it is possible to transfer submergence tolerance genes and/or QTLs to popular and elite crop cultivars. Rice was the first crop from which a successful understanding of the molecular basis of the submergence trait was gained, and cloning of some of the major genes contributing toward submergence tolerance has since been achieved. This was achieved by the cloning and characterization of the *Sub1* gene, which is a QTL controlling flooding tolerance in rice. Technical advances in molecular marker technology have unlocked rapid and accurate identification of cultivars with genes of interest, as well as their transfer into desired cultivars using several methods. The various traits that are linked with flooding tolerance have been studied thoroughly and their associated QTLs have been identified in many crops, with the most, again, being in rice. Table 10.2 gives a list of the QTLs that have been identified to contribute toward submergence/waterlogging stress in various crops. These QTLs are valuable tools in transferring the trait to genotypes that are susceptible to waterlogging stress.

10.6 Conclusions

Submergence and waterlogging is one of the major abiotic stresses that adversely affects crop growth. Flooding limits the availability of molecular oxygen to the roots as well as aerial parts. Oxygen is required to drive the aerobic metabolism that is needed to generate energy for plant survival. Flooding causes various morphological, anatomical, physiological, and metabolic disturbances in plants that might lead to slow growth or even growth arrest. Wetland plants have developed various adaptive strategies such as escape from the stress, or a slowing of growth and quiescence. Plants achieve these through efficient use of metabolic resources and by effectively tackling the toxic environment in the flooded soil. Various physiological and molecular mechanisms are involved in plant survival under flooding stress. There has been huge recent progress in understanding these mechanisms which take

Table 10.2. RNA sequencing for identifying flooding tolerance genes in plants.

Crop name	References	Crop name	References
Oryza sativa	Lasanthi-Kudahettige *et al* [183] Locke *et al* [234] Minami *et al* [235] Narsai *et al* [236]	***Arabidopsis thaliana***	[246–248]
Wild halophytic rice	Garg *et al* [237]		
Sesame	Dossa *et al* [238] Wang *et al* [239]	**Oil palm**	[249]
Sesbania cannabina	Ren *et al* [97]	***Brassica napus***	[250–252], [253, 254]
Soybean	Lin *et al* [149] Chen *et al* [240]		
Kiwi	Li *et al* [241]		
Maize	Rajhi *et al* [151] Campbell *et al* [211] Zhai *et al* [242]		
Cucumber	Xu *et al* [243]	Chinese wingnut (*Pterocarya stenoptera*)	[255]
Salix variegata	Zhang *et al* [138]	Bermuda grass (*Cynodon dactylon*)	[256]
Tomato	Safavi-Rizi *et al* [244]	*Porteresia coarctata*	[257]
Brachypodium distachyon	Rivera-Contreras *et al* [245]	*Triarrhena sacchariflora*	[115]

place at the cellular and molecular levels. Several genes and QTLs that are involved in submergence tolerance have been identified and characterized, and many have been used in breeding programmes to develop resistant genotypes.

Objective-type questions

1. Flooding causes conversion of organic nitrogen of soil to …
2. The rates of diffusion of oxygen, CO_2, and ethylene are high in water in comparison to those in air. True or false?
3. Which is the major anaerobic respiratory pathway adopted by plants during flooding stress?
4. Name the phytohormone which plays a major role in stomatal regulation?
5. The *Sub1A* gene is involved in which type of adaptation strategy for flooding tolerance?

Answers: 1 (ammonia); **2** (false); **3** (ethanolic fermentation); **4** (abscisic acid or ABA); **5** (LOQS).

Descriptive-type questions

1. What are the main chemical changes undergone by soil during flooding?
2. Describe the different adaptive strategies of plants to submergence.
3. What are the main morphological adaptations of plants during submergence/waterlogging stress?
4. Name a few important genes identified to have roles in submergence tolerance.
5. Elaborate on the role of the *Sub*1 gene in flooding tolerance and the significance of identification of this gene in crop improvement for submergence tolerance.

References

[1] FAO 2021 *The Impact of Disasters and Crises on Agriculture and Food Security: 2021* (Rome: FAO)

[2] Saldaña-Zorrilla S O 2008 Stakeholders' views in reducing rural vulnerability to natural disasters in southern Mexico: hazard exposure and coping and adaptive capacity *Glob. Environ. Change* **18** 583–97

[3] Kreft H and Jetz W 2007 Global patterns and determinants of vascular plant diversity *Proc. Natl. Acad. Sci.* **104** 5925–30

[4] Voesenek L and Bailey-Serres J 2015 Flood adaptive traits and processes: an overview *New Phytol.* **206** 57–73

[5] Kim Y, Shahzad R and Lee I-J 2020 Regulation of flood stress in plants *Plant Life under Changing Environment* (New York: Academic) pp 157–73

[6] Sasidharan R, Hartman S, Liu Z, Martopawiro S, Sajeev N, van Veen H and Voesenek L A C J 2017 Signal dynamics and interactions during flooding stress *Plant Physiol.* **176** 1106–17

[7] Robinson G W 1949 Soils, their origin, constitution, and classification *Soils* 3rd edn (London: Murby)

[8] Meetei T T, Devi Y, Chanu T and Upadhyaya H 2020 Electro-chemical properties of a submerged soil: a review *Plant Cell Biotechnol. Mol. Biol.* **21** 155–62

[9] Ponnamperuma F N 1972 The chemistry of submerged soils *Advances in Agronomy* vol 24 (Los Banos, CA: IRRI Los Banos) pp 29–96

[10] Kristensen K J and Enoch H 1964 Soil air composition and oxygen diffusion rate in soil columns at different heights above a water table *Trans. 8th Int. Cong. Soil Sci.* **2** 159–70

[11] Lee K-W, Chen P-W, Lu C-A, Chen S, Ho T-H and Yu S-M 2009 Coordinated responses to oxygen and sugar deficiency allow rice seedlings to tolerate flooding *Sci. Signal.* **2** ra61

[12] Armstrong W and Beckett P M 2011 Experimental and modelling data contradict the idea of respiratory down-regulation in plant tissues at an internal [O_2] substantially above the critical oxygen pressure for cytochrome oxidase *New Phytol.* **190** 431–41

[13] Dexter A R 1988 Advances in characterization of soil structure *Soil Tillage Res.* **11** 199–235

[14] Cook F and Knight J 2003 Oxygen transport to plant roots: modelling for physical understanding of soil aeration *Soil Sci. Soc. Am. J.* **67** 20–31

[15] Tete E, Viaud V and Walter C 2015 Organic carbon and nitrogen mineralization in a poorly drained mineral soil under transient waterlogged conditions: an incubation experiment *Eur. J. Soil Sci.* **66** 427–37

[16] Irfan M, Hayat S, Hayat Q, Afroz S and Ahmad A 2010 Physiological and biochemical changes in plants under waterlogging *Protoplasma* **241** 3–17

[17] Ponnamperuma F N 1965 *The Mineral Nutrition of the Rice Plant* (Baltimore, MD: Johns Hopkins Press) pp 295–328

[18] IRRI 1966 *Annu Rep. Int. Rice Res. Inst.* Los Bafios, Philippines

[19] Ponnamperuma F N 1955 *PhD Thesis* (Ithaca, NY: Cornell University)

[20] Matin N H and Jalali M 2017 The effect of waterlogging on electrochemical properties and soluble nutrients in paddy soils *Paddy Water Environ.* **15** 443–55

[21] Unger I M, Kennedy A C and Muzika R-M 2009 Flooding effects on soil microbial communities *Appl. Soil Ecol.* **42** 1–8

[22] Kirk G J D 2004 *The Biogeochemistry of Submerged Soils* (Chichester: Wiley)

[23] Nagarajah S, Neue H U and Alberto M C R 1989 Effect of Sesbania, Azolla and rice straw incorporation on the kinetics of N H, K, Fe, Mn, Zn, and P in some flooded rice soils *Plant Soil* **116** 37–48

[24] Saleh J, Najafi N, Oustan S, Aliasgharzad N and Ghassemi-Golezani K 2012 Effect of silicon, salinity and water logging on the extractable Zn, Cu, K and Na in a sandy loam soil *Proc. of the 8th Int. Soil Science Congress (Cesme and Izmir, Turkey)*

[25] IRRI 1965 *Annu. Rep. Int. Rice Res,* Inst. Los Bafios, Philippines

[26] Seago J L, Marsh L C, Stevens K J, Soukup A, Vortubová O and Enstone D E 2005 A re–examination of the root cortex in wetland flowering plants with respect to aerenchyma *Ann. Bot.* **96** 565–79

[27] Chen T, Yuan F, Song J and Wang B 2015 Nitric oxide participates in waterlogging tolerance through enhanced adventitious root formation in the euhalophyte *Suaeda salsa Funct. Plant Biol.* **43** 244–53

[28] Colmer T D 2003a Long–distance transport of gases in plants: a perspective on internal aeration and radial oxygen loss from roots *Plant Cell Environ.* **26** 17–36

[29] Colmer T D 2003b Aerenchyma and an inducible barrier to radial oxygen loss facilitate root aeration in upland, paddy and deepwater rice (*Oryza sativa* L.) *Ann. Bot.* **91** 301–9

[30] Colmer T D and Greenway H 2005 Oxygen transport, respiration, and anaerobic carbohydrate catabolism in roots in flooded soils *Plant Respiration: From Cell to Ecosystem* ed H Lambers and M Rivas-Carbo (Dordrecht: Springer) pp 137–58

[31] Li C, Liu D, Lin Z, Guan B, Liu D, Yang L, Deng X, Mei F and Zhou Z 2019 Histone acetylation modification affects cell wall degradation and aerenchyma formation in wheat seminal roots under waterlogging *Plant Growth Regul.* **87** 149–63

[32] Sauter M 2013 Root responses to flooding *Curr. Opin. Plant Biol.* **16** 282–6

[33] Shiono K, Ejiri M, Shimizu K and Yamada S 2019 Improved waterlogging tolerance of barley (*Hordeum vulgare*) by pretreatment with ethephon *Plant Prod. Sci.* **22** 285–95

[34] Takahashi H, Yamauchi T, Colmer T D and Nakazono M 2014 Aerenchyma formation in plants *Low-oxygen Stress in Plants: Oxygen Sensing and Adaptive Responses to Hypoxia* ed J van Dongen and F Licausi (Vienna: Springer) pp 247–65

[35] Yamauchi T, Colmer T D, Pedersen O and Nakazono M 2018 Regulation of root traits for internal aeration and tolerance to soil waterlogging-flooding stress *Plant Physiol.* **176** 1118–30

[36] Sou H D, Masumori M, Kurokochi H and Tange T 2019 Histological observation of primary and secondary aerenchyma formation in adventitious roots of *Syzygium kunstleri* (King) Bahadur and R. C. Gaur grown in hypoxic medium *Forests* **10** 137

[37] Sou H D, Masumori M, Yamanoshita T and Tange T 2021 Primary and secondary aerenchyma oxygen transportation pathways of *Syzygium kunstleri* (King) Bahadur & R. C. Gaur adventitious roots in hypoxic conditions *Nat. Sci. Rep.* **11** 4520

[38] Evans D E 2003 Aerenchyma formation *New Phytol.* **161** 35–49

[39] Visser E J W and Bögemann G M 2006 Aerenchyma formation in the wetland plant Juncus effusus is independent of ethylene *New Phytol.* **171** 305–14

[40] Striker G G, Insausti P, Grimoldi A A and León R J C 2006 Root strength and trampling tolerance in the grass *Paspalum dilatatum* and the dicot *Lotus glaber* in flooded soil *Funct. Ecol.* **20** 4–10

[41] Fan M, Bai R, Zhao X and Zhang J 2007 Aerenchyma formed under phosphorus deficiency contributes to the reduced root hydraulic conductivity in maize roots *J. Integr. Plant Bio.* **49** 598–604

[42] Wegner L H 2010 Oxygen transport in waterlogged plants *Waterlogging Signalling and Tolerance in Plants* ed S Mancuso and S Shabala (Heidelberg: Springer) pp 3–22

[43] Pedersen O, Sauter M, Colmer T D and Nakazono M 2021 Regulation of root adaptive anatomical and morphological traits during low soil oxygen *New Phytol.* **229** 42–9

[44] Rubinigg M, Stulen I, Elzenga J T M and Colmer T D 2002 Spatial patterns of radial oxygen loss and nitrate net fux along adventitious roots of rice raised in aerated or stagnant solution *Funct. Plant Biol.* **29** 1475–81

[45] Kotula L, Ranathunge K, Schreiber L and Steudle E 2009 Functional and chemical comparison of apoplastic barriers to radial oxygen loss in roots of rice (*Oryza sativa* L.) grown in aerated or deoxygenated solution *J. Exp. Bot.* **60** 2155–67

[46] Shiono K, Ogawa S, Yamazaki S, Isoda H, Fujimura T, Nakazono and Colmer T D 2011 Contrasting dynamics of radial O_2-loss barrier induction and aerenchyma formation in rice roots of two lengths *Ann.Bot.* **107** 89–99

[47] Watanabe K, Nishiuchi S, Kulichikhin K and Nakazono M 2013 Does suberin accumulation in plant roots contribute to waterlogging tolerance? *Front. Plant Sci.* **4** 178

[48] Visser E J W, Colmer T D, Blom C W P M and Voesenek L A C J 2000 Changes in growth, porosity, and radial oxygen loss from adventitious roots of selected mono- and dicotyledonous wetland species with contrasting types of aerenchyma *Plant Cell Environ.* **23** 1237–45

[49] Ejiri M and Shiono K 2019 Prevention of radial oxygen loss is associated with exodermal suberin along adventitious roots of annual wild species of *Echinochloa Front. Plant Sci.* **10** 254

[50] Ranathunge K, Lin J, Steudle E and Schreiber L 2011 Stagnant deoxygenated growth enhances root suberization and lignifications, but differentially affects water and NaCl permeabilities in rice (*Oryza sativa* L.) roots *Plant Cell Environ.* **34** 1223–40

[51] Kozlowski T T and Pallardy S G 1984 Effect of flooding on water, carbohydrate, and mineral relations *Flooding and Plant Growth* **1984** 165–93

[52] Ayi Q, Zeng B, Liu J, Li S, Van Bodegom P M and Cornelissen J H 2016 Oxygen absorption by adventitious roots promotes the survival of completely submerged terrestrial plants *Ann. Bot.* **118** 675–83

[53] Watanabe K, Takahashi H, Sato S, Nishiuchi S, Omori F, Malik A I, Colmer T D, Mano Y and Nakazono M 2017 A major locus involved in the formation of the radial oxygen loss barrier in adventitious roots of teosinte *Zea nicaraguensis* is located on the short-arm of chromosome 3 *Plant Cell Environ.* **40** 304–16
[54] Wei W, Li D, Wang L, Ding X, Zhang Y, Gao Y and Zhang X 2013 Morpho-anatomical and physiological responses to waterlogging of sesame (*Sesamum indicum* L.) *Plant Sci.* **208** 102–11
[55] Eysholdt-Derzsó E and Sauter M 2019 Hypoxia and the group VII ethylene response transcription factor HRE2 promote adventitious root elongation in *Arabidopsis Plant Biol.* **21** 103–8
[56] Liu J H and Reid D M 1992 Auxin and ethylene–stimulated adventitious rooting in relation to tissue sensitivity to auxin and ethylene production in sunflower hypocotyls *J. Exp.Bot.* **43** 1191–8
[57] Vidoz M L, Loreti E, Mensuali A, Alpi A and Perata P 2010 Hormonal interplay during adventitious root formation in flooded tomato plants *Plant J.* **63** 551–62
[58] McDonald M P and Visser E J W 2003 A study of the interaction between auxin and ethylene in wildtype and transgenic ethylene insensitive tobacco during adventitious root formation induced by stagnant root zone conditions *Plant Biol.* **5** 550–6
[59] Sharma S, Sharma J, Soni V, Kalaji H M and Elsheery N I 2021 Waterlogging tolerance: a review on regulative morphophysiological homeostasis of crop plants *J. Water Land Dev.* **49 (IV–VI)** 16–28
[60] Striker G G and Colmer T D 2017 Flooding tolerance of forage legumes *J. Exp.Bot.* **68** 1851–72
[61] Konnerup D, Islam A K and Colmer T D 2017 Evaluation of root porosity and radial oxygen loss of disomic addition lines of *Hordeum marinum* in wheat *Funct. Plant Biol.* **44** 400–9
[62] Cox M C H, Millenaar F F, van Berkel Y E M, Peeters A J M and Voesenek L A C J 2003 Plant Movement. Submergence–induced petiole elongation in *Rumex palustris* depends on hyponastic growth *Plant Physiol.* **132** 282–91
[63] Heydarian Z, Sasidharan R, Cox M C H, Pierik R, Voesenek L A C J and Peeters A J M 2010 A kinetic analysis of hyponastic growth and petiole elongation upon ethylene exposure in *Rumex palustris Ann. Bot.* **106** 429–35
[64] Vriezen W H, De Graaf B, Mariani C and Voesenek L A C J 2000 Submergence induces expansin gene expression in flooding–tolerant *Rumex palustris* and not in flooding intolerant *R. acetosa Planta* **210** 956–63
[65] Insausti P, Grimoldi A A, Chaneton E J and Vasellati V 2001 Flooding induces a suite of adaptive plastic responses in the grass *Paspalum dilatatum New Phytol.* **152** 291–9
[66] Mollard F P O, Striker G G, Ploschuk E L, Vega A S and Insausti P 2008 Flooding tolerance of *Paspalum dilatatum* (*Poaceae: Paniceae*) from upland and lowland positions in a natural grassland *Flora* **203** 548–56
[67] Mollard F P O, Striker G G, Ploschuk E L and Insausti P 2010 Subtle topographical differences along a floodplain promote different plant strategies among *Paspalum dilatatum* subspecies and populations *Austral Ecol.* **35** 189–96
[68] Rumanti I A, Sitaresmi T and Nugraha Y 2020 Rice tolerance variation to long-term stagnant flooding and germination ability under an-aerobic environment *IOP Conf.*

Series: Earth and Environmental Science. The 4th Int. Conf. on Climate Change 2019 (The 4th ICCC 2019) 18–19.11. 2019 *(Yogyakarta, Indonesia)* vol 423 012048 1–7

[69] Armstrong W, Strange M E, Cringle S and Beckett P M 1994 Microelectrode and modelling study of oxygen distribution in roots *Ann. Bot.* **74** 287–99

[70] Stevens K J, Peterson R L and Stephenson G R 1997 Morphological and anatomical responses of *Lythrum salicaria* L. (purple loosestrife) to an imposed water gradient *Int. J. Plant Sci.* **158** 172–83

[71] Striker G G, Insausti P, Grimoldi A A, Ploschuk E L and Vasellati V 2005 Physiological and anatomical basis of differential tolerance to soil flooding of *Lotus corniculatus* L. and *Lotus glaber* Mill *Plant Soil* **276** 301–11

[72] Shimamura S, Yamamoto R, Nakamura T, Shimada S and Komatsu S 2010 Stem hypertrophic lenticels and secondary aerenchyma enable oxygen transport to roots of soybean in flooded soil *Annal. Bot.* **106** 277–84

[73] Teakle N L, Armstrong J, Barrett–Lennard E G and Colmer T D 2011 Aerenchymatous phellem in hypocotyl and roots enables O_2 transport in *Melilotus siculus New Phytol.* **190** 340–50

[74] Rodríguez M E, Doffo G N, Cerrillo T and Luquez V M C 2018 Acclimation of cuttings of willow genotypes to flooding depth level *New For.* **49** 415–27

[75] Zhang F, Zhu K, Wang Y Q, Zhang Z P, Lu F, Yu H Q and Zou J Q 2019 Changes in photosynthetic and chlorophyll fluorescence characteristics of sorghum under drought and waterlogging stress *Photosynthetica.* **57** 1156–64

[76] Kozlowski T T 1997 Responses of woody plants to flooding and salinity *Tree Physiology* Monograph 1 http://heronpublishing.com/tp/monograph/kozlowski.pdf

[77] Garcia N, Da-Silva C J, Cocco K L, Pomagualli D, De Oliveira F K, Da Silva J V, De Oliveira A C and Do Amarante L 2020 Waterlogging tolerance of five soybean genotypes through different physiological and biochemical mechanism *Environ. Exp. Bot.* **172** 103975

[78] Ding J, Liang P, Wu P, Zhu M, Li C, Zhu X, Chen Y and Guo W 2020 Effects of waterlogging on grain yield and associated traits of historic wheat cultivars in the middle and lower reaches of the Yangtze River, China *Field Crops Res.* **246** 107695

[79] Nguyen T N, Tuan P A, Mukherjee S, Son S and Ayele B T 2018 Hormonal regulation in adventitious roots and during their emergence under waterlogged conditions in wheat *J. Exp. Bot.* **69** 4065–82

[80] Bailey-Serres J and Voesenek L A C J 2008 Flooding stress: acclimations and genetic diversity *Ann. Rev. Plant Biol.* **59** 313–39

[81] Bailey-Serres J and Voesenek L A C J 2010 Life in the balance: a signaling network controlling survival of flooding *Curr. Opin. Plant Biol.* **13** 489–94

[82] Stevens R M and Prior L D 1994 The effect of transient waterlogging on the growth, leaf gas exchange, and mineral composition of potted sultana grapevines *Am. J. Enol. Vitic.* **45** 285–90

[83] Pompeiano A, Reyes T H, Moles T M, Guglielminetti L and Scartazza 2020 Photosynthetic and growth responses of *Arundo donax* L. plantlets under different oxygen deficiency stresses and reoxygenation *Front. Plant Sci.* **10** 408

[84] Ahmed S, Nawata E, Hosokawa M, Domae Y and Sakuratani T 2002 Alterations in photosynthesis and some antioxidant enzymatic activities of mungbean subjected to waterlogging *Plant Sci.* **163** 117–23

[85] Colmer T D and Voesenek L A C J 2009 Flooding tolerance: suites of plant traits in variable environments *Funct. Plant Biol.* **36** 665–81
[86] Bailey-Serres J, Fukao T, Gibbs D J, Holdsworth M J and Lee S C 2012 Making sense of low oxygen sensing *Trends Plant Sci.* **17** 129–38
[87] van Dongen J T and Licausi F 2015 Oxygen sensing and signalling *Ann. Rev. Plant Biol.* **66** 345–67
[88] van Nguyen L, Takahashi R and Githiri S M *et al* 2017 Mapping quantitative trait loci for root development under hypoxia conditions in soybean (*Glycine max* L. Merr.) *Theor. Appl. Genet.* **130** 743–55
[89] Nishiuchi S, Yamauchi T, Takahashi H, Kotula L and Nakazono M 2012 Mechanisms for coping with submergence and waterlogging in rice *Rice* **5** 2
[90] Dordas C, Hasinoff B B, Igamberdiev A U, Manac'h N, Rivoal J and Hill R D 2003 Expression of a stress induced hemoglobin affects NO levels produced by alfalfa root cultures under hypoxic stress *Plant J.* **35** 763–70
[91] Igamberdiev A U and Hill R D 2004 Nitrate, NO and haemoglobin in plant adaptation to hypoxia: an alternative to classic fermentation pathways *J. Exp. Bot.* **55** 2473–82
[92] Limami A, Diab H and Lothier J 2014 Nitrogen metabolism in plants under low oxygen stress *Planta* **239** 531–41
[93] Davies D D 1980 Anaerobic metabolism and the production of organic acids *The Biochemistry of Plants* ed D D Davies vol 2 (New York: Academic) pp 581–611
[94] Kennedy R A, Rumpho M E and Fox T C 1992 Anaerobic metabolism in plants *Plant Physiol.* **84** 1204–9
[95] Oliveira H C, Salgado I and Sodek L 2013 Involvement of nitrite in the nitrate-mediated modulation of fermentative metabolism and nitric oxide production of soybean roots during hypoxia *Planta* **237** 255–64
[96] Ruperti B *et al* 2020 Flooding responses on grapevine: a physiological, transcriptional, and metabolic perspective *Front. Plant Sci.* **10** 339
[97] Ren C G, Kong cc and Yan K *et al* 2017 Elucidation of the molecular responses to waterlogging in *Sesbania cannabina* roots by transcriptome profiling *Sci. Rep.* **7** 9256
[98] Nanjo Y *et al* 2011 Transcriptional responses to flooding stress in roots including hypocotyl of soybean seedlings *Plant Mol. Biol.* **77** 129–44
[99] Oh M and Komatsu S 2015 Characterization of proteins in soybean roots under flooding and drought stresses *J. Proteomics* **114** 161–81
[100] Tamang B G, Magliozzi J O, Maroof M A S and Fukao T 2014 Physiological and transcriptomic characterization of submergence and reoxygenation responses in soybean seedlings *Plant Cell Environ.* **37** 2350–65
[101] Wu Y-S and Yang C-Y 2016 Physiological responses and expression profile of NADPH oxidase in Rice (*Oryza sativa*) seedlings under different levels of submergence *Rice* **9** 2
[102] Anee T I, Nahar K, Rahman A, Mahmud J A, Bhuiyan T F and Alam M U 2019 Oxidative damage and antioxidant defense in *Sesamum indicum* after different waterlogging durations *Plants* **8** 196
[103] Kuai L J, Liu Z, Wang Y, Meng Y, Chen B and Zhao W *et al* 2014 Waterlogging during flowering and boll forming stages affects sucrose metabolism in the leaves subtending the cotton boll and its relationship with boll weight *Plant Sci.* **223** 79–98

[104] Pierik R, de Wit M and Voesenek L A C J 2011 Growth-mediated stress escape: convergence of signal transduction pathways activated upon exposure to two different environmental stresses *New Phytol.* **189** 122–34

[105] Winkel A, Pedersen O, Ella E, Ismail A M and Colmer T D 2014 Gas film retention and underwater photosynthesis during field submergence of four contrasting rice genotypes *J. Exp. Bot.* **65** 3225–33

[106] Panda D, Sharma S G and Sarkar R K 2008 Chlorophyll fluorescence parameters, CO_2 photosynthetic rate and regeneration capacity as a result of complete submergence and subsequent re-emergence in rice (*Oryza sativa* L.) *Aquat. Bot.* **88** 127–33

[107] Panda D and Sarkar R K 2012 Leaf photosynthetic activity and antioxidant defense associated with Sub1 QTL in rice subjected to submergence and subsequent re-aeration *Rice Sci.* **19** 108–16

[108] Panda D and Sarkar R K 2013 Characterization of leaf gas exchange and anti-oxidant defense of rice (*Oryza sativa* L.) cultivars differing in submergence tolerance owing to complete submergence and consequent re-aeration *Agric. Res.* **2** 301–8

[109] Bansal R, Sharma S, Tripathi K and Kumar A 2019 Waterlogging tolerance in black gram [*Vigna mungo* (L.) Hepper] is associated with chlorophyll content and membrane integrity *Indian J. Biochem. Biophys.* **56** 81–5

[110] Jia L T, Qin X, Lyu D G, Qin S J and Zhang P 2019 ROS production and scavenging in three cherry rootstocks under short-term waterlogging conditions *Sci. Hortic.* **257** 108647

[111] Pucciariello C and Perata P 2017 New insights into reactive oxygen species and nitric oxide signalling under low oxygen in plants *Plant Cell Environ.* **40** 473–82

[112] Xu X, Wang H, Qi X, Xu Q and Chen X 2014 Waterlogging-induced increase in fermentation and related gene expression in the root of cucumber (*Cucumis sativus* L.) *Sci. Hortic.* **179** 388–95

[113] Zhang P, Lyu D, Jia L, He J and Qin S 2017 Physiological and de novo transcriptome analysis of the fermentation mechanism of *Cerasus sachalinensis* roots in response to short-term waterlogging *BMC Genomics* **18** 649

[114] Balakhnina T I, Bulak P, Matichenkov V V, Kosobryuvkhov A A and Wlodarczyk T M 2015 The influence of Si-rich mineral zeolite on the growth processes and adaptive potential of barley plants under cadmium stress *Plant Growth Regul.* **75** 557–65

[115] Wang J, Sun H, Sheng J J, Jin S R, Zhou F S, Hu Z L and Diao Y 2019 Transcriptome, physiological and biochemical analysis of *Triarrhena sacchariflora* in response to flooding stress *BMC Genet.* **20** 88

[116] Ye T T, Shi H T, Wang Y P and Chan Z L 2015 Contrasting changes caused by drought and submergence stresses in bermudagrass (*Cynodon dactylon*) *Front. Plant Sci.* **6** 951

[117] Kawano N, Ella E, Ito O, Yamauchi Y and Tanaka K 2002 Metabolic changes in rice seedlings with different submergence tolerance after desubmergence *Environ. Exp. Bot.* **47** 195–203

[118] Chugh V, Kaur N and Gupta A K 2016 Comparison of antioxidant system and anaerobic metabolism in seedlings of contrasting maize genotypes under short-term waterlogging *Int. J. Biochem. Res.* **15** 1–10

[119] Duhan S, Kumari A and Sheokand S 2017 Effect of waterlogging and salinity on antioxidative system in pigeon pea plant leaves at different stages of development *Res. Crop* **18** 559–68

[120] Zhang J, Yin D J, Fan S X, Li S G and Dong L 2019b Modulation of morphological and several physiological parameters in sedum under waterlogging and subsequent drainage *Russian J. Plant Physiol.* **66** 290–8

[121] Ella E S, Kawano N and Ito O 2003 Importance of active oxygen-scavenging system in the recovery of rice seedlings after submergence *Plant Sci.* **165** 85–93

[122] Hancock R D and Viola R 2005 Biosynthesis and catabolism of L-ascorbic acid in plants *Crit. Rev. Plant Sci.* **24** 167–88

[123] Steffens B, Kovalev A, Gorb S N and Sauter M 2012 Emerging roots alter epidermal cell fate through mechanical and reactive oxygen species signalling *Plant Cell* **24** 3296–306

[124] Najeeb U, Tan D K Y, Bange M P and Atwell B J 2018 Protecting cotton crops under elevated CO_2 from waterlogging by managing ethylene *Funct. Plant Biol.* **45** 340

[125] Zimmerman P W and Hitchcock A E 1933 Initiation and stimulation of adventitious roots caused by unsaturated hydrocarbon gases *Contrib. Boyce Thompson Inst.* **5** 351–69

[126] Wample R L and Reid D M 1979 The role of endogenous auxins and ethylene in the formation of adventitious roots and hypocotyl hypertrophy in flooded sunflower plants (*Helianthus annuus*) *Physiol. Plant.* **45** 219–26

[127] Steffens B and Sauter M 2009 Heterotrimeric G protein signalling is required for epidermal cell death in rice *Plant Physiol.* **151** 732–40

[128] Visser E J W, Blom C W P M and Voesenek L A C J 1996 Flooding-induced adventitious rooting in Rumex: morphology and development in an ecological perspective *Acta Bot. Neerl.* **45** 17–28

[129] Lorbiecke R and Sauter M 1999 Adventitious root growth and cell-cycle induction in deepwater rice *Plant Physiol.* **119** 21–30

[130] Steffens B, Geske T and Sauter M 2011 Aerenchyma formation in the rice stem and its promotion by H_2O_2 *New Phytol.* **190** 369–78

[131] Qi X H, Li Q Q, Ma X T, Qian C L, Wang H H, Ren N N, Shen C X, Huang S M, Xu X W and Xu Q *et al* 2019 Waterlogging induced adventitious root formation in cucumber is regulated by ethylene and auxin through reactive oxygen species signalling *Plant Cell Environ.* **42** 1458–70

[132] Yamauchi T, Watanabe K, Fukazawa A, Mori H, Abe F, Kawaguchi K, Oyanagi A and Nakazono M 2014 Ethylene and reactive oxygen species are involved in root aerenchyma formation and adaptation of wheat seedlings to oxygen-deficient conditions *J. Exp. Bot.* **65** 261–73

[133] Xu Q T, Yang L, Zhou Z Q, Mei F Z, Qu L H and Zhou G S 2013 Process of aerenchyma formation and reactive oxygen species induced by waterlogging in wheat seminal roots *Planta* **238** 969–82

[134] Mergemann H and Sauter M 2000 Ethylene induces epidermal cell death at the site of adventitious root emergence in rice *Plant Physiol.* **124** 609–14

[135] Zarembinski T I and Theologis A 1997 Expression characteristics of OS-ACS1 and OS-ACS2, two members of the 1-aminocyclopropane-1-carboxylate synthase gene family in rice (*Oryza sativa* L. Cv. Habiganj Aman II) during partial submergence *Plant Mol. Biol.* **33** 71–7

[136] Zhou Z, Engler J A, Rouan D, Michiels F, van Montagu M and van der Straeten D 2002 Tissue localization of a submergence-induced 1-aminocyclopropane-1-carboxylic acid synthase in rice *Plant Physiol.* **129** 72–84

[137] Fukao T, Xu K, Ronald P C and Bailey-Serres J 2006 A variable cluster of ethylene response factor-like genes regulates metabolic and developmental acclimation responses to submergence in rice *Plant Cell* **18** 2021–34

[138] Zhang Q, Tang S and Li J *et al* 2021 Integrative transcriptomic and metabolomic analyses provide insight into the long-term submergence response mechanisms of young *Salix variegata* stems *Planta* **253** 88

[139] He F, Wang H L, Li H G, Su Y, Li S, Yang Y, Feng C H, Yin W and Xia X 2018 PeCHYR1, a ubiquitin E3 ligase from *Populus euphratica*, enhances drought tolerance via ABA-induced stomatal closure by ROS production in *Populus Plant Biotechnol. J.* **16** 1514–28

[140] Shimamura S, Yoshioka T, Yamamoto R, Hiraga S, Nakamura T, Shimada S and Komatsu S 2014 Role of abscisic acid in flood induced secondary aerenchyma formation in soybean (*Glycine max*) hypocotyls *Plant Prod. Sci.* **17** 131–7

[141] Dawood T, Yang X, Visser E J, Te Beek T A, Kensche P R, Cristescu S M and Rieu I 2016 A co-opted hormonal cascade activates dormant adventitious root primordia upon flooding in *Solanum dulcamara Plant Physiol.* **170** 2351–64

[142] Zhang Y, Chen Y, Lu H, Kong X, Dai J and Li Z *et al* 2016 Growth, lint yield and changes in physiological attributes of cotton under temporal waterlogging *Field Crops Res.* **194** 83–93

[143] Nan R, Carman J G and Salisbury F B 2002 Water stress, CO_2 and photoperiod influence hormone levels in wheat *J. Plant Physiol.* **159** 307–12

[144] Lin C and Sauter M 2019 Polar auxin transport determines adventitious root emergence and growth in rice *Front. Plant Sci.* **10** 444

[145] Kim Y H, Hwang S-J, Waqas M, Khan A L, Lee J-H and Lee J-D *et al* 2015 Comparative analysis of endogenous hormones level in two soybean (*Glycine max* L.) lines differing in waterlogging tolerance *Front. Plant Sci.* **6** 714

[146] Ayano M, Kani T, Kojima M, Sakakibara H, Kitaoka T and Kuroha T 2014 Gibberellin biosynthesis and signal transduction is essential for internode elongation in deepwater rice *Plant Cell Environ.* **37** 2313–24

[147] Gibbs D J, Lee S C, Isa N M, Gramuglia S, Fukao T, Bassel G W, Correia C S, Fo C, Theodoulou F L and Bailey-Serres J 2011 Homeostatic response to hypoxia is regulated by the N-end rule pathway in plants *Nature* **479** 415–8

[148] Licausi F, Kosmacz M, Weits D A, Giuntoli B, Giorgi F M, Voesenek L A C J, Perata P and van Dongen J T 2011 Oxygen sensing in plants is mediated by an N-end rule pathway for protein destabilization *Nature* **479** 419–22

[149] Lin C S *et al* 2019 Regulatory cascade involving transcriptional and N-end rule pathways in rice under submergence *Proc. Natl. Acad. Sci. USA.* 2019 **116** 3300–9

[150] Mustroph A 2018 Improving flooding tolerance of crop plants *Agronomy* **8** 160

[151] Rajhi I, Yamauchi T, Takahashi H, Nishiuchi S, Shiono K, Watanabe R, Mliki A, Nagamura Y, Tsutsumi N and Nishizawa N K *et al* 2011 Identification of genes expressed in maize root cortical cells during lysigenous aerenchyma formation using laser micro-dissection and microarray analyses *New Phytol.* **190** 351–68

[152] Mühlenbock P, Plaszczyca M, Plaszczyca M, Mellerowicz E and Karpinski S 2007 Lysigenous aerenchyma formation in *Arabidopsis thaliana* is controlled by *LEASION SIMULATING DISEASE1 Plant Cell* **19** 3819–30

[153] Steffens B, Steffen-Heins A and Sauter M 2013 Reactive oxygen species mediate growth and death in submerged plants *Front. Plant Sci.* **4** 179

[154] Liu S, Wang J, Wang L, Wang X, Xue Y, Wu P and Shou H X 2009 Adventitious root formation in rice requires OsGNOM1 and is mediated by the OsPINs family *Cell Res.* **19** 1110–9

[155] Xu X W, Ji J, Xu Q, Qi X H, Weng Y Q and Chen X H 2018 The major-effect quantitative trait locus CsARN6.1 encodes an AAA ATPase domain-containing protein that is associated with waterlogging stress tolerance by promoting adventitious root formation *Plant J.* **93** 917–30

[156] Xu K, Xu X, Fukao T, Canlas P, Maghirang-Rodriguez R, Heuer S, Ismail A M, Bailey-Serres J, Ronald P C and Mackill D J 2006 Sub1A is an ethylene-response-factor-like gene that confers submergence tolerance to rice *Nature* **442** 705–8

[157] Ismail A M, Singh U S, Singh S, Dar M and Mackill D J 2013 The contribution of submergence-tolerant (Sub1) rice varieties to food security in flood-prone areas *Field Crops Res.* **152** 83–93

[158] Fukao T, Harris T and Bailey-Serres J 2009 Evolutionary analysis of the Sub1 gene cluster that confers submergence tolerance to domesticated rice *Ann. Bot.* **103** 143–50

[159] Bailey-Serres J, Fukao T, Ronald P C, Ismail A M, Heuer S and Mackill D J 2010 Submergence tolerant rice: SUB1 journey from landrace to modern cultivar *Rice* **3** 138–47

[160] Toojinda T, Siangliw M, Tragoonrung S and Vanavichit A 2003 Molecular genetics of submergence tolerance in rice: QTL analysis of key traits *Ann. Bot.* **91** 243–53

[161] Pucciariello C and Perata P 2013 Quiescence in rice submergence tolerance: an evolutionary hypothesis *Trends Plant Sci.* **18** 377–81

[162] Fukao T and Bailey-Serres J 2008 Submergence tolerance conferred by Sub1A is mediated by SLR1 and SLRL1 restriction of gibberellin responses in rice *Proc. Natl. Acad. Sci. USA* **105** 16814–9

[163] Neeraja C N, Rodriguez M, Pamplona A, Heuer S, Collard B C, Septiningsih E M, Vergara G, Sanchez D, Xu K and Ismail A M *et al* 2007 A marker-assisted backcross approach for developing submergence-tolerance rice cultivars *Theor. Appl. Genet.* **115** 767–76

[164] Iftekharuddaula K M, Newaz M A, Salam M A, Ahmed H U, Mahbub M A A, Septiningsih E M, Collard B C Y, Sanchez D L, Pamplona A M and Mackill D J 2011 Rapid and high-precision marker assisted backcrossing to introgress the SUB1 QTL into BR11, the rainfed lowland rice mega variety of Bangladesh *Euphytica* **178** 83–97

[165] Manivong P, Korinsak S, Siangliw J L, Vanavichit A and Toojinda T 2014 Marker-assisted selection to improve submergence tolerance, blast resistance and strong fragrance in glutinous rice *Thai J. Genet.* **7** 110–22

[166] Oladosu Y, Rafii M Y, Arolu F, Chukwu S C, Muhammad I, Kareem I, Salisu M A and Arolu I W 2020 Submergence tolerance in rice: review of mechanism, breeding and prospects *Sustainability* 2020 **12** 1632

[167] Baksh S K Y, Donde R and Kumar J 2021 Genetic relationship, population structure analysis and pheno-molecular characterization of rice (*Oryza sativa* L.) cultivars for bacterial leaf blight resistance and submergence tolerance using trait specific STS markers *Physiol Mol Biol Plants* **27** 543–62

[168] Hattori Y, Nagai K, Furukawa S, Song X J, Kawano R, Sakakibara H, Wu J Z, Matsumoto T, Yoshimura A and Kitano H *et al* 2009 The ethylene response factors SNORKEL1 and SNORKEL2 allow rice to adapt to deep water *Nature* **460** 1026–30

[169] Hattori Y, Nagai K and Ashikari M 2011 Rice growth adapting to deepwater *Curr. Opin. Plant Biol.* **14** 100–5

[170] Kuroha T, Nagai K, Gamuyao R, Wang D R, Furuta T, Nakamori M, Kitaoka T, Adachi K, Minami A and Mori Y *et al* 2018 Ethylene-gibberellin signalling underlies adaptation of rice to periodic flooding *Science* **361** 181–6

[171] Giuntoli B, Lee S C, Licausi F, Kosmacz M, Oosumi T, van Dongen J T, Bailey-Serres J and Perata P 2014 A trihelix DNA binding protein counterbalances hypoxia-responsive transcriptional activation in *Arabidopsis PLoS Biol.* **12** e1001950

[172] Tang H, Bi H, Liu B, Lou S, Yan S and Tong S *et al* 2021 WRKY33 interacts with WRKY12 protein to up-regulate RAP2.2 during submergence induced hypoxia response in *Arabidopsis thaliana New Phytol.* **229** 106–25

[173] Wei X N, Xu H J, Rong W, Ye X G and Zhang Z Y 2019 Constitutive expression of a stabilized transcription factor group VII ethylene response factor enhances waterlogging tolerance in wheat without penalizing grain yield *Plant Cell Environ.* **42** 1471–85

[174] Zhu Q G, Gong Z Y, Huang J W, Grierson D, Chen K S and Yin X R 2019 High-CO_2/hypoxia-responsive transcription factors DkERF24 and DkWRKY1 interact and activate DkPDC2 promoter *Plant Physiol.* **180** 621–33

[175] Mohanty B, Takahashi H, de los Reyes B G, Wijaya E, Nakazono M and Lee D-Y 2016 Transcriptional regulatory mechanism of alcohol dehydrogenase 1-deficient mutant of rice for cell survival under complete submergence *Rice* **9** 51

[176] Raineri J, Ribichich K F and Chan R L 2015 The sunflower transcription factor HaWRKY76 confers drought and flood tolerance to *Arabidopsis thaliana* plants without yield penalty *Plant Cell Rep.* **34** 2065–80

[177] Klok E J, Wilson I W, Wilson D, Chapman S C, Ewing R M and Somerville S C *et al* 2002 Expression profile analysis of the low-oxygen response in *Arabidopsis* root cultures *Plant Cell* **14** 2481–94

[178] Sun Y, Liu Z, Guo J, Zhu Z and Sun X 2020 WRKY33-PIF4 loop is required for the regulation of H_2O_2 homeostasis *Biochem. Biophys. Res. Commun.* **527** 9228

[179] Nakashima K and Yamaguchi-Shinozaki K 2013 ABA signalling in stress-response and seed development *Plant Cell Rep.* **32** 959–70

[180] Wang S *et al* 2021 bZIP72 promotes submerged rice seed germination and coleoptile elongation by activating ADH1 *Plant Physiol. Biochem.* **169** 112–18

[180a] Mohanty B 2021 Promoter architecture and transcriptional regulation of genes upregulated in germination and coleoptile elongation of diverse rice genotypes tolerant to submergence *Front. Genet.* **12** 639654

[181] Mustroph A, Lee S C, Oosumi T, Zanetti M E, Yang H, Ma K, Yaghoubi-Masihi A, Fukao T and Bailey-Serres J 2010 Cross-kingdom comparison of transcriptomic adjustments to low-oxygen stress highlights conserved and plant-specific responses *Plant Physiol.* **152** 1484–500

[182] Gomez-Ariza J *et al* 2019 A transcription factor coordinating internode elongation and photoperiodic signals in rice *Nat. Plants* **5** 358–62

[183] Lasanthi-Kudahettige R, Magneschi L, Loreti E, Gonzali S, Licausi F and Novi G *et al* 2007 Transcript profiling of the anoxic rice coleoptile *Plant Physiol.* **144** 218–31

[184] Narsai R, Edwards J M, Roberts T H, Whelan J, Joss G H and Atwell B J 2015 Mechanisms of growth and patterns of gene expression in oxygen deprived rice coleoptiles *Plant J.* **82** 25–40

[185] Kurokawa Y, Nagai K, Huan P D, Shimazaki K, Qu H, Mori Y, Toda Y, Kuroha T, Hayashi N and Aiga S *et al* 2018 Rice leaf hydrophobicity and gas films are conferred by a wax synthesis gene (LGF1) and contribute to flood tolerance *New Phytol.* **218** 1558–69

[186] Nagai K, Mori Y and Ishikawa S *et al* 2020 Antagonistic regulation of the gibberellic acid response during stem growth in rice *Nature* **584** 109–14

[187] Miro B and Ismail A M 2013 Tolerance of anaerobic conditions caused by flooding during germination and early growth in rice (*Oryza sativa* L.) *Front. Plant Sci.* **4** 269

[188] Xuan L, Hua J, Zhang F, Wang Z, Pei X, Yang Y, Yin Y and Creech D L 2021 Identification and functional analysis of ThADH1 and ThADH4 genes involved in tolerance to waterlogging stress in *Taxodium* hybrid 'Zhongshanshan 406' *Genes* **12** 225

[189] Tougou M, Hashiguchi A, Yukawa K, Nanjo Y, Hiraga S, Nakamura T, Nishizawa K and Komatsu S 2012 Responses to flooding stress in soybean seedlings with the alcohol dehydrogenase transgene *Plant Biotechnol.* **29** 301–5

[190] Xu L, Pan R, Zhou M, Xu Y and Zhang W 2020 Lipid remodelling plays an important role in wheat (*Triticum aestivum*) hypoxia stress *Funct. Plant Biol.* **47** 58–66

[191] Basu S, Kumari S, Kumar P, Kumar G and Rajwanshi R 2021 Redox imbalance impedes photosynthetic activity in rice by disrupting cellular membrane integrity and induces programmed cell death under submergence *Physiol. Plant.* **172** 1764–78

[192] Septiningsih E M, Sanchez D L, Singh N, Sendon P M, Pamplona A M and Heuer S 2012 Identifying novel QTLs for submergence tolerance in rice cultivars IR72 and Madabaru *Theor. Appl. Genet.* **124** 867–74

[193] Angaji S A, Septiningsih E M, Mackill D J and Ismail A M 2010 Identification of QTLs associated with tolerance of anaerobic conditions during germination in rice (*Oryza sativa* L.) *Euphytica* **172** 159–68

[194] Septiningsih E M, Ignacio J C, Sendon P M, Sanchez D L, Ismail A M and Mackill D J 2013 QTL mapping and confirmation for tolerance of anaerobic conditions during germination derived from the rice landrace Ma-Zhan Red *Theor. Appl. Genet.* **126** 1357–66

[195] Hamamura K and Kupkanchankul T 1979 Inheritance of floating ability of rice *Jpn. J. Breed.* **29** 211–6

[196] Tripathi R S and Rao M J B 1985 Inheritance studies of characters associated with floating habit and their linkage relationship in rice *Euphytica* **34** 875–81

[197] Tang D Q, Kasai Y, Miyamoto N, Ukai Y and Nemoto K 2005 Comparison of QTLs for early elongation ability between two floating rice cultivars with a different phylogenetic origin *Breed. Sci.* **55** 1–5

[198] Hattori Y, Miura K, Asano K, Yamamoto E, Mori H, Kitano H, Matsuoka M and Ashikari M 2007 A major QTL confers rapid internode elongation in response to water rise in deepwater rice *Breed. Sci.* **57** 305–14

[199] Xu K, Xu X, Ronald P C and Mackill D J 2000 A high-resolution linkage map in the vicinity of the rice submergence tolerance locus *Sub1 Mol. Gen. Genetics* **263** 681–9

[200] Suge H 1987 Physiological genetics of internode elongation under submergence in floating rice *Jpn. J. Genet.* **62** 69–80

[201] Siangliw M, Toojinda T, Tragoonrung S and Vanavichit A 2003 Thai jasmine rice carrying QTLch9 (SubQTL) is submergence tolerant *Ann. Bot.* **91** 255–61

[202] Xu K and Mackill D J 1996 A major locus for submergence tolerance mapped on rice chromosome 9 *Mol. Breed.* **2** 219–24

[203] Ramiah K and Ramaswami K 1940 Floating habit in rice *Indian J. Agric. Sci.* **11** 1–8

[204] Nandi S, Subudhi P K, Senadhira D, Manigbas N L, Sen-Mandi S and Huang N 1997 Mapping QTLs for submergence tolerance in rice by AFLP analysis and selective genotyping *Mol. Gen. Genet.* **255** 1–8

[205] Gonzaga Z J C, Carandang J, Sanchez D L, Mackill D J and Septiningsih E M 2016 Mapping additional QTLs from FR13A to increase submergence tolerance beyond SUB1 *Euphytica* **209** 627–36

[206] Jiang L, Liu S, Hou M, Tang J, Chen L, Zhai H and Wan J 2006 Analysis of QTLs for seed low temperature germinability and anoxia germinability in rice (*Oryza sativa* L.) *Field Crops Res.* **98** 68–75

[207] Gill M B, Zeng F, Shabala L, Zhang G, Fan Y, Shabala S and Zhou M 2017 Cell-based phenotyping reveals QTL for membrane potential maintenance associated with hypoxia and salinity stress tolerance in barley *Front. Plant Sci.* **8** 1941

[208] Manangkil O E, Vu H T T, Mori N, Yoshida S and Nakamura C 2013 Mapping of quantitative trait loci controlling seedling vigor in rice (*Oryza sativa* L.) under submergence *Euphytica* **192** 63–75

[209] Mano Y, Muraki M and Takamizo T 2006 Identification of QTL controlling flooding tolerance in reducing soil conditions in maize (*Zea mays* L.) seedlings *Plant Prod. Sci.* **9** 176–81

[210] Mano Y and Omori F 2013 Flooding tolerance in interspecific introgression lines containing chromosome segments from teosinte (*Zea nicaraguensis*) in maize (*Zea mays* subsp. *mays*) *Ann. Bot.* **112** 1125–39

[211] Campbell M T, Proctor C A, Dou Y, Schmitz A J, Phansak P and Kruger G R 2015 Genetic and molecular characterization of submergence response identifies *Subtol6* as a major submergence tolerance locus in maize *PLoS One* **10** e0120385

[212] Mano Y, Muraki M, Fujimori M, Takamizo T and Kindiger B 2005 Identification of QTL controlling adventitious root formation during flooding conditions in teosinte (*Zea mays* ssp. *huehuetenangensis*) seedlings *Euphytica* **142** 33–42

[213] Mano Y, Omori F, Muraki M and Takamizo T 2005a QTL mapping of adventitious root formation under flooding conditions in tropical maize (*Zea mays* L.) seedlings *Breed. Sci.* **255** 343–7

[214] Mano Y and Omori F 2009 High-density linkage map around the root aerenchyma locus Qaer1.06 in the backcross populations of maize Mi29 × teosinte *Zea nicaraguensis Breed. Sci.* **59** 427–33

[215] Zhang X, Tang B, Yu F, Li L, Wang M, Xue Y, Zhang Z, Yan J, Yue B and Zheng Y *et al* 2013 Identification of major QTL for waterlogging tolerance using genome-wide association and linkage mapping of maize seedlings *Plant Mol. Biol. Rep.* **31** 594–606

[216] Mano Y, Omori F, Takamizo T, Kindiger B, Bird R M, Loaisiga C H and Takahashi H 2007 QTL mapping of root aerenchyma formation in seedlings of a maize × rare teosinte *Zea nicaraguensis* cross *Plant Soil* **295** 103–13

[217] Mano Y and Omori F 2008 Verification of QTL controlling root aerenchyma formation in a maize × teosinte *Zea nicaraguensis* advanced backcross population *Breed. Sci.* **58** 217–23

[218] Mano Y and Omori F 2009 High-density linkage map around the root aerenchyma locus Qaer1.06 in the backcross populations of maize Mi29 × teosinte *Zea nicaraguensis Breed. Sci.* **59** 427–33
[219] Zaidi P H, Rafique S and Singh N N 2003 Response of maize (*Zea mays* L.) genotypes to excess soil moisture stress: morpho-physiological effects and basis of tolerance *Eur. J. Agron.* **19** 383–99
[220] Zhou M 2011 Accurate phenotyping reveals better QTL for waterlogging tolerance in barley: QTL for waterlogging tolerance in barley *Plant Breed.* **130** 203–08
[221] Zhou M, Johnson P, Zhou G, Li C and Lance R 2012 Quantitative trait loci for waterlogging tolerance in a barley cross of Franklin × YuYaoXiangTian Erleng and the relationship between waterlogging and salinity tolerance *Crop Sci.* **52** 2082–8
[222] Broughton S, Zhou G, Teakle N L, Matsuda R, Zhou M, O'Leary R A, Colmer T D and Li C 2015 Waterlogging tolerance is associated with root porosity in barley (*Hordeum vulgare* L.) *Mol. Breed.* **35** 27
[223] Bertholdsson N-O, Holefors A, Macaulay M and Crespo-Herrera L A 2015 QTL for chlorophyll fluorescence of barley plants grown at low oxygen concentration in hydroponics to simulate waterlogging *Euphytica* **201** 357–65
[224] Osman K A, Tang B, Wang Y, Chen J, Yu F, Li L, Han X, Zhang Z, Yan J and Zheng Y *et al* 2013 Dynamic QTL analysis and candidate gene mapping for waterlogging tolerance at maize seedling stage *PLoS One* **8** e79305
[225] Githiri S M, Watanabe S, Harada K and Takahashi R 2006 QTL analysis of flooding tolerance in soybean at an early vegetative growth stage *Plant Breed.* **125** 613–8
[226] vanToai T T, St. Martin S K, Chase K, Boru G, Schnipke V, Schmitthenner A F and Lark K G 2001 Identification of a QTL associated with tolerance of soybean to soil waterlogging *Crop Sci.* **41** 1247–252
[227] Ye H, Song L, Chen H, Valliyodan B, Cheng P, Ali L, Vuong T, Wu C, Orlowski J and Buckley B *et al* 2018 A major natural genetic variation associated with root system architecture and plasticity improves waterlogging tolerance and yield in soybean: waterlogging tolerance and root system plasticity in soybean *Plant Cell Environ.* **41** 2169–82
[228] Cornelious B, Chen P, Chen Y, de Leon N, Shannon J G and Wang D 2005 Identification of QTLs underlying water-logging tolerance in soybean *Mol. Breed.* **16** 103–12
[229] Valliyodan B, Ye H, Song L, Murphy M, Shannon J G and Nguyen H T 2017 Genetic diversity and genomic strategies for improving drought and waterlogging tolerance in soybeans *J. Exp. Bot.* **68** 1835–49
[230] Nguyen V T, Vuong T D, VanToai T, Lee J D, Wu X, Mian M A R, Dorrance A E, Shannon J G and Nguyen H T 2012 Mapping of quantitative trait loci associated with resistance to and flooding tolerance in soybean *Crop Sci.* **52** 2481–93
[231] Yeboah M A, Xuehao C, Feng C R, Alfandi M, Liang G and Gu M 2008 Mapping quantitative trait loci for waterlogging tolerance in cucumber using SRAP and ISSR markers *Biotech.* **7** 157–67
[232] Ballesteros D C, Mason R E, Addison C K, Andrea Acuña M, Nelly Arguello M, Subramanian N, Miller R G, Sater H, Gbur E E and Miller D *et al* 2015 Tolerance of wheat to vegetative stage soil waterlogging is conditioned by both constitutive and adaptive QTL *Euphytica* **201** 329–43
[233] Yu M, Mao S, Chen G, Liu Y, Li W, Wei Y, Liu C and Zheng Y 2014 QTLs for waterlogging tolerance at germination and seedling stages in population of recombinant

inbred Lines derived from a cross between synthetic and cultivated wheat genotypes *J. Integr. Agric.* **13** 31–9

[234] Locke A M, Barding G A Jr, Sathnur S, Larive C K and Bailey-Serres J 2018 Rice SUB1A constrains remodelling of the transcriptome and metabolome during submergence to facilitate post-submergence recovery *Plant Cell Environ.* **41** 721–36

[235] Minami A, Yano K, Gamuyao R, Nagai K, Kuroha T and Ayano M *et al* 2018 Time-course transcriptomics analysis reveals key responses of submerged deepwater rice to flooding *Plant Physiol.* **176** 3081–102

[236] Narsai R, Howell K A, Carroll A, Ivanova A, Millar A H and Whelan J 2009 Defining core metabolic and transcriptomic responses to oxygen availability in rice embryos and young seedlings *Plant Physiol.* **151** 306–22

[237] Garg R, Verma M, Agrawal S, Shankar R, Majee M and Jain M 2014 Deep transcriptome sequencing of wild halophyte rice, *Porteresia coarctata*, provides novel insights into the salinity and submergence tolerance factors *DNA Res.* **21** 69–84

[238] Dossa K *et al* 2019 Transcriptomic profiling of sesame during waterlogging and recovery *Sci. Data.* **6** 204

[239] Wang L, Li D, Zhang Y, Gao Y, Yu J, Wei X and Zhang X 2016 Tolerant and susceptible sesame genotypes reveal waterlogging stress response patterns *PLoS One* **11** e0149912

[240] Chen W *et al* 2016 Identification and comparative analysis of differential gene expression in soybean leaf tissue under drought and flooding stress revealed by RNA-Seq *Front. Plant Sci.* **7** 1044

[241] Li Z, Zhong Y, Bai D, Abid M, Zhang Y, Lin M, Qi X, Hu C and Fang J 2020 Comparative transcriptome analysis of two contrasting kiwifruit (*Actinidia*) genotypes under waterlogging stress *Preprints* https://doi.org/10.20944/preprints202007.0395.v1

[242] Zhai L, Liu Z, Zou X, Jiang Y, Qiu F, Zheng Y and Zhang Z 2013 Genome-wide identification and analysis of microRNA responding to long-term waterlogging in crown roots of maize seedlings *Physiol. Plant.* **147** 181–93

[243] Xu X, Chen M, Ji J, Xu Q, Qi X and Chen X 2017 Comparative RNA-seq based transcriptome profiling of waterlogging response in cucumber hypocotyls reveals novel insights into the de novo adventitious root primordia initiation *BMC Plant Biol.* **17** 129

[244] Safavi-Rizi V, Herde M and Stöhr C 2020 RNA-Seq reveals novel genes and pathways associated with hypoxia duration and tolerance in tomato root *Sci. Rep.* **10** 1692

[245] Rivera-Contreras I, Zamora-Hernández T and Huerta-Heredia A *et al* 2016 Transcriptomic analysis of submergence-tolerant and sensitive *Brachypodium distachyon* ecotypes reveals oxidative stress as a major tolerance factor *Sci. Rep.* **6** 27686

[246] Van Veen H, Vashisht D, Akman M, Girke T, Mustroph A, Reinen E, Hartman S, Kooiker M, van Tienderen P and Schranz M E *et al* 2016 Transcriptomes of eight *Arabidopsis thaliana* accessions reveal core conserved, genotype- and organ-specific responses to flooding stress *Plant Physiol.* **172** 668–89

[247] Yeung E, van Veen H, Vashisht D, Sobral Paiva A L, Hummel M and Rankenberg T *et al* 2018 A stress recovery signaling network for enhanced flooding tolerance in *Arabidopsis thaliana Proc. Natl Acad. Sci. USA* **115** E6085–94

[248] van Dongen J T, Fröhlich A, Ramírez-Aguilar S J, Schauer N, Fernie A R, Erban A, Kopka J, Clark J, Langer A and Geigenberger P 2009 Transcript and metabolite profiling of the adaptive response to mild decreases in oxygen concentration in the roots of *Arabidopsis* plants *Ann. Bot.* **103** 269–80

[249] Nuanlaong S, Wuthisuthimethavee S, Mekanawakuland M and Suraninpong P 2020 Transcriptome analysis of oil palm (*Elaeis guineensis* Jacq.) roots under waterlogging stress *Plant Omics J.* **13** 46–56

[250] Zou X L, Zeng L, Lu G Y, Cheng Y, Xu J S and Zhang X K 2015 Comparison of transcriptomes undergoing waterlogging at the seedling stage between tolerant and sensitive varieties of *Brassica napus* L. *J. Integr. Agric.* **14** 1723–34

[251] Wittig P R, Stefanie A, Müller J T, Bammer B, Álvarez-Cansino L, Konnerup D, Pedersen O and Mustroph A 2020 Two *Brassica napus* cultivars differ in gene expression, but not in their response to submergence *Physiol. Plant.* **171** 400–15

[252] Zou X L, Tan X Y, Hu C W, Zeng L, Lu G Y and Fu G P 2013b The transcriptome of *Brassica napus* L. roots under waterlogging at the seedling stage *Int. J. Mol. Sci.* **14** 2637–51

[253] Li J, Iqbal S, Zhang Y, Chen Y, Tan Z, Ali U and Guo L 2021a Transcriptome analysis reveals genes of flooding-tolerant and flooding sensitive rapeseeds differentially respond to flooding at the germination stage *Plants* **10** 693

[254] Lee Y-H, Kim K-S, Hwang J-H, Jang Y-S, Lee D-H and Choi I-H 2014 Global gene expression responses to waterlogging in leaves of rape seedlings *Plant Cell Rep.* **33** 289–99

[255] Li Y, Shi L-C, Yang J, Qian Z-H, He Y-X and Li M-W 2021b Physiological and transcriptional changes provide insights into the effect of root waterlogging on the aboveground part of *Pterocarya stenoptera Genomics* **113** 2583–90

[256] Yuan Z, Ni X, Arif M, Dong Z, Zhang L, Tan X, Li J and Li C 2021 Transcriptomic analysis of the photosynthetic, respiration, and aerenchyma adaptation strategies in Bermudagrass (*Cynodon dactylon*) under different submergence stress *Int. J. Mol. Sci.* **22** 7905

[257] Rohini G, Mohit V, Shashank A and Rama S *et al* 2013 Deep transcriptome sequencing of wild halophyte rice, *Porteresia coarctata*, provides novel insights into the salinity and submergence tolerance factors *DNA Res.* **21** 69–84

IOP Publishing

Molecular Response and Genetic Engineering for Stress in Plants, Volume 1
Abiotic stress
Pawan Shukla, Anirudh Kumar, Rakesh Kumar and Manish K Pandey

Chapter 11

Understanding nitric oxide signaling: plant abiotic stress perspective

Mallikarjuna Rao Puli[1*]

[1]*Tomato Genomics Research, Department of Plant Sciences, University of Hyderabad, Hyderabad 500046, India*

Plants face challenges from several abiotic sources during their lifetime and, consequently, obstacles to their growth, physiology, and development. To counter them effectively, plants follow different strategies, including generating nitric oxide (NO). Upon sensing abiotic stress, plants generate NO directly or through plant hormones from various sources. Nitric oxide synthase (NOS)-like activity and nitrate reductases contribute to maximum NO biogenesis in plants. NO can induce short-term changes, like prevention of water loss through inducing stomatal closure via activating signaling cascades, and long-term changes by regulating transcription, post-translational modifications, and epigenetic changes. S-nitrosoglutathione acts as an NO reservoir and can be transported to various tissues and facilitate *in planta* NO signaling. Applying synthetic NO donors or transient ectopic expression of NOS from cyanobacteria, green alga, and mammalian sources can enhance plant tolerance towards abiotic stresses. This chapter summarizes the consequences of abiotic stresses in plants, NO production by various abiotic sources, modes of NO manipulation and detection, mechanisms of NO action, and the consequences of NO modulation against abiotic stresses in plants.

11.1 Introduction

Nitric oxide (NO), a small gaseous molecule, was initially identified as an air pollutant. However, systematic studies revealed its function as a signaling molecule across biological systems and proposed it to play an important role in signaling networks. NO has emerged as an essential component in plants' defenses against

* Address for correspondence Email: mallikarjunpuli@gmail.com.

doi:10.1088/978-0-7503-4921-5ch11

abiotic and biotic stresses. NO controls numerous critical physiological and developmental processes such as stomatal conductance, photosynthesis, respiration, seed germination, root growth, flowering, and adaptive responses to biotic and abiotic stresses, as well as iron homeostasis [1–4].

Plants encounter continuous challenges from climatic and geochemical fluctuations throughout their life cycle, either alone or in combination [5]. Alterations in water content (drought, submergence), temperature (radiation, heat, cold), gas composition (through pollutants), soil chemical composition (salts, heavy metals), and physical challenges constitute abiotic stresses. These abiotic stresses cause a severe challenge to plant survival and productivity [5–9]. Drought and salinity are two major abiotic factors that cause severe water scarcity to plants through osmotic stress. In contrast, flooding and submergence (due to heavy rains and waterlogging) form an opposite challenge [10, 11]. Water scarcity and flooding cause survival and development defects in crops, resulting in losses to productivity [11–13]. Yield losses concerning drought and salinity are mainly associated with their detrimental effects on flower formation, maintenance, pollen viability, seed number, and seed maturation [12]. Contrarily, submergence of plant shoots and roots causes hypoxic and diminished light conditions. Plants adapted for terrestrial habitats cannot tolerate continuous submergence. Flooding drastically affects photosynthesis and aerobic respiration in submerged plant parts, which can however survive up to a certain extent under anaerobic conditions [10]. Global climatic change and air pollutants have been considered an additive contribution to the complexity of such abiotic stresses [5, 14–16]. Air pollutants carrying nitrogen, sulfur, and carbon oxides and ozone have a substantial effect on plant survival [15, 16]. Heavy metals such as cadmium (Cd), lead (Pb), and arsenic (As), and other metallic micronutrients such as manganese, (Mn), copper (Cu), zinc (Zn), and selenium (Se) at higher concentrations cause severe effects on growth, photosynthesis, membrane lipid composition, and water content in crop plants [17–22].

Plants can detect fluctuations in abiotic factors through their sensory mechanisms, either directly or through plant hormones, and cope with them by adjusting their physiology and development [9, 23]. Plants can sense direct signals from alterations in quality of light and temperature through specific receptors such as phototropins, phytochromes, ultraviolet B (UV-B) receptors, and cryptochromes, and coordinate their developmental plasticity and physiology to adapt accordingly [24–27]. However, other abiotic stressors incur their effects on the plants through alterations in plant hormone compositions [28]. The plant hormone abscisic acid (ABA) plays a central role in regulating several abiotic stresses in plants [29, 30]. Plants' tolerances towards abiotic stress are not limited to those involving ABA only: there also exist ABA-independent mechanisms and cross-talk between them [31–33]. Gibberellic acids (GAs), jasmonic acid (JA), ethylene, auxins, brassinosteroids (BRs), strigolactones, phytochromes, karrikins, and cytokinins have been reported to participate to a certain extent in various abiotic stresses [10, 34–38]. For instance, synergistic interaction between ABA and BRs plays a critical role in salt tolerance in rice [39].

The generation of NO has been reported as an essential signal upon plant exposure to various abiotic stressors, and it plays a pivotal role in the plant's protection against them [40]. This chapter summarizes the current knowledge and updates on NO biosynthesis, signaling, and cross-talk with stress regulators, emphasizing abiotic stresses and how chemical or genetic manipulations of NO can lead to crop protection against them. An account of the current approaches used to detect and characterize NO is also given.

11.2 Nitric oxide production: an evident outcome of multiple abiotic stresses

NO generation has emerged as an important early signal that appears as part of abiotic stress adjustment in several plants. An array of abiotic stressors, such as drought, salinity, heat, low temperature, heavy metals, and ozone, has been proposed to involve NO signaling in plants [40, 41]. Drought- or dehydration-induced direct elevation of NO was reported in the root apices of cucumber and nitrate- or ammonium-treated rice seedlings [42, 43]. However, drought-related NO generation was reported through ABA signaling in guard cells during stomatal closure [44–46]. Salt (NaCl) exposure induced NO production in roots of red kidney beans (*Phaseolus vulgaris*), calluses of reed (*Phragmites communis* Trin.), and *Populus euphratica* [47–49]. Heat stress was shown to elevate NO levels in *Arabidopsis thaliana* roots, *Medicago sativa*, and *Nicotiana tabacum* leaf peels and cell suspensions while reducing it in *Pisum sativum* leaves, *Chrysanthemum morifolium* seedlings, and guard cell protoplasts of *N. glauca* [41, 50]. In *Arabidopsis*, *hot5* (which encodes a GSNO oxidoreductase, GSNOR) null alleles displayed higher NO, and other reactive nitrogen species were defective in thermotolerance and reproductive development. Removal of NO rescued these phenotypes [51]. Heavy metal antimony toxicity induced NO generation in roots of *Dittrichia viscosa* [52], while Se reduced internal NO in wheat leaves and roots [22]. Simulated acid rains, by applying sulfate and nitrate mixtures at low pH, reduced NO content in mung beans [53]. UV-B radiation induced NO production in excised leaves of kidney beans [54].

11.3 Nitric oxide biosynthesis in plants: molecular perspective

11.3.1 Sources of NO production in the plant kingdom

Biosynthesis of NO has been proposed to happen through various roots in plants that include oxidative and reductive means by both enzymatic and nonenzymatic methods (figure 11.1). NOS-like activities, nitrate reductases (NRs), the mitochondrial electron transport system, and xanthine dehydrogenase/oxidase have been reported to produce NO in plants [40, 55]. Nonenzymatic chemical degradation of nitrite to NO at acidic pH contributes to plant NO production [56]. Polyamines and hydroxylamines also contribute to NO formation in plants during developmental or stress responses [55].

The principal oxidative means of NO production was proposed through nitric oxide synthase (NOS)-like activity in plants (figure 11.1(a)). Mammalian tissues produce NO as a byproduct while oxidizing arginine to citrulline through three defined NOS enzymes, namely endothelial (eNOS), inducible (iNOS), and neuronal

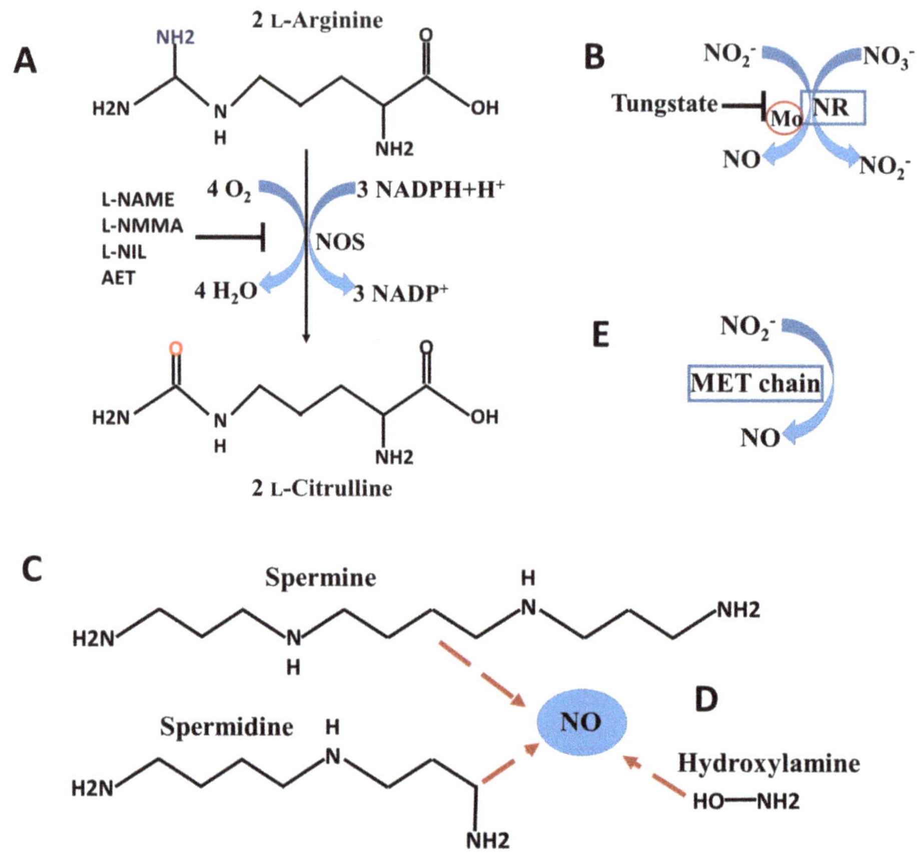

Figure 11.1. NO biosynthesis in plants. (a) NOS enzymes generate NO through the conversion of L-arginine to L-citrulline; (b) nitrate reductase enzymes generate NO through reduction of nitrite; (c) polyamines spermine and spermidine; (d) hydroxylamine generates NO through as yet unknown mechanisms; and (e) the mitochondrial electron transport chain (probably cytochrome C) generates NO through reduction of nitrite.

(nNOS) nitric oxide synthases [57, 58]. Molecular oxygen accepts electrons from NADPH+H^+ during this reaction [57–59].

NO generation by plant extracts via conversion of arginine to citrulline, inhibition of this process by mammalian NOS inhibitors, and detection of peptides by NOS antibodies suggest NOS-related activity in plants [60–63]. Surprisingly, characterizing the defined plant-specific NOS enzymes by genome sequence analysis has not so far been successful in terrestrial plants [2, 40, 55]. In *Arabidopsis*, initially, the AtNOS1/NO-associated protein-1 (AtNOA1) was proposed as a plant NOS enzyme capable of NO production [64]. Later, it was characterized as a GTPase that participates in ribosomal biogenesis and probably indirectly contributes to NO elevation [65]. However, a defined NOS-like enzyme was identified from a marine green alga *Ostreococcus tauri* (OtNOS) carrying 45% sequence similarities with human eNOS and the capability of utilizing L-arginine as a substrate for NO production [66]. A sequence-based transcriptomic search for NOS enzymes in 1300 species of the plant kingdom, including 1087 land plants and 265 algal species, suggested that only 15 species of algae (though none of the land plants) possess the defined NOS-like sequences [67]. However, this study did not rule out the presence of a multipeptide noncanonical NOS in higher plants.

Although the molecular mechanisms have not yet been established, polyamines and hydroxylamine generate NO via oxidative roots. The elevation of polyamines spermine and spermidine induce NO release in plants (figure 11.1(c)). Hydroxylamine-mediated NO production involves the interaction of hydroxylamine with reactive oxygen intermediates [40, 55] (figure 11.1(d)).

Nitrate reductases are well-established plant proteins that generate NO via reductive pathways (figure 11.1(b)). Even though NO production's oxidative roots were not well established, NO generation via nitrite reduction (NO_2^-) by NRs were successfully characterized in land plants [45, 68, 69]. During NO generation, plants are frequently susceptible to NR inhibitors such as tungstate [45, 46, 70]. Similarly, an NR-deficient double mutant of *nia1nia2* was impaired in the accumulation of NO or NO mediated responses in both roots and stomata of *Arabidopsis* [45, 71]. NRs preferentially catalyze nitrate to nitrite reduction utilizing a NAD(P)H under typical cellular conditions, where typical cellular nitrite levels are up to 100 times higher than nitrite. However, NR uses nitrite as a substrate instead of nitrate during certain situations and converts it to NO and its reactive product, peroxynitrite ($ONOO^-$) [44, 45]. Despite both NIA1 and NIA2 being capable of reducing nitrate to nitrite and nitrite to NO, NIA1 prefers nitrite reduction while NIA2 prioritizes nitrate reduction in isoform-specific studies by ectopic expression [72]. A plasma membrane-bound nitrite, NO-reductase (NI-NOR), was found to produce apoplastic NO at slightly acidic pH (6.0) in roots of *Nicotiana tabacum* L. cv. Samsun [73]. An additional source of NO generation was reported from tobacco via the mitochondrial electron transport system, probably through cytochrome C using nitrite as a substrate [74] (figure 11.1(e)). NO generation through this process was found to participate in pathogenesis by *Pseudomonas syringae* in *Arabidopsis* [75].

11.3.2 Plant hormones that regulate NO biosynthesis in plants during abiotic stress

Despite several plant hormones contributing to plant resistance against abiotic stresses, the chief plant hormone regulating abiotic stresses is ABA. The perception of abiotic stress by plants leads to ABA biosynthesis, which leads to both short- and long-term effects on plant physiology and development. The immediate response after ABA accumulation by dehydration stress is induction of stomatal closure and thereby prevention of water loss through guard cells. ABA also alters the root vascular differentiation and protoxylem formation through miRNA expression in endodermal initials [76]. ABA or ABA analog pyrabactin elevate NO production in guard cells while inducing stomatal closure via both NR and NOS-like activities [44–46, 70], suggesting that NO biosynthesis is part of ABA signaling and plays a role downstream of it during stomatal closure. *Arabidopsis* NR double mutants *nia1nia2* and *Atnoa1* were impaired in ABA-induced NO production and activation of ion channels in guard cells leading to impaired stomatal closure, suggesting that NO deficiency leads to ABA hyposensitivity [44, 64, 71] and supposed stress susceptibility. However, in contrast, maximal uncoupling of NO biosynthesis through a triple knockout of *nia1nia2Atnoa1* was hypersensitive to ABA during stomatal closure and seed germination and resistant to drought stress [77]. Besides ABA, other plant hormones that can induce stomatal closure also produce NO in guard cells. For example, salicylic acid (SA) induced NO production in guard cells of *Arabidopsis* while inducing stomatal closure. The SA-induced generation of NO and stomatal closure was impaired in *nia1* and *nia2* single mutants and *nia1nia2* double mutant, and susceptible to both NR and NOS inhibitors [78]. Ethylene donor 1-aminocyclopropane-1-carboxylic acid induced NO production while inducing stomatal closure in *Arabidopsis* [79]. Methyl jasmonate, a methyl ester of JA, induced NO production in *Taxus* guard cells during stomatal closure and cell cultures while inducing defense responses and secondary metabolism [80, 81]. Synthetic auxins like indole butyric acid also induce NO production while inducing later root primordia in *Arabidopsis* [82].

In contrast to ABA-induced NO production, NO can induce ABA biosynthesis while protecting plants from abiotic stressors. NO mediates salt-induced ABA biosynthesis in maize roots, and external application of sodium nitroprusside (SNP) promotes ABA biosynthesis [83]. NO mediates the biosynthesis of ABA through the *vp14* gene by BRs while protecting maize plants from polyethylene glycol (PEG)-induced oxidative stress [84]. Polyamine-mediated NO production has been proposed to be involved in root development and embryogenesis [85], cadmium toxicity [17], and drought stress [42].

11.3.3 Tools for detecting and manipulating NO in plants

11.3.3.1 Synthetic compounds in NO research

Several pharmacological compounds successfully tested in mammalian research were used to donate, absorb, and inhibit NO biosynthesis in plants. The major NO donors that release NO include SNP (figure 11.2(a)), S-nitro-N-acetylpenicillamine

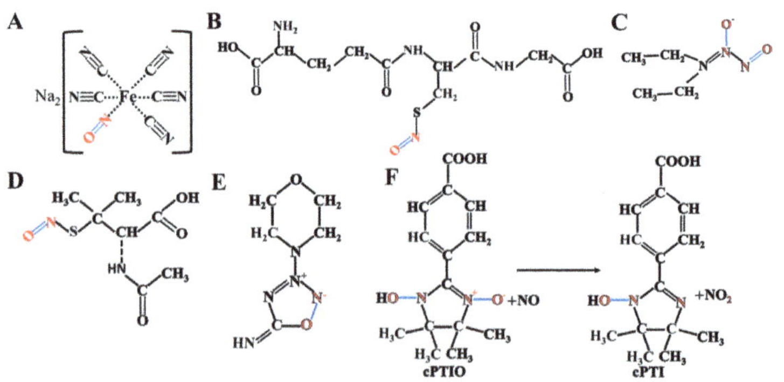

Figure 11.2. NO donors and scavengers. (a) SNP, (b) GSNO, (c) NONOates, (d) SNAP, (e) SIN-1; and (f) cPTIO and its reaction with NO leads to the formation of cPTI and NO_2. PTIO does not carry the carboxyl group and differs slightly from cPTIO and forms PTI after reaction with NO.

(SNAP) (figure 11.2(d)), S-nitroglutathione (GSNO) (figure 11.2(b)), NONOates (figure 11.2(c)), 3-morpholinosydnonimine (SIN-1) (figure 11.2(e)), and organic nitrates. SNP, SNAP, and GSNO have been widely used in plants as NO donors. In solution, SNP displayed a slow rate of NO release with a prolonged half-life (12 hr) compared to GSNO (7 hr) and SNAP (3 h)r, and SNAP is capable of producing NO at a higher rate compared to GSNO and SNP [86]. The most widely used NO scavenger in plants is 2-phenyl-4,4,5-tetramethylimidazole-1-oxyl 3-oxide (PTIO) or its carboxy derivative, cPTIO, and trimethylammonio-PTIO. PTIO reacts with NO to form PTI and NO_2 (figure 11.2(f)). Though there were questions over the assurance of the effects of PTIO or cPTIO working through NO quenching alone, these NO scavengers were shown to play a pivotal role in the confirmation of NO signaling in plants [87].

Along with NO donors and scavengers, several studies have utilized inhibitors of NO biosynthesis through the competitive inhibition of NOS enzyme activity and inhibitors of NRs. Structural analogs of L-arginine such as N(G)-nitro-L-arginine methyl ester (L-NAME), N(G)-monomethyl-L-arginine acetate (L-NMMA), L-N6-(1-Iminoethyl)lysine, and 2-aminoethyl-isothiourea are competitive inhibitors of mammalian NOS enzymes. L-NAME and L-NMMA have been effectively applied in plant research to confirm the presence of NOS activity in plants. Plant NRs use molybdenum (Mo) as a cofactor. Tungstate can replace Mo as a cofactor and acts as an inhibitor of Mo-cofactor-containing enzymes, including NRs, and is used accordingly as an inhibitor of NRs in plant research. Although tungstate is not a specific inhibitor of NR, a wide array of NO research utilizes it as a source for NR function, particularly with regards to stomatal and ABA signaling pathways [46, 68–70]. Although all the above chemicals are not extremely specific to the purpose and have limitations, they have played a significant role in NO research.

11.3.3.2 Detection of NO in plants

Detection of precise levels of NO in plants is challenging and capricious due to its unique properties such as short half-life, free mobility across membranes, and ubiquitous presence at submicromolar concentrations in plant tissues, as well as it responding to various developmental and physiological signals [86, 88]. Various detection methods have been employed in plants depending on the nature of the studies and tissue types. Such methods have primarily utilized spectrophotometric, fluorescence, and luminescence detection methods [88]. *In planta* NO localization is detected through the following widely used techniques.

The oxyhemoglobin (HbO_2) method. An *in planta* assay relays the interaction between NO and HbO, which generates methemoglobin (MetHb) and nitrate, leading to a shift in absorbance from 415–421 to 401 [89]. The HbO_2 method is a tried-and-tested and robust assay that can detect NO up to 1.3 to 2.8 nM. It has limitations due to the nonspecificity of HbO_2 to NO and H_2O_2 and the demand for the generation of fresh HbO_2.

Greiss method and assay kits. This method utilizes the technique of azo dye formation through NO interaction with sulphanilic acid and α-naphthylamine. In this assay, NO interacts with CrO_3 and rapidly oxidizes to NO_2. NO_2 then interacts with sulfanilic acid to form an intermediate diazonium salt, which further interacts with N-(1-naphthyl) ethylenediamine (NED) to form the azo dye (λmax 540 nm), a stable water-soluble compound. The Griess assay has limitations in detecting NO at submicromolar concentrations. However, a modified method of Greiss reaction was generated that can detect the generation of NO up to the nanomolar range [86, 88, 90].

Fluorescence dyes. The most widely used method of NO detection for *in planta* and intracellular localizations are the diaminofluorescence (DAF) dyes 4, 5-diaminofluorescence diacetate (DAF-2DA) and 4-Amino-5-methylamino-2',7'-difluorofluorescein diacetate (DAF-FM DA) [86, 88]. Both DAF-2DA and DAF-FM DA are cell permeable, more efficient, and can detect NO up to as low as 5 nM and 3 nM, respectively [86, 89]. Upon their loading, intracellular esterases will activate and remove their esters to form nonfluorescent DAF-2 and DAF-FM, respectively. In living cells, NO rapidly interacts with molecular oxygen to form N2O3, which converts DAF-2 and DAF-FM into the highly fluorescent stable compounds DAF-2T and DAF-FM T, respectively [90, 91]. To confirm NO detection through DAF dyes, particularly DAF-2DA, cPTIO has been widely used in plant cells, as cPTIO rapidly interacts with NO to form an NO_2 radical and protects it from oxidization. However, in a study cPTIO did not mask the NO-related fluorescence when DAF-FM was used as a detector [90]. DAF-based dyes also have limitations such as that they are light-sensitive and prone to react with dehydroascorbic acid and ascorbic acid at higher concentrations (>5 mM) that give rise to fluorescent compounds with similar detection ranges. As DAF dyes can only detect N_2O_3 (the formation of which requires molecular oxygen) but not NO directly, their practical application in plant cells surviving under anaerobic conditions is debatable [86, 88].

Spin-trapping method. Electron paramagnetic resonance (EPR), also known as electron spin resonance or electron magnetic resonance, utilizes the trapping and detection of transient and reactive free radicals on a highly concentrated diamagnetic spin trap [86, 92]. As the steady-state level is lower for EPR detection, it requires more stable spin traps like iron-dithiocarbamates to properly detect NO through EPR [86, 88, 92]. Despite NO detection through EPR having several advantages, plant biology has less received attention using this method due to issues of cost effectiveness and technical requirements (table 11.1).

Chemiluminescence method. Extracellular NO detection in plants has been mainly achieved through the use of the chemiluminescence method. In this method, NO interacts with ozone to form an exciting high-energy state NO_2 radicle that emits light at >600 nm while reaching the ground state. A photomultiplier tube absorbs this light and detects the NO level based on the intensity of emitted light [86, 88]. The chemiluminescence method can be employed to measure NO both in gaseous and dissolved states but cannot detect NO's oxidized forms.

NO *in planta* detection can also be employed through NO-specific electrodes and mass spectroscopy [88].

Laser-based infrared spectroscopic methods have also been used to detect NO emission from plants; however, they have attracted less attention. Despite the several methods available for NO detection, DAF-based fluorescence methods have

Table 11.1. Effects of external application of NO donors on plant abiotic stresses caused by drought, heat, salinity, and heavy metals.

Method of NO manipulation	Plant species	Abiotic stress affected	References
SNP (200 μM) and GSH (200 μM) in Hyponex hydroponics solution	Rice (*Oryza sativa* L.)	Protected seedlings from uptaking Cu and Cu-induced toxicity on growth, biomass, water content, and photosynthesis pigments.	[123]
SNP (100 μM) in Hoagland's solution	Tomato (*Solanum lycopersicum* L.)	Protected plants from uptaking Cd and Cd-induced toxicity on growth, biomass, water content, and photosynthesis pigments.	[110]
SNP (500 μM) foliar spray	Bean (*Phaseolus vulgaris* L.)	Protected from Mn toxicity on leaf area, leaf water content, chlorosis, necrosis spots, antioxidant content, and membrane lipid composition.	[20]
SNP (100 μM) in nutrient solution	Wheat (*Triticum aestivum* L.)	Protected plants from selenium toxicity and internal NO required for spermine-induced protection from selenium toxicity on plant growth, photosynthesis, water content, and antioxidants.	[22]
SNP (100 μM) in nutrient solution	Luffa (*Luffa acutangula* (L.) Roxb.)	Protected plants from arsenic toxicity on photosynthesis and oxidative stress.	[18]
SNP (25 μM) foliar spray	Safflower (*Carthamus tinctorius* L.)	Protected plants from drought-induced inhibition of growth and photosynthesis.	[124]
SNP (200 μM) foliar spray	Chickpea (*Cicer arietinum* L.)	Protected plants from salt-induced changes in water content, lipid peroxidation, and electrolyte leakage.	[117]

(*Continued*)

Table 11.1. *(Continued)*

Method of NO manipulation	Plant species	Abiotic stress affected	References
SNP (50 μM) added to nutrient solution	Chickpea (Cicer arietinum L.)	Protected plants from salt-induced changes in growth, biomass, water content, chlorophylls, antioxidants, and electrolyte leakage.	[118]
SNP (100 μM) foliar spray	Mustard (*Brassica juncea* L.)	Protected plants from salt-induced changes in water content, chloroplast organization, photosynthesis, and growth performance.	[119]
SNP (50–100 μM) added to Murashige and Skoog medium growth medium	Raspberry (*Rubus idaeus* var. Danehdrosht.)	Protected plants from salt-induced changes in morphophysiological characters and reduction in water content, membrane stability, soluble sugars, and antioxidants.	[120]
SNP (100 μM) combined with PGPR biofilms	Sugarcane (*Saccharum officinarum* L.)	Enhanced biofilm formation and root colonization of PGPR. Protected plants from salt-stress-induced growth reduction, oxidative stress, and ion balance.	[121]
SNP (0.25 mM to 0.75 mM) in Hoagland's solution	Mung bean (*Vigna radiata*)	Protected from simulated-acid-rain-induced toxicity and reduction in antioxidants.	[53]
SNP (100 μM) in nutrient solution	Alfalfa (*Medicago sativa* L.)	Protected from iron deficiency.	[125]
GSNO (10–100 μM) foliar spray/chitin nanoparticles embedded GSNO application	Sugarcane (*Saccharum* spp.) cv. IACSP94–2094	Protected from drought-induced changes in water content, photosynthesis, and antioxidants.	[113–115]
Excess nitrate supply	Sugarcane (*Saccharum* spp.) cv. IACSP95–5000	Protected from drought-induced oxidative damage and changes in leaf gas exchange and photosynthesis.	[116]

SNP (100 μM) and GSNO (100 μM) in water and roots immersed for 12 h	Cucumber (*Cucumis sativus* cv. Dar)	Protected plants from progressive-dehydration-induced changes in membrane permeability, water content, and lipid peroxidation.	[42]
SNP (300 μM) foliar spray	Apple rootstocks (*Malus hupehensis* (Pamp.) Rehd. and *Malus sieversii* (Ledeb.) M.Roem.)	Protected plants from drought-induced changes in membrane permeability, melondialdehyde content, water potential, water content, photosynthesis, and antioxidants.	[112]

attracted significant attention in plants, mainly to measure NO levels in guard cells and root apices.

11.4 Nitric oxide signaling: mode of action and cross-talk with co-signaling components

Nitric oxide regulates plant processes through modulation in transcription, post-translational, and epigenetic regulations [93, 94]. NO upregulates the genes related to kinases, biosynthetic genes of plant hormones and cell walls, secondary metabolites, transcription factors, glutathione S-transferase, ATP-binding cassette (ABC) transporters, and defense-related genes [95]. NO mediates post-translational modifications such as S-nitrosylation, the reversible addition of NO to cysteine residues of targeted proteins, to generate S-nitrosothiols in plants. S-nitrosothiols such as GSNO, generated by reaction between the antioxidant glutathione and NO, act as an intracellular store for NO and facilitate its transfer through phloem in plants [93]. GSNO reductase (GSNOR) breaks down GSNO, leading to the prevention of S-nitrosylation of proteins and NO signaling. Reactive oxygen species (ROS) inhibit GSNOR activity, prevent GSNO degradation, and facilitate NO protection of plants [96]. The feedback regulation between ROS and NO through GSNO levels facilitates plant survival under stressful conditions. NO activates epigenetic regulation through the regulation of histone-acetyltransferases during stress responses. Application of GSNO induces the histone-acetylation activation by deactivating histone-deacetylases in *Arabidopsis* [97]. Application of GSNO activated histone-acetylation, whereas cPTIO prevented it, suggesting the epigenetic regulation of targeted proteins by NO signaling [94]. NO protects plants from the deleterious effects of ROS produced by various stressors by activating antioxidant enzymes [98]. NO modulates stomatal closure through several signaling components, including cyclic guanosine monophosphate, mitogen-activated protein kinase, sphingosine kinases and their derivatives, ion channels, and Ca^{2+} pathways [98, 99].

11.5 Modulating nitric oxide biosynthesis in plants for engineering abiotic stress tolerance

Understanding the molecular mechanisms underlying plant responses against abiotic stresses will be critical for the rational design of crops to counter them successfully. Being a chief regulator, ABA biosynthesis, perception, and signaling was traditionally targeted for improving abiotic stress tolarance in several plants [28, 29, 100–102]. ABA or ABA analogs were able to induce stomatal closure and their foliar spray enhanced abiotic stress resistance in different plants [70, 103–105]. Application of BRs increased tolerance to cadmium toxicity in tomato and cold-stress-induced damage in cucumber [106–108]. External application of karrikins improved salt- and mannitol-induced osmotic stress in oil plant *Sapium sebiferum* during seed germination. External application of JA and GA3 alleviated lead and Cd toxicity in tomato and chickpea [19, 109, 110].

Therefore, being a principal abiotic stress mediator, manipulation of the NO metabolism would also seem to be a potential tool for crop improvement. Henceforth, recent studies tested NO modulation via an external application, which successfully alleviated abiotic stress in several plants [111]. NO's exogenous foliar application protected crop plants from oxidative stress and lipid peroxidation caused by various abiotic stressors. External application of SNP and GSNO as root immersion/foliar spray protected cucumber roots and apple rootstocks from dehydration and drought [42, 112]. External NO through foliar spray of GSNO/chitin nanoparticles embedded in GSNO application, or excess nitrate supply, protected sugarcane from drought effects such as elevated antioxidants on the shoots and roots, preventing altered stomatal function and photosynthesis and increased Rubisco activity [113–116]. External application of NO as SNP/SNAP through foliar spray, mixing with growth medium, or PGPR biofilms enhanced crop protection against salt stress in chickpea, mustard, raspberry, and sugarcane [117–121]. NO alleviated salt-stress-induced damage in plant–water relation, photosynthesis, electrolyte leakage, and oxidative damage. In mung beans, NO mediated the methane prevention of osmotic stress by PEG during seed germination [122]. Similarly, application of NO also gave protection against heavy metal toxicity in crop plants. External NO application recovered luffa, rice, tomatoes, wheat, and beans from heavy metals As, Cu, Se, Cd, and Mn induced reduction in photosynthesis, growth, lipid content, and composition [18–22, 110, 123]. In tomato, foliar application of SNP enhanced tolerance to glyphosate by alleviating herbicide-induced redox disorders and stimulating pathways related to herbicide detoxification [111].

Despite the external foliar application of ABA, ABA analogs, and NO donors successfully alleviating abiotic stresses in certain plants, their application on crops has practical limitations. Chemicals like ABA and SNP are light-sensitive and have short life spans. Their repeated administration might not be cost effective. Henceforth, attempts have been made to engineer abiotic stress tolerance by genetic manipulation. ABA biosynthesis, perception, and signaling components were targeted to achieve this purpose in a wide range of plants [100, 102, 126–128]. In addition, loss-of-function or gain-of-function mutants related to other plant hormone signaling pathways have also exhibited abiotic stress tolerance in plants. Ectopic expression of maize phytochrome interacting factor *ZmPIF3* in rice and *Arabidopsis* enhanced abiotic stress tolerance through ABA hypersensitivity [36]. Modulation of BR signaling genes through loss of function in GSK3-like kinase BIN2 or gain of function in transcription factors BZR-1 and BES-1 resulted in cold resistance in *Arabidopsis* [129].

Studies into genetic manipulation of NO signaling through ectopic expression or genome editing of NRs and NOS-related enzymes have not received much attention. Individual loss-of-function alleles of *Atnoa1*, *nia1*, or *nia2*, or *nia1nia2* double mutants were more susceptible to abiotic stresses [130, 131]. Though triple knockout mutants of *nia1nia2Atnoa1* were ABA hypersensitive and resistant to drought, they have severe defects in plant morphology, particularly the plant size and biomass production [77]. Heat-shock-induced NO generation and heat tolerance was rescued when *Arabidopsis* plants were overexpressed with *AtNOA1* or *NIA2* under a cauliflower mosaic virus 35S constitutively active promoter under a cyclic nucleotide-gated ion channel 6 mutation background [50].

Although nothing has been reported on the presence of a defined plant-specific NOS so far, there have been attempts to investigate the transient ectopic expression of algal or mammalian counterparts to validate its significance in plant adaptation towards abiotic stresses. Transient ectopic expression of NOS from green alga *Ostreococcus tauri* (OtNOS) promoted stomatal development and abiotic stress tolerance in *Arabidopsis* [66]. Ectopic expression of rat nNOS enhanced abiotic stress tolerance in *Arabidopsis* and rice while enhancing resistance to microbes in tobacco [132–135]. Transient ectopic expression of SyNOS from cyanobacteria *Synechococcus* PCC 7335 enhanced nitrogen use efficiency, nitrogen assimilation, and nitogen deficiency tolerance, and thereby yield, in *Arabidopsis* [136].

11.6 Conclusions

Protection of crop plants against abiotic stresses has been highly reported by external application of NO in many plants. Engineering plants with enhanced NO by genetic means could produce plants with enhanced tolerance against various abiotic stressors. Hence, novel CRISPR-Cas tools might help generate lines with high NO by targeting promoters and conserved sites of AtNOA, NIA1, and NIA2 homologs in crop plants with elevated expression or hyperactivation. CRISPR-Cas editing will open the way to making genetic changes in plants with segregation out of transgenes and cultivation of plants without fear of horizontal gene pollution.

Acknowledgments

M.R.P. thanks Prof. Rameshwar P Sharma and Dr Sreelaksmi Yellamaraju, Repository of Tomato Genomics Resources, Dept. Plant Sciences, University of Hyderabad, Hyderabad, India for permission to write this chapter during their stay at RTGR.

Multiple choice questions

1. A gaseous molecule identified as an air pollutant but which plays a significant role in plant defense:
(a) Carbon oxides;
(b) Nitric oxide;
(c) Sulfur;
(d) Nitrogen gas.

2. A widely used compound used as donor to detect nitric oxide signaling in plants:
(a) Sodium nitroprusside;
(b) Iodine;
(c) Hydroxylamine;
(d) Phenol.

3. Biosynthesis of nitric oxide in plants is through …
(a) Enzymatic method;
(b) Nonenzymatic method;
(c) Both enzymatic and nonsenzymatic methods;
(d) None of the above.

4. Detection method of nitric oxide in plants:
(a) Oxyhemoglobin (HbO_2) method;
(b) Diaminofluorescence (DAF) dye method;
(c) Spin-trapping method;
(d) All the above.

5. The foliar application of pyrabactin results in …
(a) An increase of nitric oxide production;
(b) Promotion of stomatal closure;
(c) An increase nitrate reductases activity;
(d) All of the above.

Answers: 1 (b); 2 (a); 3 (c); 4 (d); 5 (d).

Descriptive-type questions

1. Define the biosynthetic pathway of nitric oxide in plants.
2. What is the functional role of nitric oxide in plants?
3. What are the methods to detect the nitric oxide in plants?
4. Describe the mechanism by which nitric oxide provides abiotic stress tolerance to plants.
5. What are the important metabolic pathways engineered to improve abiotic stress tolerance?
6. Write short notes on the genetic manipulation of the nitric oxide metabolism in plants.

References

[1] Astier J, Jeandroz S and Wendehenne D 2018 Nitric oxide synthase in plants: the surprise from algae *Plant Sci.* **268** 64–6

[2] Imran Q M, Hussain A, Lee S-U, Mun B-G, Falak N, Loake G J and Yun B-W 2018 Transcriptome profile of NO-induced *Arabidopsis* transcription factor genes suggests their putative regulatory role in multiple biological processes *Sci. Rep.* **8** 1–14

[3] Sánchez-Vicente I, Lechón T, Fernández-Marcos M, Sanz L and Lorenzo O 2021 Nitric oxide alters the pattern of auxin maxima and PIN-FORMED1 during shoot development *Front. Plant Sci.* **12** 630792

[4] Tewari R K, Horemans N and Watanabe M 2021 Evidence for a role of nitric oxide in iron homeostasis in plants *J. Exp. Bot.* **72** 990–1006

[5] Rivero R M, Mittler R, Blumwald E and Zandalinas S I 2021 Developing climate-resilient crops: improving plant tolerance to stress combination *Plant J.* **109** 373–89

[6] Hatfield J L, Boote K J, Kimball B A, Ziska L, Izaurralde R C, Ort D, Thomson A M and Wolfe D 2011 Climate impacts on agriculture: implications for crop production *Agronomy Journal* **103** 351–70

[7] Lobell D B and Gourdji S M 2012 The influence of climate change on global crop productivity *Plant Physiol.* **160** 1686–97

[8] Emamverdian A, Ding Y, Mokhberdoran F and Xie Y 2015 Heavy metal stress and some mechanisms of plant defense response *Sci. World J.* **2015** 756120

[9] Bücker-Neto L, Paiva A L S, Machado R D, Arenhart R A and Margis-Pinheiro M 2017 Interactions between plant hormones and heavy metals responses *Genet. Mol. Biol.* **40** 373–86

[10] Yeung E, Bailey-Serres J and Sasidharan R 2019 After the deluge: plant revival post-flooding *Trends Plant Sci.* **24** 443–54

[11] Ma X, Sukiran N L, Ma H and Su Z 2014 Moderate drought causes dramatic floral transcriptomic reprogramming to ensure successful reproductive development in *Arabidopsis BMC Plant Biol.* **14** 1–16

[12] Su Z, Ma X, Guo H, Sukiran N L, Guo B, Assmann S M and Ma H 2013 Flower development under drought stress: morphological and transcriptomic analyses reveal acute responses and long-term acclimation in *Arabidopsis Plant Cell* **25** 3785–807

[13] Shavrukov Y, Kurishbayev A, Jatayev S, Shvidchenko V, Zotova L, Koekemoer F, de Groot S, Soole K and Langridge P 2017 Early flowering as a drought escape mechanism in plants: How can it aid wheat production? *Front. Plant Sci.* **8** 1950

[14] Lobell D B, Schlenker W and Costa-Roberts J 2011 Climate trends and global crop production since 1980 *Science* **333** 616–20

[15] Devrajani S K, Qureshi M, Imran U and Nisa T U 2020 Impact of gaseous air pollutants on agricultural crops in developing countries: a review *J. Environ. Sci. Public Health* **4** 71–82

[16] Hong C, Mueller N D, Burney J A, Zhang Y, AghaKouchak A, Moore F C, Qin Y, Tong D and Davis S J 2020 Impacts of ozone and climate change on yields of perennial crops in California *Nat. Food* **1** 166–72

[17] Groppa M, Ianuzzo M, Tomaro M and Benavides M 2007 Polyamine metabolism in sunflower plants under long-term cadmium or copper stress *Amino Acids* **32** 265–75

[18] Singh V P, Srivastava P K and Prasad S M 2013 Nitric oxide alleviates arsenic-induced toxic effects in ridged Luffa seedlings *Plant Physiol. Biochem.* **71** 155–63

[19] Ahmad P, Raja V, Ashraf M, Wijaya L, Bajguz A and Alyemeni M N 2021 Jasmonic acid (JA) and gibberellic acid (GA3) mitigated Cd-toxicity in chickpea plants through restricted cd uptake and oxidative stress management *Sci. Rep.* **11** 1–17

[20] Mahjoubi Y, Rzigui T, Kharbech O, Mohamed S N, Abaza L, Chaoui A, Nouairi I and Djebali W 2021 Exogenous nitric oxide alleviates manganese toxicity in bean plants by modulating photosynthesis in relation to leaf lipid composition *Protoplasma* **259** 949–64

[21] Angulo-Bejarano P I, Puente-Rivera J and Cruz-Ortega R 2021 Metal and metalloid toxicity in plants: an overview on molecular aspects *Plants* **10** 635

[22] Hasan M, Alharbi B M, Alhaithloul H A S, Abdulmajeed A M, Alghanem S M, Al-Mushhin A A, Jahan M S, Corpas F J, Fang X-W and Soliman M H 2021 Spermine-mediated tolerance to selenium toxicity in wheat (*Triticum aestivum* L.) depends on endogenous nitric oxide synthesis *Antioxidants* **10** 1835

[23] Zhang H, Zhao Y and Zhu J-K 2020 Thriving under stress: how plants balance growth and the stress response *Dev. Cell* **55** 529–43

[24] Jung J-H, Domijan M, Klose C, Biswas S, Ezer D, Gao M, Khattak A K, Box M S, Charoensawan V and Cortijo S 2016 Phytochromes function as thermosensors in *Arabidopsis Science* **354** 886–9

[25] Legris M, Klose C, Burgie E S, Rojas C C R, Neme M, Hiltbrunner A, Wigge P A, Schäfer E, Vierstra R D and Casal J J 2016 Phytochrome B integrates light and temperature signals in *Arabidopsis Science* **354** 897–900

[26] Legris M, Nieto C, Sellaro R, Prat S and Casal J J 2017 Perception and signalling of light and temperature cues in plants *Plant J.* **90** 683–97

[27] Casal J J and Qüesta J I 2018 Light and temperature cues: multitasking receptors and transcriptional integrators *New Phytol.* **217** 1029–34

[28] Maksymiec W 2007 Signaling responses in plants to heavy metal stress *Acta Physiol. Plant.* **29** 177

[29] Sah S K, Reddy K R and Li J 2016 Abscisic acid and abiotic stress tolerance in crop plants *Front. Plant Sci.* **7** 571

[30] Vishwakarma K, Upadhyay N, Kumar N, Yadav G, Singh J, Mishra R K, Kumar V, Verma R, Upadhyay R and Pandey M 2017 Abscisic acid signaling and abiotic stress tolerance in plants: a review on current knowledge and future prospects *Front. Plant Sci.* **8** 161

[31] Nakashima K, Yamaguchi-Shinozaki K and Shinozaki K 2014 The transcriptional regulatory network in the drought response and its crosstalk in abiotic stress responses including drought, cold, and heat *Front. Plant Sci.* **5** 170

[32] Yoshida T, Mogami J and Yamaguchi-Shinozaki K 2014 ABA-dependent and ABA-independent signaling in response to osmotic stress in plants *Curr. Opin. Plant Biol.* **21** 133–9

[33] Soma F, Takahashi F, Yamaguchi-Shinozaki K and Shinozaki K 2021 Cellular phosphorylation signaling and gene expression in drought stress responses: ABA-dependent and ABA-independent regulatory systems *Plants* **10** 756

[34] Colebrook E H, Thomas S G, Phillips A L and Hedden P 2014 The role of gibberellin signalling in plant responses to abiotic stress *J. Exp. Biol.* **217** 67–75

[35] Du H, Huang F, Wu N, Li X, Hu H and Xiong L 2018 Integrative regulation of drought escape through ABA-dependent and-independent pathways in rice *Mol. Plant* **11** 584–97

[36] Gao Y, Wu M, Zhang M, Jiang W, Liang E, Zhang D, Zhang C, Xiao N and Chen J 2018 Roles of a maize phytochrome-interacting factors protein ZmPIF3 in regulation of drought stress responses by controlling stomatal closure in transgenic rice without yield penalty *Plant Mol. Biol.* **97** 311–23

[37] Nolan T M, Vukašinović N, Liu D, Russinova E and Yin Y 2020 Brassinosteroids: multidimensional regulators of plant growth, development, and stress responses *Plant Cell* **32** 295–318

[38] Li S-M, Zheng H-X, Zhang X-S and Sui N 2020 Cytokinins as central regulators during plant growth and stress response *Plant Cell Rep.* **40** 271–82

[39] Li Q, Xu F, Chen Z, Teng Z, Sun K, Li X, Yu J, Zhang G, Liang Y and Huang X 2021 Synergistic interplay of ABA and BR signal in regulating plant growth and adaptation *Nat. Plants* **7** 1108–18

[40] Fancy N N, Bahlmann A K and Loake G J 2017 Nitric oxide function in plant abiotic stress *Plant Cell Environ.* **40** 462–72

[41] Parankusam S, Adimulam S S, Bhatnagar-Mathur P and Sharma K K 2017 Nitric oxide (NO) in plant heat stress tolerance: current knowledge and perspectives *Front. Plant Sci.* **8** 1582

[42] Arasimowicz-Jelonek M, Floryszak-Wieczorek J and Kubiś J 2009 Involvement of nitric oxide in water stress-induced responses of cucumber roots *Plant Sci.* **177** 682–90

[43] Cao X, Zhu C, Zhong C, Zhang J, Wu L, Jin Q and Ma Q 2019 Nitric oxide synthase-mediated early nitric oxide burst alleviates water stress-induced oxidative damage in ammonium-supplied rice roots *BMC Plant Biol.* **19** 1–14

[44] Bright J, Desikan R, Hancock J T, Weir I S and Neill S J 2006 ABA-induced NO generation and stomatal closure in *Arabidopsis* are dependent on H_2O_2 synthesis *Plant J.* **45** 113–22

[45] Desikan R, Griffiths R, Hancock J and Neill S 2002 A new role for an old enzyme: nitrate reductase-mediated nitric oxide generation is required for abscisic acid-induced stomatal closure in *Arabidopsis thaliana Proc. Natl. Acad. Sci. USA* **99** 16314–8

[46] Gonugunta V K, Srivastava N, Puli M R and Raghavendra A S 2008 Nitric oxide production occurs after cytosolic alkalinization during stomatal closure induced by abscisic acid *Plant Cell Environ.* **31** 1717–24

[47] Zhao L, Zhang F, Guo J, Yang Y, Li B and Zhang L 2004 Nitric oxide functions as a signal in salt resistance in the calluses from two ecotypes of reed *Plant Physiol.* **134** 849–57

[48] Liu Y, Wu R, Wan Q, Xie G and Bi Y 2007 Glucose-6-phosphate dehydrogenase plays a pivotal role in nitric oxide-involved defense against oxidative stress under salt stress in red kidney bean roots *Plant Cell Physiol.* **48** 511–22

[49] Zhang F, Wang Y, Yang Y, Wu H, Wang D and Liu J 2007 Involvement of hydrogen peroxide and nitric oxide in salt resistance in the calluses from *Populus euphratica Plant Cell Environ.* **30** 775–85

[50] Peng X, Zhang X, Li B and Zhao L 2019 Cyclic nucleotide-gated ion channel 6 mediates thermotolerance in *Arabidopsis* seedlings by regulating nitric oxide production via cytosolic calcium ions *BMC Plant Biol.* **19** 1–13

[51] Lee U, Wie C, Fernandez B O, Feelisch M and Vierling E 2008 Modulation of nitrosative stress by S-nitrosoglutathione reductase is critical for thermotolerance and plant growth in *Arabidopsis Plant Cell* **20** 786–802

[52] Espinosa-Vellarino F L, Garrido I, Ortega A, Casimiro I and Espinosa F 2021 Response to antimony toxicity in *Dittrichia viscosa* plants: ROS, NO, H_2S, and the antioxidant system *Antioxidants* **10** 1698

[53] Jiao R, Zhang M, Wei Z, Xu J and Zhang H 2021 Alleviative effects of nitric oxide on Vigna radiata seedlings under acidic rain stress *Mol. Biol. Rep.* **48** 2243–51

[54] Zhang L and Zhao L 2008 Production of nitric oxide under ultraviolet-B irradiation is mediated by hydrogen peroxide through activation of nitric oxide synthase *J. Plant Biol.* **51** 395–400

[55] Yu M, Lamattina L, Spoel S H and Loake G J 2014 Nitric oxide function in plant biology: a redox cue in deconvolution *New Phytol.* **202** 1142–56

[56] Wendehenne D, Pugin A, Klessig D F and Durner J 2001 Nitric oxide: comparative synthesis and signaling in animal and plant cells *Trends Plant Sci.* **6** 177–83

[57] Alderton W K, Cooper C E and Knowles R G 2001 Nitric oxide synthases: structure, function and inhibition *Biochem. J* **357** 593–615

[58] Förstermann U and Sessa W C 2012 Nitric oxide synthases: regulation and function *Eur. Heart J.* **33** 829–37

[59] Feng C 2012 Mechanism of nitric oxide synthase regulation: electron transfer and interdomain interactions *Coord. Chem. Rev.* **256** 393–411

[60] Cueto M, Hernández-Perera O, Martín R, Bentura M L, Rodrigo J, Lamas S and Golvano M P 1996 Presence of nitric oxide synthase activity in roots and nodules of *Lupinus albus FEBS Lett.* **398** 159–64

[61] Durner J, Wendehenne D and Klessig D F 1998 Defense gene induction in tobacco by nitric oxide, cyclic GMP, and cyclic ADP-ribose *Proc. Natl. Acad. Sci.* **95** 10328–33

[62] Ribeiro E A Jr, Cunha F Q, Tamashiro W M and Martins I S 1999 Growth phase-dependent subcellular localization of nitric oxide synthase in maize cells *FEBS Lett.* **445** 283–6

[63] Guo F-Q and Crawford N M 2005 Arabidopsis nitric oxide synthasel is targeted to mitochondria and protects against oxidative damage and dark-induced senescence *Plant Cell* **17** 3436–50

[64] Guo F-Q, Okamoto M and Crawford N M 2003 Identification of a plant nitric oxide synthase gene involved in hormonal signaling *Science* **302** 100–3

[65] Moreau M, Lee G I, Wang Y, Crane B R and Klessig D F 2008 AtNOS/AtNOA1 is a functional *Arabidopsis thaliana* cGTPase and not a nitric-oxide synthase *J. Biol. Chem.* **283** 32957–67

[66] Foresi N, Mayta M L, Lodeyro A F, Scuffi D, Correa-Aragunde N, García-Mata C, Casalongué C, Carrillo N and Lamattina L 2015 Expression of the tetrahydrofolate-dependent nitric oxide synthase from the green alga *Ostreococcus tauri* increases tolerance to abiotic stresses and influences stomatal development in *Arabidopsis Plant J.* **82** 806–21

[67] Jeandroz S, Wipf D, Stuehr D J, Lamattina L, Melkonian M, Tian Z, Zhu Y, Carpenter E J, Wong G K-S and Wendehenne D 2016 Occurrence, structure, and evolution of nitric oxide synthase–like proteins in the plant kingdom *Sci. Signal.* **9** re2

[68] Xiong J, Fu G, Yang Y, Zhu C and Tao L 2012 Tungstate: is it really a specific nitrate reductase inhibitor in plant nitric oxide research? *J. Exp. Bot.* **63** 33–41

[69] Chamizo-Ampudia A, Sanz-Luque E, Llamas A, Galvan A and Fernandez E 2017 Nitrate reductase regulates plant nitric oxide homeostasis *Trends Plant Sci.* **22** 163–74

[70] Puli M R and Raghavendra A S 2012 Pyrabactin, an ABA agonist, induced stomatal closure and changes in signalling components of guard cells in abaxial epidermis of *Pisum sativum* *J. Exp. Bot.* **63** 1349–56

[71] Chen Z H, Wang Y, Wang J W, Babla M, Zhao C, García-Mata C, Sani E, Differ C, Mak M and Hills A 2016 Nitrate reductase mutation alters potassium nutrition as well as nitric oxide-mediated control of guard cell ion channels in *Arabidopsis New Phytol.* **209** 1456–69

[72] Mohn M A, Thaqi B and Fischer-Schrader K 2019 Isoform-specific NO synthesis by *Arabidopsis thaliana* nitrate reductase *Plants* **8** 67

[73] Stöhr C, Strube F, Marx G, Ullrich W R and Rockel P 2001 A plasma membrane-bound enzyme of tobacco roots catalyses the formation of nitric oxide from nitrite *Planta* **212** 835–41

[74] Planchet E, Jagadis Gupta K, Sonoda M and Kaiser W M 2005 Nitric oxide emission from tobacco leaves and cell suspensions: rate limiting factors and evidence for the involvement of mitochondrial electron transport *Plant J.* **41** 732–43

[75] Modolo L V, Augusto O, Almeida I M, Magalhaes J R and Salgado I 2005 Nitrite as the major source of nitric oxide production by *Arabidopsis thaliana* in response to *Pseudomonas syringae FEBS Lett.* **579** 3814–20

[76] Bloch D, Puli M R, Mosquna A and Yalovsky S 2019 Abiotic stress modulates root patterning via ABA-regulated microRNA expression in the endodermis initials *Development* **146** dev177097

[77] Lozano-Juste J and Leo̧n J 2010 Enhanced abscisic acid-mediated responses in nia1nia2-noa1–2 triple mutant impaired in NIA/NR-and AtNOA1-dependent nitric oxide biosynthesis in *Arabidopsis Plant Physiol.* **152** 891–903

[78] Hao F, Zhao S, Dong H, Zhang H, Sun L and Miao C 2010 Nia1 and Nia2 are involved in exogenous salicylic acid-induced nitric oxide generation and stomatal closure in *Arabidopsis J. Integr. Plant Biol.* **52** 298–307

[79] Zhang T Y, Li Z Q, Zhao Y D, Shen W J, Chen M S, Gao H Q, Ge X M, Wang H Q, Li X and He J M 2021 Ethylene-induced stomatal closure is mediated via MKK1/3–MPK3/6 cascade to EIN2 and EIN3 *J. Integr. Plant Biol.* **63** 1324–40

[80] Saito N, Munemasa S, Nakamura Y, Shimoishi Y, Mori I C and Murata Y 2008 Roles of RCN1, regulatory A subunit of protein phosphatase 2A, in methyl jasmonate signaling and signal crosstalk between methyl jasmonate and abscisic acid *Plant Cell Physiol.* **49** 1396–401

[81] Wang J W and Wu J Y 2005 Nitric oxide is involved in methyl jasmonate-induced defense responses and secondary metabolism activities of *Taxus* cells *Plant Cell Physiol.* **46** 923–30

[82] Kolbert Z, Bartha B and Erdei L 2008 Exogenous auxin-induced NO synthesis is nitrate reductase-associated in *Arabidopsis thaliana* root primordia *J. Plant Physiol.* **165** 967–75

[83] Chen K, Li J, Tang J, Zhao F-G and Liu X 2006 Involvement of nitric oxide in regulation of salt stress-induced ABA accumulation in maize seedling *J. Plant Physiol. Mol. Biol.* **32** 577–82

[84] Zhang A, Zhang J, Zhang J, Ye N, Zhang H, Tan M and Jiang M 2011 Nitric oxide mediates brassinosteroid-induced ABA biosynthesis involved in oxidative stress tolerance in maize leaves *Plant Cell Physiol.* **52** 181–92

[85] Tun N N, Holk A and Scherer G F 2001 Rapid increase of NO release in plant cell cultures induced by cytokinin *FEBS Lett.* **509** 174–6

[86] Lau S-E, Hamdan M F, Pua T-L, Saidi N B and Tan B C 2021 Plant nitric oxide signaling under drought stress *Plants* **10** 360
[87] Planchet E and Kaiser W M 2006 Nitric oxide production in plants: facts and fictions *Plant Signal. Behav.* **1** 46–51
[88] Mur L A, Mandon J, Cristescu S M, Harren F J and Prats E 2011 Methods of nitric oxide detection in plants: a commentary *Plant Sci.* **181** 509–19
[89] Murphy M E and Noack E 1994 Nitric oxide assay using hemoglobin method *Methods Enzymol.* **233** 240–50
[90] Vitecek J, Reinohl V and Jones R L 2008 Measuring NO production by plant tissues and suspension cultured cells *Mol. Plant* **1** 270–84
[91] Kojima H, Nakatsubo N, Kikuchi K, Kawahara S, Kirino Y, Nagoshi H, Hirata Y and Nagano T 1998 Detection and imaging of nitric oxide with novel fluorescent indicators: diaminofluoresceins *Anal. Chem.* **70** 2446–53
[92] Csonka C, Páli T, Bencsik P, Görbe A, Ferdinandy P and Csont T 2014 Measurement of nitric oxide in biological samples *Br. J. Pharmacol.* **172** 1620–32
[93] Begara-Morales J C, Mata-Pérez C, Padilla M N, Chaki M, Valderrama R, Aranda-Caño L and Barroso J B 2021 Role of electrophilic nitrated fatty acids during development and response to abiotic stress processes in plants *J. Exp. Bot.* **72** 917–27
[94] Ageeva-Kieferle A, Georgii E, Winkler B, Ghirardo A, Albert A, Hüther P, Mengel A, Becker C, Schnitzler J-P and Durner J 2021 Nitric oxide coordinates growth, development, and stress response via histone modification and gene expression *Plant Physiol.* **187** 336–60
[95] Parani M, Rudrabhatla S, Myers R, Weirich H, Smith B, Leaman D W and Goldman S L 2004 Microarray analysis of nitric oxide responsive transcripts in *Arabidopsis Plant Biotechnol. J.* **2** 359–66
[96] Kovacs I, Holzmeister C, Wirtz M, Geerlof A, Fröhlich T, Römling G, Kuruthukulangarakoola G T, Linster E, Hell R and Arnold G J 2016 ROS-mediated inhibition of S-nitrosoglutathione reductase contributes to the activation of anti-oxidative mechanisms *Front. Plant Sci.* **7** 1669
[97] Mengel A, Ageeva A, Georgii E, Bernhardt J, Wu K, Durner J and Lindermayr C 2017 Nitric oxide modulates histone acetylation at stress genes by inhibition of histone deacetylases *Plant Physiol.* **173** 1434–52
[98] Wani K I, Naeem M, Castroverde C D M, Kalaji H M, Albaqami M and Aftab T 2021 Molecular mechanisms of nitric oxide (NO) signaling and reactive oxygen species (ROS) homeostasis during abiotic stresses in plants *Int. J. Mol. Sci.* **22** 9656
[99] Puli M R, Rajsheel P, Aswani V, Agurla S, Kuchitsu K and Raghavendra A S 2016 Stomatal closure induced by phytosphingosine-1-phosphate and sphingosine-1-phosphate depends on nitric oxide and pH of guard cells in *Pisum sativum Planta* **244** 831–41
[100] Park S-Y, Peterson F C, Mosquna A, Yao J, Volkman B F and Cutler S R 2015 Agrochemical control of plant water use using engineered abscisic acid receptors *Nature* **520** 545–8
[101] Park S-I, Kim J-J, Shin S-Y, Kim Y-S and Yoon H-S 2020 ASR enhances environmental stress tolerance and improves grain yield by modulating stomatal closure in rice *Front. Plant Sci.* **10** 1752
[102] Hu Q, Ao C, Wang X, Wu Y and Du X 2021 GhWRKY1-like, a WRKY transcription factor, mediates drought tolerance in *Arabidopsis* via modulating ABA biosynthesis *BMC Plant Biol.* **21** 1–13

[103] Okamoto M, Peterson F C, Defries A, Park S-Y, Endo A, Nambara E, Volkman B F and Cutler S R 2013 Activation of dimeric ABA receptors elicits guard cell closure, ABA-regulated gene expression, and drought tolerance *Proc. Natl Acad. Sci.* **110** 12132–7

[104] Helander J D, Vaidya A S and Cutler S R 2016 Chemical manipulation of plant water use *Bioorg. Med. Chem.* **24** 493–500

[105] Vaidya A S, Peterson F C, Yarmolinsky D, Merilo E, Verstraeten I, Park S-Y, Elzinga D, Kaundal A, Helander J and Lozano-Juste J 2017 A rationally designed agonist defines subfamily IIIA abscisic acid receptors as critical targets for manipulating transpiration *ACS Chem. Biol.* **12** 2842–8

[106] Hayat S, Hasan S A, Hayat Q and Ahmad A 2010 Brassinosteroids protect *Lycopersicon esculentum* from cadmium toxicity applied as shotgun approach *Protoplasma* **239** 3–14

[107] Hasan S A, Hayat S and Ahmad A 2011 Brassinosteroids protect photosynthetic machinery against the cadmium induced oxidative stress in two tomato cultivars *Chemosphere* **84** 1446–51

[108] Jiang Y P, Huang L F, Cheng F, Zhou Y H, Xia X J, Mao W H, Shi K and Yu J Q 2013 Brassinosteroids accelerate recovery of photosynthetic apparatus from cold stress by balancing the electron partitioning, carboxylation and redox homeostasis in cucumber *Physiol. Plant.* **148** 133–45

[109] Bali S, Kaur P, Kohli S K, Ohri P, Thukral A K, Bhardwaj R, Wijaya L, Alyemeni M N and Ahmad P 2018 Jasmonic acid induced changes in physio-biochemical attributes and ascorbate-glutathione pathway in *Lycopersicon esculentum* under lead stress at different growth stages *Sci. Total Environ.* **645** 1344–60

[110] Ahmad P, Ahanger M A, Alyemeni M N, Wijaya L and Alam P 2018 Exogenous application of nitric oxide modulates osmolyte metabolism, antioxidants, enzymes of ascorbate-glutathione cycle and promotes growth under cadmium stress in tomato *Protoplasma* **255** 79–93

[111] Soares C, Rodrigues F, Sousa B, Pinto E, Ferreira I M, Pereira R and Fidalgo F 2021 Foliar application of sodium nitroprusside boosts *Solanum lycopersicum* L. tolerance to glyphosate by preventing redox disorders and stimulating herbicide detoxification pathways *Plants* **10** 1862

[112] Zhang L, Li X, Li X, Wei Z, Han M, Zhang L and Li B 2016 Exogenous nitric oxide protects against drought-induced oxidativestress in *Malus* rootstocks *Turkish J. Bot.* **40** 17–27

[113] Silveira N M, Frungillo L, Marcos F C, Pelegrino M T, Miranda M T, Seabra A B, Salgado I, Machado E C and Ribeiro R V 2016 Exogenous nitric oxide improves sugarcane growth and photosynthesis under water deficit *Planta* **244** 181–90

[114] Silveira N M, Marcos F C, Frungillo L, Moura B B, Seabra A B, Salgado I, Machado E C, Hancock J T and Ribeiro R V 2017 S-nitrosoglutathione spraying improves stomatal conductance, Rubisco activity and antioxidant defense in both leaves and roots of sugarcane plants under water deficit *Physiol. Plant.* **160** 383–95

[115] Silveira N M, Seabra A B, Marcos F C, Pelegrino M T, Machado E C and Ribeiro R V 2019 Encapsulation of S-nitrosoglutathione into chitosan nanoparticles improves drought tolerance of sugarcane plants *Nitric Oxide* **84** 38–44

[116] Pissolato M D, Silveira N M, Prataviera P J C, Machado E C, Seabra A B, Pelegrino M T, Sodek L and Ribeiro R V 2020 Enhanced nitric oxide synthesis through nitrate supply improves drought tolerance of sugarcane plants *Front. Plant Sci.* **11** 970

[117] Sheokand S, Kumari A and Sawhney V 2008 Effect of nitric oxide and putrescine on antioxidative responses under NaCl stress in chickpea plants *Physiol. Mol. Biol. Plants* **14** 355–62
[118] Ahmad P, Abdel Latef A A, Hashem A, Abd_Allah E F, Gucel S and Tran L-S P 2016 Nitric oxide mitigates salt stress by regulating levels of osmolytes and antioxidant enzymes in chickpea *Front. Plant Sci.* **7** 347
[119] Fatma M, Masood A, Per T S and Khan N A 2016 Nitric oxide alleviates salt stress inhibited photosynthetic performance by interacting with sulfur assimilation in mustard *Front. Plant Sci.* **7** 521
[120] Ghadakchiasl A, Mozafari A-a and Ghaderi N 2017 Mitigation by sodium nitroprusside of the effects of salinity on the morpho-physiological and biochemical characteristics of *Rubus idaeus* under *in vitro* conditions *Physiol. Mol. Biol. Plants* **23** 73–83
[121] Sharma A, Singh R K, Singh P, Vaishnav A, Guo D-J, Verma K K, Li D-P, Song X-P, Malviya M K and Khan N 2021 Insights into the bacterial and nitric oxide-induced salt tolerance in sugarcane and their growth-promoting abilities *Microorganisms* **9** 2203
[122] Zhang Y, Su J, Cheng D, Wang R, Mei Y, Hu H, Shen W and Zhang Y 2018 Nitric oxide contributes to methane-induced osmotic stress tolerance in mung bean *BMC Plant Biol.* **18** 1–12
[123] Mostofa M G, Seraj Z I and Fujita M 2014 Exogenous sodium nitroprusside and glutathione alleviate copper toxicity by reducing copper uptake and oxidative damage in rice (*Oryza sativa* L.) seedlings *Protoplasma* **251** 1373–86
[124] Chavoushi M, Najafi F, Salimi A and Angaji S A 2020 Effect of salicylic acid and sodium nitroprusside on growth parameters, photosynthetic pigments and secondary metabolites of safflower under drought stressSci. Hortic. **259** 108823
[125] Rahman M A, Kabir A H, Song Y, Lee S-H, Hasanuzzaman M and Lee K-W 2021 Nitric oxide prevents Fe deficiency-induced photosynthetic disturbance, and oxidative stress in alfalfa by regulating Fe acquisition and antioxidant defense *Antioxidants* **10** 1556
[126] Dey A, Samanta M K, Gayen S and Maiti M K 2016 The sucrose non-fermenting 1-related kinase 2 gene SAPK9 improves drought tolerance and grain yield in rice by modulating cellular osmotic potential, stomatal closure and stress-responsive gene expression *BMC Plant Biol.* **16** 1–20
[127] Li J, Li Y, Yin Z, Jiang J, Zhang M, Guo X, Ye Z, Zhao Y, Xiong H and Zhang Z 2017 Os ASR 5 enhances drought tolerance through a stomatal closure pathway associated with ABA and H2O2 signalling in rice *Plant Biotechnol. J.* **15** 183–96
[128] Martínez-Andújar C, Martínez-Pérez A, Albacete A, Martínez-Melgarejo P A, Dodd I C, Thompson A J, Mohareb F, Estelles-Lopez L, Kevei Z and Ferrández-Ayela A 2021 Overproduction of ABA in rootstocks alleviates salinity stress in tomato shoots *Plant, Cell Environ* **44** 2966–86
[129] Li H, Ye K, Shi Y, Cheng J, Zhang X and Yang S 2017 BZR1 positively regulates freezing tolerance via CBF-dependent and CBF-independent pathways in *Arabidopsis Mol. Plant* **10** 545–59
[130] Zhao M-G, Tian Q-Y and Zhang W-H 2007 Nitric oxide synthase-dependent nitric oxide production is associated with salt tolerance in *Arabidopsis Plant Physiol.* **144** 206–17
[131] Xuan Y, Zhou S, Wang L, Cheng Y and Zhao L 2010 Nitric oxide functions as a signal and acts upstream of AtCaM3 in thermotolerance in *Arabidopsis* seedlings *Plant Physiol.* **153** 1895–906

[132] Shi H-T, Li R-J, Cai W, Liu W, Wang C-L and Lu Y-T 2012 Increasing nitric oxide content in *Arabidopsis thaliana* by expressing rat neuronal nitric oxide synthase resulted in enhanced stress tolerance *Plant Cell Physiol.* **53** 344–57

[133] Shi H, Ye T, Zhu J-K and Chan Z 2014 Constitutive production of nitric oxide leads to enhanced drought stress resistance and extensive transcriptional reprogramming in *Arabidopsis J. Exp. Bot.* **65** 4119–31

[134] Cai W, Liu W, Wang W-S, Fu Z-W, Han T-T and Lu Y-T 2015 Overexpression of rat neurons nitric oxide synthase in rice enhances drought and salt tolerance *PLoS One* **10** e0131599

[135] Chun H J, Park H C, Koo S C, Lee J H, Park C Y, Choi M S, Kang C H, Baek D, Cheong Y H and Yun D-J 2012 Constitutive expression of mammalian nitric oxide synthase in tobacco plants triggers disease resistance to pathogens *Mol. Cells* **34** 463–71

[136] Del Castello F, Foresi N, Nejamkin A, Lindermayr C, Buegger F, Lamattina L and Correa-Aragunde N 2021 Cyanobacterial NOS expression improves nitrogen use efficiency, nitrogen-deficiency tolerance and yield in *Arabidopsis Plant Sci.* **307** 110860

Chapter 12

Possible role of osmolytes in enhancing abiotic stress tolerance in plants

Gulab Khan Rohela[1*], Pawan Saini[1], Pawan Shukla[2]

[1]*Central Sericultural Research and Training Institute, Central Silk Board, Pampore 192 121, Jammu & Kashmir, India*
[2]*Seri-Biotech Research Laboratory (SBRL), Carmelram Post, Kodathi, Bangalore 560035, India*

In plants, abiotic stress leads to the production of reactive oxygen species (ROS), which in turn causes oxidative stress and damage to the organelle membranes and cell membranes. In order to protect itself from the oxidative stress of ROS, the defense mechanisms of tolerant plant species produce certain compatible solute compounds, known as osmolytes, in higher concentrations as compared to susceptible plant species. Among various types, the main osmolyte compounds produced in plants are glycine and glycine-derived compounds, proline and proline-derived compounds, polyamines, and sugar alcohols. Among the different types of abiotic stress, salinity, low temperature, and drought stresses majorly affect the growth and productivity of crops. To counter these abiotic stresses, tolerant plant species sense the stress signals and activate a cascade of molecular activities, leading to enhanced gene expression for production of osmolytes and providing abiotic stress tolerance. This book chapter reviews the possible role of osmolytes in enhancing the abiotic stress tolerance in plants.

12.1 Introduction

Food security is a major concern for feeding the growing human population in the present and future situations [1]. Most humans depend on plant crops of wheat, rice, maize, sorghum, pulses, and vegetable crops for their food intake [2]. Plants are exposed to environmental stresses, which are caused by either natural (low rainfall, high or low temperatures, flooding, soil erosion) or man-made (excess usage of

* Author to whom any correspondence should be addressed Email: gulab_biotech@yahoo.co.in.

doi:10.1088/978-0-7503-4921-5ch12

pesticides and chemical fertilizers, water and soil pollution, rough agricultural practices) factors [3]. Due to rapid global climatic changes, the intensity of the aforementioned abiotic stresses has been increasing over the years, which has resulted in a reduction of viable cultivation area for agricultural crops worldwide [4]. Food security for present and future generations can be maintained by protecting agricultural fields and improving the stress-adaptive traits in crop plant species. Thus, to have sufficient plant-based food products, it is essential to enable and encourage stress-resistant characteristics in plants against abiotic and biotic stresses.

Plants' adaption to stress conditions occurs at different levels, ranging from the molecular, biochemical, organellar, cellular, tissue, and organ (root, leaf, stem, flower, and fruit) level [5]. Overall, the main stress adaptions of a plant species lie at the molecular level, which triggers stress-adaptive genes to undergo upregulation during stressful conditions. The upregulation of stress-adaptive genes results in reestablishing homeostasis, protection of vital metabolic- and growth-related proteins, and protection of cell and organelle membranes [6]. So, the stress -adaptive response of plants starts at the molecular level and proceeds through biochemical, cellular, and organ levels; the overall result provides the plants with resistance against abiotic stresses [7, 8].

With regard to abiotic-stress-responsive genes, there are several hundreds of genes distributed in the plant genome [9]. All these are categorized into three groups of genes [10]. One group encodes transcription factors, such as abscisic acid responsive element binding factors (ABFs), basic leucine zippers (bZIPs), core binding factors (CBFs), dehydration responsive element binding factors (DREBs), MYB, (B-myc) MYC (C-myc), and the NAM, ATAF and CUC (NAC) transcription factor families, etc, which are involved in the transcription regulation of stress-related genes [11, 12]. The second group of genes encodes the production of free radical scavengers, osmoprotectants, chaperones, heat shock proteins (Hsps), and late embryogenesis abundant proteins (LEAs), which are involved in protecting the vital proteins of plants and their membranes [13, 14]. The third group of genes encodes the proteins of membrane transporters and ion channels for easy ion and water uptake during stressful environmental conditions, ensuring the plant system carries out its essential physiological activities at optimum levels and enabling them to survive harsh conditions [15].

During different types of abiotic stress conditions (drought, salinity, light, high and low temperatures, ultraviolet radiation, and heavy metal stress), the plant faces extreme levels of oxidative stress from the increased levels of reactive oxygen species (ROS), which have a high potential to damage the membranes [16]. To counter the increased levels of ROS and for protection of cells' osmotic levels, certain stress-adaptive genes encode the production and accumulation of solute compounds (compatible solutes), which play a protective role in reestablishing homeostasis and making the plant cell turgid for the growth and development of the plant during stressful environmental conditions [17, 18]. As the compatible solutes fulfill the function of osmotic adjustment of a plant cell during stressful conditions, they are commonly known as osmolytes [19]. Osmolytes are chiefly produced in response to stress conditions and confer abiotic stress tolerance by protecting the cellular

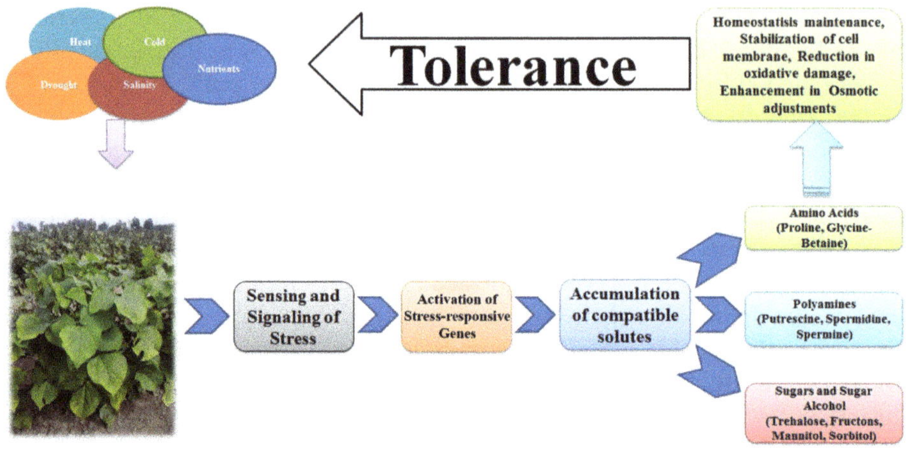

Figure 12.1. An overview of the roles of osmolytes during abiotic stress.

machinery, such as vital proteins and membranes, from the oxidative effects of stress-generated oxygen free radicals [20].

Biochemically, osmolytes are sugars, sugar alcohols (mannitol, sorbitol), polyamines, secondary metabolites, amino acids (glycine, proline), and their derivatives [21]. Osmolyte compounds have been reported to confer abiotic stress tolerance in plants against abiotic stresses of cold, drought, salinity, and light stress (figure 12.1). Osmolyte compounds induce abiotic stress tolerance in plants through complex mechanisms of molecular and biochemical pathways involving the upregulation of stress-adaptive genes and enhanced production of compatible solutes [22]. For instance, several major studies of salinity-tolerant plant species (halophytes) reported the accumulation of high amounts of proline, which carries the vital function of membrane stabilization, and protection of vital proteins and subcellular structures during salinity stress [23, 24]. Like proline, other osmolyte compounds and their mechanisms of action have been reported in several crop species [25–27]. Molecular-level work has been carried in several stress-tolerant plant species to identify the specific genes encoding the biosynthesis of osmolytes. Gene-transfer techniques and genome-editing approaches have also been utilized to raise stress-tolerant crop species [28]. Through the present book chapter, we will present an overview of these various types of osmolytes and their possible roles in providing abiotic stress tolerance to plants.

12.2 Proline

In response to abiotic stress, a plant's defense mechanism produces various biomolecules to counter and protect its cellular machinery from the generated oxidative stress. Among the different type of osmolytes, proline plays a pivotal role in providing resistance against the different types of abiotic stress (figure 12.2) [29]. Proline was reported to maintain the water relations of plants in drought stress conditions. Similarly, higher amounts of proline accumulation were reported in *Pisum* spp. and *Petunia hybrida* plants under drought conditions [30, 31]. Lower-molecular-weight

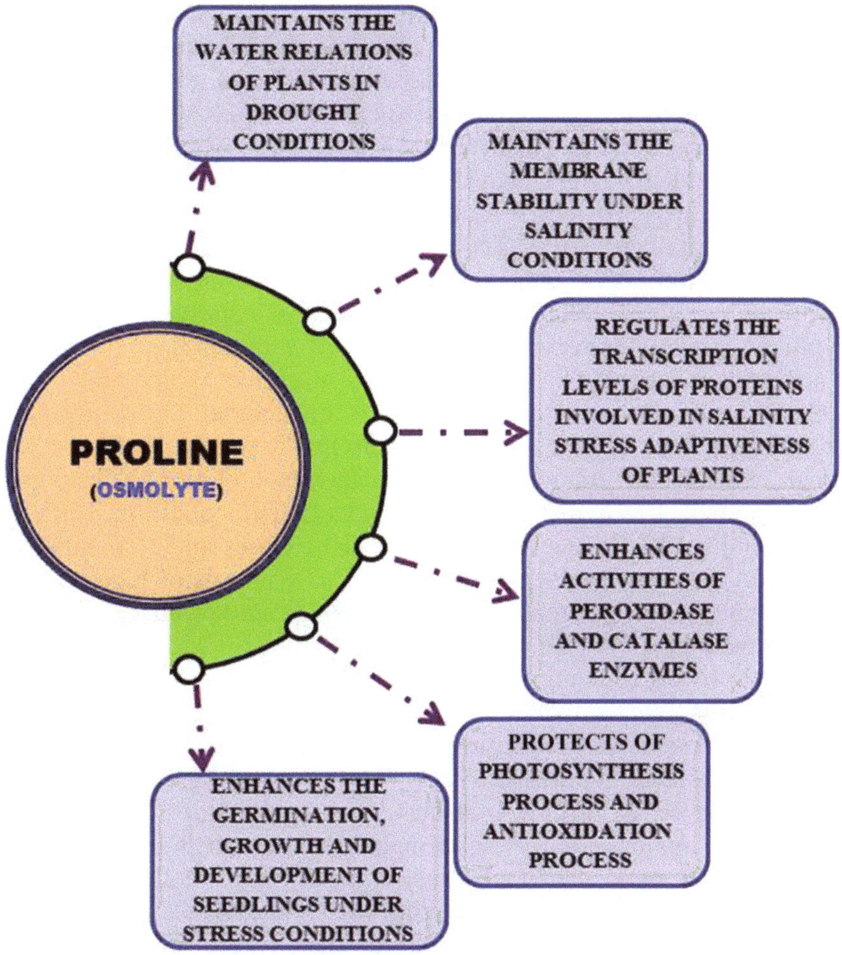

Figure 12.2. Roles of proline in conferring abiotic stress tolerance in plants.

osmolytes, such as glycine betaine, proline, and polyamines, have been reported to enhance the stress tolerance capacity of plants by maintaining membrane stability under the stress of high salinity conditions. Proline was also reported to enhance the percentage of germination, growth, and development of seedlings under stress conditions [32]. Accumulation of higher amounts of proline co-relates with better stress adaptiveness of plants under drought/salinity stress conditions [33, 34]. Exogenous application of proline was shown to provide salt tolerance in germinating seeds by regulating the uptake of sodium ions in *Arabidopsis altissima* [35].

Proline was also reported to regulate the transcription levels of proteins involved in the salinity stress adaptation of plants [36]. In *Panacratium maritimum* stress tolerance levels were correlated with enhanced activities of peroxidase and catalase enzymes under the influence of higher amounts of proline [36]. In tobacco, the proline enhances the activities of the enzymes which carry antioxidant function under oxidative stresses [37]. Proline accumulation is also correlated with increased levels of cold stress tolerance in plants [38]; hence, external application of exogenous proline is also recommended to provide cold tolerance to plants during low-temperature periods such as the winter season [39, 40]. To counter metallicity stress (copper, chromium) in plants, external/exogenous application of brassinosteroids has been recommended, which enhances the production and accumulation of proline content, which in turn provides resistance to various abiotic stresses including the aforementioned metallicity stress [41, 42]. Apart from the endogenous content, the exogenous application of a combination of 24-epibrassinolide and proline was reported to enhance the abiotic stress tolerance against salinity stress with increased photosynthetic rate and enhanced yield in *Brassica juncea* [43].

The accumulation of proline in response to salt stress is not only reported in plants but also in marine organisms such as eubacteria, algae, protozoa, and invertebrates [44]. This clearly suggests that proline is widely distributed among a wide range of plant and animal species and has its evolutionary background in providing tolerance for living organisms under harsh conditions such as high salinity. In conditions of water shortage, the plant senses and responds by undertaking a series of molecular events which leads to synthesis of higher amounts of important osmolytes such as trehalose, glycine betaine, and proline. These osmolytes carry out the osmotic adjustment of plant cells in harsh and adverse environmental conditions by lowering the osmotic potential of the cell sap [45]. Proline-compatible solutes can also function as hydroxyl free radical scavengers in plants under stressed conditions [46]. Exogenous application of proline treatment to seeds has shown a good response in alleviating drought stress in wheat crops [47].

As enhanced levels of proline synthesis have a correlation with abiotic stress tolerance in plants against salinity and drought stress, scientists and researchers have carried out several molecular and genetic engineering studies to increase the expression levels of proline by upregulating the synthesis of enzymes involved in its biosynthesis. In a study transgenic tobacco plants were raised by carrying the overexpression of 1-pyrroline-5-carboxylate synthase, which increases proline production and confers tolerance under drought stress conditions [48]. The overexpression of 1-pyrroline-5-carboxylate synthase enzyme leads to the catalytic conversion of glutamate to pyrroline-5-carboxylate, which finally reduces to synthesize proline.

Similarly, in another study transgenic lines of *Panicum virgatum* (switch grass) were raised by incorporating a *Lolium perenne* L.-based gene encoding Δ1-pyrroline-5-carboxylate synthase, which is the rate-limiting enzyme of proline synthesis. Among the raised transgenic lines, some transgenic switch grass showed a high proline content under stress conditions and thereby ensured high-biomass production as compared to the control line [49]. In transgenic sorghum (*Sorghum bicolor*), enhanced levels of proline synthesis and accumulation resulted in protection of the photosynthesis process and antioxidation process that plays a main role in alleviating salinity stress conditions [50]. Silencing of the proline catabolizing enzyme, i.e. proline dehydrogenase, has been shown to confer abiotic stress tolerance in transgenic tobacco [51].

12.3 Glycine betaine

Glycine betaine is a quaternary ammonium compound produced naturally in plant species in high amounts under stressful environmental conditions such as salinity, drought, freezing, flooding, high UV light radiation, heavy metal stress, and under dehydrated states [52]. Glycine betaine is found in large amounts in the chloroplasts of stress-tolerant plant species; it is essentially required to maintain the stability of thylakoid membranes under oxidative stress conditions and to ensure the plants survive under harsh environments via enabling of the chloroplasts to perform optimum levels of photosynthesis (figure 12.3) [53]. Glycine betaine protects the

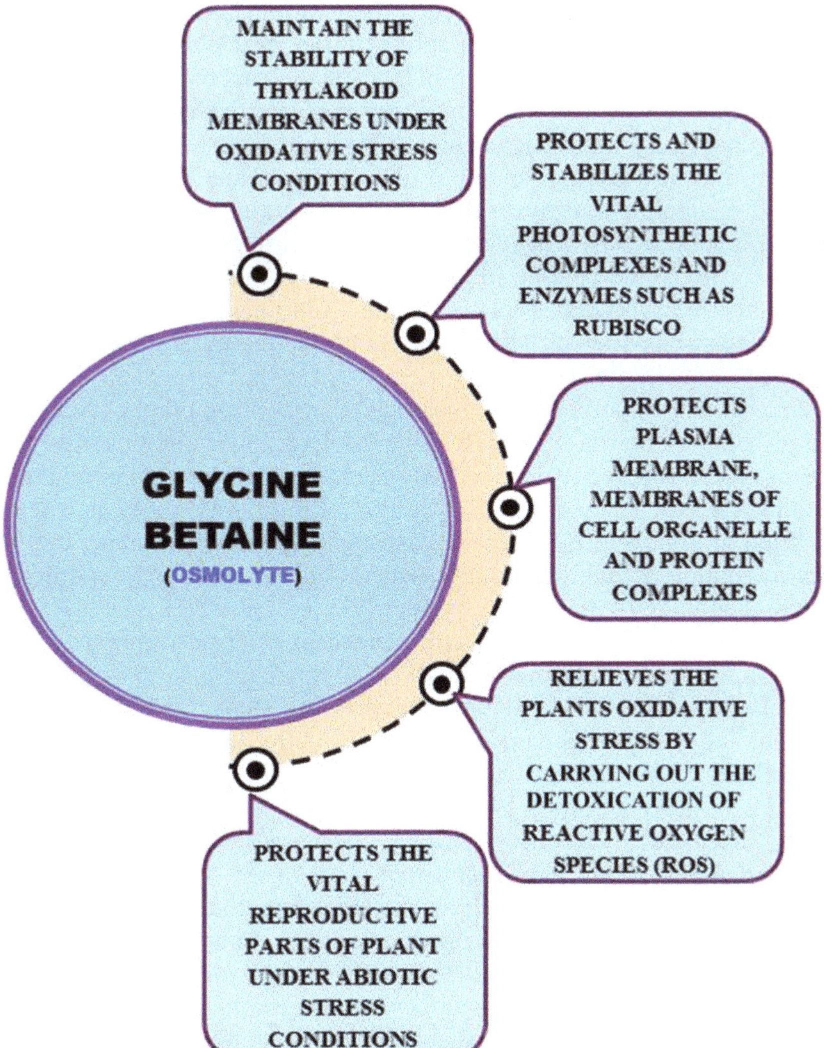

Figure 12.3. Role of glycine betaine in conferring abiotic stress tolerance in plants.

main components of plants such as the plasma membranes, membranes of cell organelles, and protein complexes [54]. Glycine betaine biosynthesis occurs by utilizing the serine amino acid as a precursor through the intermediates of ethanolamine, choline, and betaine aldehyde. Compatible solutes of proline and glycine betaine have a role in efflux of potassium ions (K^+) in plants under stressful conditions [55].

Glycine betaine is an osmoprotective compound and plays a role in minimizing the damage of abiotic stress in several important crop species, which includes wheat, rice, sorghum, vegetables, and pulse crops [56–58]. It relieves the plants from the oxidative stress of oxygen free radicals by causing the nonaccumulation of ROS through a detoxification mechanism [59]. Glycine betaine also acts as a chaperone in plant systems under stressful environmental conditions and protects vital proteins for survival [60]. It also protects and stabilizes the vital photosynthetic complexes and enzymes such as ribulose 1, 5-bisphosphate carboxylase (also known as Rubisco) and photosystem-II [61]. It further protects the vital reproductive parts of the plant under cold and water-deficit stress conditions [59].

The biosynthesis of glycine betaine occurs from choline as a precursor. Choline gets oxidized by choline monooxygenase to form betaine aldehyde (BA). In a final step, betaine aldehyde dehydrogenase (BADH) carries the incorporation of NAD+ into BA to form glycine betaine [62]. Alternatively, in some plant species glycine betaine can also be synthesized using glycine amino acid as a precursor involving methyl transferase enzymes such as glycine sarcosine methyl transferase or dimethyl glycine methyl transferase [10].

If glycine betaine is synthesized in insufficient amounts in plants, it can be applied exogenously on crop plants as a stress protector to counter abiotic stress conditions [63]. There exist several reports on exogenous/external application of glycine betaine in providing stress tolerance against drought and salinity stress in various crop species, listed as follows: tolerance against chilling stress in tomato [64]; drought tolerance in sunflower [65], cotton [66], wheat [56], and rice [67]; salinity stress in tomato [68], canola [69], maize [70], pepper [71], ryegrass [72], rice [63], soya bean [73], and snap bean [74]; heavy metal stress in wheat [75] is also similarly induced by exogenous application of glycine betaine.

Genetic engineering of plants with overexpression of glycine-betaine-synthesizing enzymes (choline monooxygenase and BADH, methyl tranferases) plays a crucial role in making crop plants which are able to tolerate stress conditions by accumulation of high amounts of glycine betaine [76, 77]. Transgenic plants of *Arabidopsis thaliana, Eucalyptus globulus, Nicotiana tobbacum, Oryzae sativa, Triticum aestivum,* and *Solanum tuberosum* were raised by transferring the genes of glycine betaine biosynthesis, which resulted in accumulation of higher amounts of glycine betaine and conferring tolerance towards salinity, drought, and chilling stresses [62, 78–84].

The expression of the transgenes, synthesis and accumulation of glycine betaine in leaves, and conferring of abiotic stress tolerance in transgenic plants majorly depends on the availability of the precursor molecules, i.e. choline, the specific gene transferred, and the promoters which are clubbed with the transgene [85–88].

As glycine betaine also plays a crucial role in the growth and development of the reproductive parts of plants under stress conditions, in some reports of transgenic plants with higher accumulation of glycine betaine, along with the properties of tolerance towards abiotic stress they also displayed superior development of flowers and fruits when compared with control plants [62, 89, 90]. This is due to higher amounts of glycine betaine accumulation in the reproductive parts than leaves, which results in, for example, bigger size tomatoes [90].

12.4 Polyamines

Polyamines (PAs) are the oldest organic substances in the history of biochemistry, originally derived from amino acids when they were discovered by Anton van Leeuwenhook [91] in the year 1678 in the crystalline form of human semen. PAs are present in all biological organisms (plants, animals, microorganisms, humans) [92]. They are are positively charged organic substances at physiological pH [93]. There are three most common types of PA, e.g. spermine (Spm), $[NH_2(CH_2)_3NH(CH_2)_4NH(CH_2)_4NH_2]$, spermidine (Spd) $[NH_2(CH_2)_3NH(CH_2)_4NH_2]$, and putrescine (Put) $[NH2(CH2)_4NH_2]$, that are present in all cellular compartments of plants [94]. The PAs are present in free, soluble conjugated, and insoluble bound forms [95]. These PA substances are also considered growth substances as they are involved in growth, development, physiological, and metabolic activities. PAs are further involved in various important cellular processes such as DNA replication, RNA transcription, protein synthesis, cell division, differentiation, proliferation, phytohormone synthesis, stabilization of cellular membrane, enzymatic activities, as well as developmental processes such as embryogenesis, reproductive organs, root growth and their network, tuberization, floral initiation, fruit development, ripening and senescence [92, 96]. PAs play diverse functions, e.g. as signaling molecules [97, 98], and in abiotic stress tolerance responses such as cold tolerance [99–101], drought resistance [102, 103], salinity [98, 104–106], oxidative stress [107–109], and metal toxicity and deficiency [110, 111].

The concentration of PAs within the cellular compartments depends upon the anabolic, catabolic, and their conjugated reactions in plants. The cascade of enzymatic reactions leads to biosynthesis of Put, Spm, Spd, Cad, etc, PAs from their precursor compounds. The nomenclature of such PAs is based upon their source of origin, i.e. putrescine is isolated from putrefying meat, cadaverine from dead human bodies [112], and spermidine and spermine from semen [113, 114]. The three most commonly present PAs (Put, Spm, Spd) are derived from L-arginine (via L-ornithine) and L-methionine amino acids [115]. The biosynthesis of PAs commences with the formation of diamine PA, i.e. Put, from L-ornithine amino acids through a decarboxylation process catalyzed by the ornithine decarboxylase (ODC) enzyme. This process is common in mammals and fungi, whereas in plants and bacteria, Put synthesis is catalyzed by arginine decarboxylase (ADC). Further, the addition of aminopropyl moiety (decarboxylated S-adenosylmethionine; dcSAM) through spermidine synthase (SPDS) and spermine synthase (SPMS) leads to synthesis of the spermidine and spermine PAs [116, 117]. It has been reported that when different types of stresses occur, these PAs are synthesized through the ADC pathway [118, 119], while when a plants is undergoing its active growth and development phase, the ODC pathway will take the lead for Pas biosynthesis.

A decrease in potassium concentration resulted in ninhydrin positive spots in barley plants, which indicated the enhancement of the Put PA [120]. Later, it was found that K+ deficient shoots with L-^{14}C-arginine application enhanced the level of PAs in comparison to ornithine [121]. The connection of the ADC pathway with plant response towards abiotic stress tolerance was established by Galston and coworkers at Yale University [122]. It has been observed that, under varied environmental conditions, the accumulation of PAs increases [123, 124], which revealed that the level of PAs in plant organs can be regulated by PA catabolic enzymes. An increase in the level of PAs during stress conditions leads toward production of ROS, which upregulates abiotic-stress-responsive proteins [125].

Salinity affects the levels of PAs in several plants, e.g. *Lactuca sativa, Capsicum annum, Spinacia oleracea, Beta vulgaris, Brassica oleraceae,* and *Lycopersiucm esculentum* [123]. Salt stress resulted in reduced levels of PAs (Put, Spd, and Spm) [126, 127]. A model plant of *A. thaliana* has provided new opportunities for understanding the role of PAs in abiotic stress tolerance. It was found that activation of *ADC2* with stimulation of oxidase amine has a significant role in salinity stress. Thus, the enhanced expression of ADC2 and SPMS was linked with salt stress in *Arabidopsis* [99, 128]. The overexpression of PAs has been shown to enhance salinity tolerance in different plants such as tobacco, rice, and *Arabidopsis* [124, 129–131]. In apple and lupine, Put and Spd PAs accreted when salt and water stress conditions were induced [132, 133]. Under saline conditions, an increase in Spd and Spm was observed in maize seedlings and flowers and in *Arabidopsis* flowers [134]. A similar response was observed in *Arabidopsis* flowers by Tassoni *et al* [135]. An exogenous treatment of Spd or Spm resulted in an improvement in salt tolerance in three *indica* rice varieties under different salt concentrations [136].

The level of PAs also increases under osmotic stress in plants [124]. Osmolytes such as sorbitol, mannitol, sucrose, proline, and glycine betaine also enhanced the level of PAs in plants. In oats under water stress conditions the level of PAs was seen to accumulate in protoplasts and other cells [122]. During stress conditions the ADC transcript level increases, which imparts osmotic stress tolerance in plants [137]. In *Arabidopsis*, it was observed that overexpression of homologous *ADC2* increased the level of PAs, which led to closure of stomata and ultimately protected the plants against drought stress [105]. Abscisic acid (ABA), which is well known as a stress hormone, positively impacts the PA metabolism and enhances the expression of ADC2, SPDS1, and SPMS during osmotic stress conditions [105]. In *Arabidopsis*, the upregulation of S-adenosylmethionine decarboxylase (SAMDC) encoding gene favors an increase in the level of Spm, which subsequently triggers *NCED3,* the major gene of ABA biosynthesis. The mutant versions of PA biosynthesis genes, as well as ABA biosynthesis and ABA deficiency, also showed enhanced water and salt stress tolerance in plants. The three PAs, i.e. Put, Spm, and Spd, control the closing of the stomatal aperture during water stress [97, 138]; Put is also involved in the ABA biosynthesis metabolism in response to abiotic stress. The enhanced expression

Table 12.1. List of polyamine biosynthesis genes expressed in different plants through transgenic technology.

#	Gene	Crop plant	Tolerance	References
1	*ADC*	*Oryza sativa*	Salt	[129]
2	*ADC*	*Brassica juncea*	Chilling and salt	[139]
3	*ADC1, ADC2*	*Oryza sativa*	Drought	[140]
4	*ADC1, ADC2*	*Arabidopsis thaliana*	Freezing	[100]
5	*ADC*	*Malus domestica*	Chilling, salt, and dehydration	[141]
6	*ADC*	*Oryza sativa*	Chilling	[142]
7	*AtADC2*	*Arabidopsis thaliana*	Salt	[143]
8	*ADC*	*Solanum melongena*	Chilling, salt, and dehydration	[144]
9	*ADC*	*Zea mays*	Salt	[134]
10	*MdADC*	*Malus sylvestris* (L.) Mill var. *domestica*	Salt	[133]
11	*PaADC2*	*Pringlea antiscorbutica*	Chilling and salt	[145]
12	*Mouse ODC cDNA*	*Populous nigra*	Stress	[146]
		Nicotiana tabaccum	Salt	[131]
13	*MdSAMDC2*	*Malus sylvestris*	Cold and salt	[141]
		Pyrus communis	Salt	[147]
14	*SPDS cDNA* from *Cucurbita ficifolia*	*Arabidopsis thaliana*	Chilling, salinity, and drought	[148]
		Sweet Potato	Increase Spd	[149]
15	*MdSPDS*1	*Pyrus communis*	Salt, heavy metal, and osmotic stress	[150]
16	*SAMDC*	*Oryza sativa*	Salt	[130]
17	*ACC synthase*	*Nicotiana tabaccum*	Broad spectrum	[151]
18	*ACC oxidase*			
19	*SAMDC*		Salt and osmotic	[152]
			Broad spectrum	[153]
20	*SAMDC1*	*Arabidopsis thaliana*	Broad spectrum	[105]
21	*ADC1*		Freezing	[154]
22	*SAMDC*	*Solanum lycopersicum*	Heat	[155]
23	*SPDS*	*Pyrus communis*	Broad spectrum	[150]
24	*ADC2*	*Arabidopsis thaliana*	Drought	[96]

of PA biosynthesis genes as well as accretion of Put during osmotic stress is controlled by an ABA-mediated mechanism. Several PA biosynthesis genes have been transformed through transgenic technology to impart abiotic stress tolerance in crop plants (table 12.1).

Under low temperatures, i.e. chilling and freezing, the PA biosynthesis pathways respond very well and provide tolerance under cold stress in different crop plants [121, 156, 157]. In *Arabidopsis*, a higher level of Put imparts freezing tolerance [100] due to activation of ADC genes. The level of Put increases during cold stress while the levels of Spm and Spd remain the same or may decrease [96]. Under high-temperature situations, also, the level of free Put has been seen to increase and impart heat tolerance. The increased PAs under heat stress are related again with the ADC genes. Transcript profiling of the ADC1, ADC2, and SAMDC2 genes showed increased cold tolerance [100, 101]. Exposure to cold stress mainly activates the ADC1 gene [158], while ADC2 is activated under all types of abiotic stresses [99, 159].

12.5 Sugars

Plants are autotrophic in nature and produce as well as consume carbohydrates. Carbohydrates are produced through the process of photosynthesis, which utilizes sunlight, water, and carbon dioxide. Glucose is the major carbohydrate produced through photosynthesis. Other types of carbohydrates are also produced at different steps of the photosynthesis cycle. These carbohydrates produced by plants are also known as sugars. Sugars are the biochemically derived products involved in important growth and development activities of plants such as seed germination, respiration, photosynthesis, reproduction, senescence, root formation, cell division, etc. Sugars are a source of energy, which is necessary for biological organisms. With our current understanding, we see that sugars and their derivatives, sugar alcohols/polyols, play important roles in abiotic stress management in plants. The accumulation of sugars (trehalose, raffinose, sucrose, fructons) and polyols (sorbitol, mannitol, inositols) results in diverse functions during abiotic stress in plants [160]. These sugars and polyols are designated as compatible solutes or osmoprotectants as they impart stress tolerance in plants. Under normal and stress conditions, sugars play multiple roles such as sensing and signaling molecules, receptors, and osmoprotectants [161]. Besides their role in growth and development, sugar and sugar alcohols/polyols are involved in imparting abiotic stress tolerance in crop plants (table 12.2). As such, the genes linked with enhancement of sugars and polyols have also been transformed in crop plants using transgenic technology (table 12.3).

When plants are exposed to a stress, a series of reactions leads to the activation of genes against the stress condition(s), i.e. drought, salt, cold, heat, etc [200]. Exposure to salt stress leads to a disruption of the ionic balance in cellular compartments and damages organelles like chloroplasts and mitochondria, which leads to repression of metabolic processes such as photosynthesis, respiration, nitrogen assimilation, osmotic balance, etc [201–203]. Salinity stress in plants leads to an accumulation of sugars and sugar alcohols, which enhances the salinity tolerance [204, 205] by increasing the osmotic adjustment of the cell. Increased concentrations of trehalose have been observed under salinity stress. Ashraf and Tufail [206] have reported an enhancement in soluble sugars under salt stress in sunflower. Salinity-tolerant plants, i.e. *Eruca sativa, Melilotusindica*, etc, have higher levels of compatible sugars than salt-susceptible plants [160]. The application of trehalose externally to rice seedlings led to recovery from salt stress [207]. According to Bonhert [208], polyols are also

Table 12.2. Role of different sugars and sugar alcohols during abiotic stress in crop plants (adopted from Bhattacharya and Kundu 2020) [160]. John Wiley & Sons. © 2020 John Wiley & Sons Ltd.

#	Compatible solutes	Response towards stress	References
		Sugars	
1	Trehalose	Stabilizes cell membrane, protects cell from oxidative damage	[162]
2	Sucrose	Osmolyte, maintains turgor pressure	[163]
3	Fructons	Stabilizes cell membrane, protects cell from oxidative damage	[164]
4	Raffinose	Osmotic tolerance, alleviates oxidative stress	[165]
		Sugar alcohols	
5	Mannitol	Reduces ROS species, stabilizes biological macromolecules	[166]
6	Sorbitol	Osmotic adjustments, maintains redox homeostatis	[167]
7	Myo-inositol	Osmotic adjustments, scavenges ROS	[168]
8	Ononitol	Osmotic adjustments, maintains redox homeostatis	[169]
9	Pinitol		

involved in osmotic regulation and salt tolerance in plants. It has been found that an increase in polyols content in halophytic plants helps against osmotic stress. Mannitol accumulation in celery results in salt tolerance [193]. Mannitol also protects the antioxidant enzymes that are involved in the ROS-scavenging mechanism [190].

Glucose, which provides energy for physiological activities and is generated through photosynthesis, provides drought stress tolerance by closing stomata, reducing photosynthesis, maintaining leaf water content, and by preventing cell membrane oxidation [209–211]. The concentration of sugar alcohol mannitol, soluble proteins, and other osmolytes increased significantly when drought-sensitive wheat varieties were exposed to drought stress during the stage of anthesis and enhanced their growth and development [212]. Sugars like trehalose have been shown to be the most effective osmolyte during drought stress, as the accumulation of trehalose helps in maintaining the integrity of cellular membranes [213]. Other sugar molecules such as raffinose, stachyose, verbascose, etc, help in avoiding dehydration conditions in plants [214]. During temperature stress, the accumulation of raffinose, galactinol, etc, helps in overexpressing the Hsps in *A. thaliana* and imparts tolerance towards chilling and salinity stress [215]. During low-temperature conditions, sugars and polyols help in stabilization of cellular membranes. In rice, also, trehalose accumulation leads to chilling stress tolerance [172]. The transport and storage carbohydrate known as sucrose plays the role of a signal molecule during heat stress in low concentrations, while when the level of sucrose is elevated it helps in mitigating oxidative stress. The heat-tolerant tomato responds towards heat stress by accumulating high concentrations of sucrose in young fruits [216]. The low concentration of soluble sugars in sorghum pollen grains provides tolerance to heat stress [217].

Table 12.3. Genes associated with sugars and sugar alcohols engineered for abiotic stress tolerance in plants.

#	Compatible solutes	Gene	Crop plant	Tolerance	References
			Sugars		
1	Trehalose	*ScTPS1*	Tobacco	Drought	[162]
			Potato	Drought	[170]
			Tomato	Drought, salinity, and oxidative stress	[171]
		OtsA + *OtsB*	Rice	Stress tolerance, photo-oxidative damage tolerance	[172]
		TPS		Salinity	[173]
		ScTPS1 + *ScTPS2*	*Arabidopsis*	Dehydration tolerance	[174]
		TPS	Tobacco	Drought	[175]
2	Fructons	*SacB*	Tobacco	Drought	[176]
			Sugar beet	Dehydration tolerance	[177]
		1-SST + *6-SFT*	Ryegrass	Cold	[178]
		1-SST	Tobacco		[164]
			Rice	Freezing	[179]
		Levansucrase	Tobacco		[180]
3	Raffinose family oligosaccharides (RFO)	*GolS*	Tobacco	Low temperature	[181]
				Desiccation	[182]
			Arabidopsis	Drought	[183]
		α-galactosidase	*Petunia* spp.	Freezing	[184]
		UDP-glucose 4-epimerase	*Arabidopsis*	Drought, freezing	[185]
			Sugar alcohols		
4	Mannitol	*mtlD*	Tobacco	Salt	[166]
			Sorghum		[186]
			Potato		[187]
			Wheat		[188]
		HVA1 + *mtlD*	Maize		[189]
		Mannitol-1-phosphate dehydrogenase	Tobacco	Oxidative stress	[190]
			Rice	Drought, salinity	[191]
			Pinustaeda	Salinity	[192]
			Arabidopsis		[193]
5	Sorbitol	*SP6DH*	Tobacco	Salt	[164]
			Persimmon		[194]
			Pinustaeda	Salinity	[192]
					[195]

(*Continued*)

Table 12.3. (*Continued*)

#	Compatible solutes	Gene	Crop plant	Tolerance	References
			Sugars		
		Glucitol-6-phosphate dehydrogenase	*Diospyros kaki*		
6	Myo-inositol	*MIPS*	*Arabidopsis*	Salinity	[168]
		PcINO1	Tobacco	Salt	[196]
		PcINO1 + *McIMT1*			[200]
7	Pinitol	*MIPS* + *IMT1*	Tobacco	Salt	[197]
8	Ononitol	*McIMT1*	Tobacco	Salt	[167]
9	Galactinol	Galactinol synthase	*Arabidopsis*	Drought, salinity, oxidative damage	[165]
10	Invertase	Apoplastic invertase		Salinity	[198]
11	Aldehyde reductase	*MsALR*	Tobacco	Chemical	[199]

12.6 Conclusions

In the present scenario of rapid global climate and environmental change, there is much need to raise climate-resilient crops through genetic engineering approaches. These crop plants should be designed to cope with the abiotic stresses which are occurring naturally or created by humans. Osmolytes have a huge role in making plant species tolerant towards abiotic/environmental stresses. The biosynthetic pathways of different osmolytes has been well documented, along with the identification of genes coding for important enzymes of osmolytes' biosynthetic pathway. This chapter has discussed the production of different osmolytes in plants in response to different abiotic stresses. Several genetically engineered, stress-resistant crop plant species have been raised either through gene-transfer or genome-editing technologies, in which osmolytes were overexpressed through strong promoters or stress-inducible promoters and the resulting transgenic plants displayed enhanced stress tolerance. However, further investigations are needed to develop climate-smart crops through advance omics tools to help farmers and agriculturists cultivate crops with enhanced agricultural productivity, by means of which food security can be provided for the growing human population under continued changing environmental conditions.

Multiple choice questions

1. Most of the osmolytes naturally synthesized in plants in response to abiotic stress are …

(a) Low-molecular-weight compounds such as amino acids, derivatives of amino acids, sugar alcohols, and polyamines;
(b) High-molecular-weight compounds such as complex polysaccharides and aromatic compounds;
(c) Both low-molecular-weight and high-molecular-weight compounds;
(d) None of the above.

2. Which amino acid acts as a most vital osmolyte in plants and provides osmoregulation during stressful conditions such as salinity, drought, cold, and flooding stress?
(a) Alanine;
(b) Glutamic acid;
(c) Asparagine;
(d) Proline.

3. Carbohydrates which can act as an osmolyte are mostly …
(a) Monosaccharides (glucose, ribose) and disaccharides (lactose, cellobiose);
(b) Reducing sugars (trehalose, raffinose, sucrose) and sugar alcohols (mannitol, sorbitol, myo-inositol, pinitol, galactinol);
(c) Polysaccharides (starch, cellulose, glycogen);
(d) Glycopolysaccharides.

4. Which of the following is correct about proline in functioning as an osmolyte?
(a) It enhances the percentage of germination, growth, and development of seedlings under stress conditions;
(b) Proline regulates the transcription levels of proteins involved in salinity stress adaptiveness of plants;
(c) Accumulation of higher amounts of proline co-relates with better stress adaptiveness of plants under drought/salinity stress conditions;
(d) All of the above.

5. Polyamines are …
(a) Derivatives of fatty acids;
(b) Derivatives of amino acids;
(c) Derivatives of simple sugars;
(d) Derivatives of polyethene.

Answers: 1 (a); 2 (d); 3 (b); 4 (d); 5 (b).

Descriptive-type questions

1. What are the compatible solutes and their roles in plants?
2. Describe the evolutionary role of proline in plants.
3. Write short notes on proline, glycine betaine, and polyamines.

4. How has genetic engineering been used to engineer the osmolyte production pathways in plants?
5. Describe the role of polyamines during abiotic stress.

References

[1] Godfray H C J, Beddington J R, Crute I R, Haddad L, Lawrence D, Muir J F, Pretty J, Robinson S, Thomas S M and Toulmin C 2010 Food security: the challenge of feeding 9 billion people *Science* **327** 812–8

[2] Medek D E, Schwartz J and Myers S S 2017 Estimated effects of future atmospheric CO_2 concentrations on protein intake and the risk of protein deficiency by country and region *Environ. Health Perspect.* **125** 087002

[3] Hussain S, Huang J, Huang J, Ahmad S, Nanda S, Anwar S, Shakoor A, Zhu C, Zhu L and Cao X 2020 Rice production under climate change: adaptations and mitigating strategies *Environment, Climate, Plant and Vegetation Growth* (Berlin: Springer) pp 659–86

[4] Ramesh P, Panwar N, Singh A, Ramana S, Yadav S K, Shrivastava R and Rao A S 2010 Status of organic farming in India *Curr. Sci.* **98** 1190–4

[5] Hasanuzzaman M, Nahar K, Alam M, Roychowdhury R and Fujita M 2013 Physiological, biochemical, and molecular mechanisms of heat stress tolerance in plants *Int. J. Mol. Sci.* **14** 9643–84

[6] Wang X, Zenda T, Liu S, Liu G, Jin H, Dai L, Dong A, Yang Y and Duan H 2019 Comparative proteomics and physiological analyses reveal important maize filling-kernel drought-responsive genes and metabolic pathways *Int. J. Mol. Sci.* **20** 3743

[7] Bravo R, Parra V, Gatica D, Rodriguez A E, Torrealba N, Paredes F, Wang Z V, Zorzano A, Hill J A and Jaimovich E 2013 Endoplasmic reticulum and the unfolded protein response: dynamics and metabolic integration *Int. Rev. Cell Mol. Biol.* **301** 215–90

[8] Ahmed I, Shukla P and Gautam R 2021 *Omics Technologies for Sustainable Agriculture and Global Food Security* vol. 1 (Berlin: Springer) pp 169–89

[9] Deshmukh R, Sonah H, Patil G, Chen W, Prince S, Mutava R, Vuong T, Valliyodan B and Nguyen H T 2014 Integrating omic approaches for abiotic stress tolerance in soybean *Front. Plant Sci.* **5** 244

[10] Khan M S, Ahmad D and Khan M A 2015 Utilization of genes encoding osmoprotectants in transgenic plants for enhanced abiotic stress tolerance *Electron. J. Biotechnol.* **18** 257–66

[11] Frank S R, Schroeder M, Fernandez P, Taubert S and Amati B 2001 Binding of c-Myc to chromatin mediates mitogen-induced acetylation of histone H4 and gene activation *Genes Develop.* **15** 2069–82

[12] Shinozaki K, Yamaguchi-Shinozaki K and Seki M 2003 Regulatory network of gene expression in the drought and cold stress responses *Curr. Opin. Plant Biol.* **6** 410–7

[13] Bohnert H J and Sheveleva E 1998 Plant stress adaptations—making metabolism move *Curr. Opin. Plant Biol.* **1** 267–74

[14] Buchanan B B, Gruissem W and Jones R L 2015 *Biochemistry and Molecular Biology of Plants* (New York: Wiley)

[15] Blumwald E 2000 Sodium transport and salt tolerance in plants *Curr. Opin. Cell Biol.* **12** 431–4

[16] Das K and Roychoudhury A 2014 Reactive oxygen species (ROS) and response of antioxidants as ROS-scavengers during environmental stress in plants *Front. Environ. Sci.* **2** 53

[17] Soliman M, Elkelish A, Souad T, Alhaithloul H and Farooq M 2020 Brassinosteroid seed priming with nitrogen supplementation improves salt tolerance in soybean *Physiol. Mol. Biol. Plants* **26** 501–11

[18] Jia H, Wang L, Li J, Sun P, Lu M and Hu J 2020 Physiological and metabolic responses of *Salix sinopurpurea* and *Salix suchowensis* to drought stress *Trees* **34** 563–77

[19] Gupta N, Thind S K and Bains N S 2014 Glycine betaine application modifies biochemical attributes of osmotic adjustment in drought stressed wheat *Plant Growth Regul.* **72** 221–8

[20] Sharma A, Shahzad B, Kumar V, Kohli S K, Sidhu G P S, Bali A S, Handa N, Kapoor D, Bhardwaj R and Zheng B 2019 Phytohormones regulate accumulation of osmolytes under abiotic stress *Biomolecules* **9** 285

[21] Singh M, Kumar J, Singh S, Singh V P and Prasad S M 2015 Roles of osmoprotectants in improving salinity and drought tolerance in plants: a review *Rev. Environ. Sci. Bio/Technol.* **14** 407–26

[22] Suprasanna P, Nikalje G and Rai A 2016 *Osmolytes and Plants Acclimation to Changing Environment: Emerging Omics Technologies* (Berlin: Springer) pp 1–12

[23] Meena M, Divyanshu K, Kumar S, Swapnil P, Zehra A, Shukla V, Yadav M and Upadhyay R 2019 Regulation of L-proline biosynthesis, signal transduction, transport, accumulation and its vital role in plants during variable environmental conditions *Heliyon* **5** e02952

[24] Shafi A, Zahoor I and Mushtaq U 2019 *Salt Stress, Microbes, and Plant Interactions: Mechanisms and Molecular Approaches* (Berlin: Springer) pp 269–300

[25] Majumder A, Sengupta S and Goswami L 2010 Osmolyte regulation in abiotic stress *Abiotic Stress Adaptation in Plants: Physiological, Molecular and Genomic Foundation* ed A Pareek, S K Sopory, H J Bohnert and Govindjee (Dordrecht: Springer) pp 349–70

[26] Slama I, Abdelly C, Bouchereau A, Flowers T and Savouré A 2015 Diversity, distribution and roles of osmoprotective compounds accumulated in halophytes under abiotic stress *Annal. Bot.* **115** 433–47

[27] Hasanuzzaman M, Anee T I, Bhuiyan T F, Nahar K and Fujita M 2019 *Advances in Rice Research for Abiotic Stress Tolerance* (Amsterdam: Elsevier) pp 677–708

[28] Kumar R, Nizampatnam N R, Alam M, Thakur T K and Kumar A 2021 Genome editing technologies for plant improvement: advances, applications and challenges ed A Kumar, R Kumar, P Shukla and M K Pandey *Omics Technologies for Sustainable Agriculture and Global Food Security* vol 1 (Singapore: Springer) pp 213–40

[29] Anjum S A, Ashraf U, Tanveer M, Khan I, Hussain S, Shahzad B, Zohaib A, Abbas F, Saleem M F and Ali I 2017 Drought induced changes in growth, osmolyte accumulation and antioxidant metabolism of three maize hybrids *Front. Plant Sci.* **8** 69

[30] Alexieva V, Sergiev I, Mapelli S and Karanov E 2001 The effect of drought and ultraviolet radiation on growth and stress markers in pea and wheat *Plant Cell Environ.* **24** 1337–44

[31] Yamada M, Morishita H, Urano K, Shiozaki N, Yamaguchi-Shinozaki K, Shinozaki K and Yoshiba Y 2005 Effects of free proline accumulation in petunias under drought stress *J. Exp. Bot.* **56** 1975–81

[32] Sudhakar C, Lakshmi A and Giridarakumar S 2001 Changes in the antioxidant enzyme efficacy in two high yielding genotypes of mulberry (*Morus alba* L.) under NaCl salinity *Plant Sci.* **161** 613–9

[33] Ahmad P, Jaleel C A, Salem M A, Nabi G and Sharma S 2010 Roles of enzymatic and nonenzymatic antioxidants in plants during abiotic stress *Crit. Rev. Biotechnol.* **30** 161–75

[34] Hasanuzzaman M, Nahar K and Fujita M 2013 *Ecophysiology and Responses of Plants Under Salt Stress* (Berlin: Springer) pp 25–87
[35] Nounjan N, Nghia P T and Theerakulpisut P 2012 Exogenous proline and trehalose promote recovery of rice seedlings from salt-stress and differentially modulate antioxidant enzymes and expression of related genes *J. Plant Physiol.* **169** 596–604
[36] Khedr A H A, Abbas M A, Wahid A A A, Quick W P and Abogadallah G M 2003 Proline induces the expression of salt-stress-responsive proteins and may improve the adaptation of *Pancratium maritimum* L. to salt-stress *J. Exp. Bot.* **54** 2553–62
[37] Hoque M A, Banu M N A, Nakamura Y, Shimoishi Y and Murata Y 2008 Proline and glycinebetaine enhance antioxidant defense and methylglyoxal detoxification systems and reduce NaCl-induced damage in cultured tobacco cells *J. Plant Physiol.* **165** 813–24
[38] Kishor P K, Sangam S, Amrutha R, Laxmi P S, Naidu K, Rao K S, Rao S, Reddy K, Theriappan P and Sreenivasulu N 2005 Regulation of proline biosynthesis, degradation, uptake and transport in higher plants: its implications in plant growth and abiotic stress tolerance *Curr. Sci.* **88** 424–38
[39] Hayat S, Hayat Q, Alyemeni M N, Wani A S, Pichtel J and Ahmad A 2012 Role of proline under changing environments: a review *Plant Signal. Behav.* **7** 1456–66
[40] Jonytiene V, Burbulis N, Kupriene R and Blinstrubiene A 2012 Effect of exogenous proline and de-acclimation treatment on cold tolerance in *Brassica napus* shoots cultured *in vitro J. Food Agric. Environ.* **10** 327–30
[41] Verbruggen N and Hermans C 2008 Proline accumulation in plants: a review *Amino Acids* **35** 753–9
[42] Kido E A, Neto J R F, Silva R L, Belarmino L C, Neto J P B, Soares-Cavalcanti N M, Pandolfi V, Silva M D, Nepomuceno A L and Benko-Iseppon A M 2013 Expression dynamics and genome distribution of osmoprotectants in soybean: identifying important components to face abiotic stress *BMC Bioinf.* **14** 1–11
[43] Wani A S, Ahmad A, Hayat S and Tahir I 2019 Epibrassinolide and proline alleviate the photosynthetic and yield inhibition under salt stress by acting on antioxidant system in mustard *Plant Physiol. Biochem.* **135** 385–94
[44] Roosens N H, Al Bitar F, Loenders K, Angenon G and Jacobs M 2002 Overexpression of ornithine-δ-aminotransferase increases proline biosynthesis and confers osmotolerance in transgenic plants *Mol. Breed.* **9** 73–80
[45] Lazar T 2003 Plant physiology (book review) *Annal. Bot.* **91** 750–1
[46] Fahad S, Nie L, Chen Y, Wu C, Xiong D, Saud S, Hongyan L, Cui K and Huang J 2015 Crop plant hormones and environmental stress *Sustain. Agri. Rev.* **2015** 371–400
[47] Kamran M, Shahbaz M, Ashraf M and Akram N A 2009 Alleviation of drought-induced adverse effects in spring wheat (*Triticum aestivum* L.) using proline as a pre-sowing seed treatment *Pak. J. Bot.* **41** 621–32
[48] Kishore P, Hong Z, Miao G and DPS V 2007 Overexpression of 1-pyrroline-5-car boxylate synthase increases proline production and confers osmotolerence in transgenic plants *Plant Physiol.* **108** 842–7
[49] Guan C, Huang Y-H, Cen H-F, Cui X, Tian D-Y and Zhang Y-W 2019 Overexpression of the *Lolium perenne* L. delta1-pyrroline 5-carboxylate synthase (LpP5CS) gene results in morphological alterations and salinity tolerance in switchgrass (*Panicum virgatum* L.) *PLoS One* **14** e0219669

[50] Reddy P S, Jogeswar G, Rasineni G K, Maheswari M, Reddy A R, Varshney R K and Kishor P K 2015 Proline over-accumulation alleviates salt stress and protects photosynthetic and antioxidant enzyme activities in transgenic sorghum [*Sorghum bicolor* (L.) Moench] *Plant Physiol. Biochem.* **94** 104–13

[51] Tateishi Y, Nakagawa T and Esaka M 2005 Osmotolerance and growth stimulation of transgenic tobacco cells accumulating free proline by silencing proline dehydrogenase expression with double-stranded RNA interference technique *Physiol. Plant.* **125** 224–34

[52] Mohanty A, Kathuria H, Ferjani A, Sakamoto A, Mohanty P, Murata N and Tyagi A 2002 Transgenics of an elite indica rice variety Pusa Basmati 1 harbouring the codA gene are highly tolerant to salt stress *Theor. Appl. Genetics* **106** 51–7

[53] Yang W J, Rich P J, Axtell J D, Wood K V, Bonham C C, Ejeta G, Mickelbart M V and Rhodes D 2003 Genotypic variation for glycinebetaine in sorghum *Crop Sci.* **43** 162–9

[54] Ashraf M and Foolad M R 2007 Roles of glycine betaine and proline in improving plant abiotic stress resistance *Environ. Exp. Bot.* **59** 206–16

[55] Cuin T, Zhou M, Parsons D and Shabala S 2012 Genetic behaviour of physiological traits conferring cytosolic K+/Na+ homeostasis in wheat *Plant Biol.* **14** 438–46

[56] Raza M, Saleem M, Shah G, Khan I and Raza A 2014 Exogenous application of glycinebetaine and potassium for improving water relations and grain yield of wheat under drought *J. Soil Sci. Plant Nutr.* **14** 348–64

[57] Ali Rezaei M, Jokar I, Ghorbanli M, Kaviani B and Kharabian-Masouleh A 2012 Morpho-physiological improving effects of exogenous glycine betaine on tomato (‘*Lycopersicum esculentum*’ Mill.) cv. PS under drought stress conditions *Plant Omics* **5** 79–86

[58] Sakamoto A and Murata N 2001 The use of bacterial choline oxidase, a glycinebetaine-synthesizing enzyme, to create stress-resistant transgenic plants *Plant Physiol.* **125** 180–8

[59] Chen T H and Murata N 2008 Glycinebetaine: an effective protectant against abiotic stress in plants *Trends Plant Sci.* **13** 499–505

[60] Diamant S, Rosenthal D, Azem A, Eliahu N, Ben-Zvi A P and Goloubinoff P 2003 Dicarboxylic amino acids and glycine-betaine regulate chaperone-mediated protein-disaggregation under stress *Mol. Microbiol.* **49** 401–10

[61] Yang X, Liang Z, Wen X and Lu C 2008 Genetic engineering of the biosynthesis of glycinebetaine leads to increased tolerance of photosynthesis to salt stress in transgenic tobacco plants *Plant Mol. Biol.* **66** 73

[62] Sulpice R, Tsukaya H, Nonaka H, Mustardy L, Chen T H and Murata N 2003 Enhanced formation of flowers in salt-stressed *Arabidopsis* after genetic engineering of the synthesis of glycine betaine *Plant J.* **36** 165–76

[63] Hasanuzzaman M, Alam M, Rahman A, Hasanuzzaman M, Nahar K and Fujita M 2014 Exogenous proline and glycine betaine mediated upregulation of antioxidant defense and glyoxalase systems provides better protection against salt-induced oxidative stress in two rice (*Oryza sativa* L.) varieties *BioMed Res. Int.* **2014** 757219

[64] Park E-J, Jeknic Z and Chen T H 2006 Exogenous application of glycinebetaine increases chilling tolerance in tomato plants *Plant Cell Physiol.* **47** 706–14

[65] Hussain M, Malik M, Farooq M, Ashraf M and Cheema M 2008 Improving drought tolerance by exogenous application of glycinebetaine and salicylic acid in sunflower *J. Agron. Crop Sci.* **194** 193–9

[66] Ahmad S, Raza I, Ali H, Shahzad A N and Sarwar N 2014 Response of cotton crop to exogenous application of glycinebetaine under sufficient and scarce water conditions *Brazil. J. Bot.* **37** 407–15
[67] Tisarum R, Theerawitaya C, Samphumphung T, Takabe T and Cha-um S 2019 Exogenous foliar application of glycine betaine to alleviate water deficit tolerance in two Indica rice genotypes under greenhouse conditions *Agronomy* **9** 138
[68] Heuer B 2003 Influence of exogenous application of proline and glycinebetaine on growth of salt-stressed tomato plants *Plant Sci.* **165** 693–9
[69] Wahid A and Jamil A 2009 Inducing salt tolerance in canola (*Brassica napus* L.) by exogenous application of glycinebetaine and proline: response at the initial growth stages *Pak. J. Bot.* **41** 1311–9
[70] Nawaz K and Ashraf M 2010 Exogenous application of glycinebetaine modulates activities of antioxidants in maize plants subjected to salt stress *J. Agron. Crop Sci.* **196** 28–37
[71] Korkmaz A and Şirikçi R 2011 Improving salinity tolerance of germinating seeds by exogenous application of glycinebetaine in pepper *Seed Sci. Technol.* **39** 377–88
[72] Hu L, Hu T, Zhang X, Pang H and Fu J 2012 Exogenous glycine betaine ameliorates the adverse effect of salt stress on perennial ryegrass *J. Am. Soc. Hortic. Sci.* **137** 38–46
[73] Malekzadeh P 2015 Influence of exogenous application of glycinebetaine on antioxidative system and growth of salt-stressed soybean seedlings (*Glycine max* L.) *Physiol. Mol. Biol. Plants* **21** 225–32
[74] Desoky E, Ibrahim S A and Merwad A 2019 Mitigation of salinity stress effects on growth, physio-chemical parameters and yield of snapbean (*Phaseolus vulgaris* L.) by exogenous application of glycine betaine *Int. Lett. Nat. Sci.* **76** 60–71
[75] Bhatti K H, Anwar S, Nawaz K, Hussain K, Siddiqi E, Sharif R, Talat A and Khalid A 2013 Effect of exogenous application of glycinebetaine on wheat (*Triticum aestivum* L.) under heavy metal stress *Middle-East J. Sci. Res.* **14** 130–7
[76] McCue K F and Hanson A D 1990 Drought and salt tolerance: towards understanding and application *Trends Biotechnol.* **8** 358–62
[77] Hussain Wani S, Brajendra Singh N, Haribhushan A and Iqbal Mir J 2013 Compatible solute engineering in plants for abiotic stress tolerance-role of glycine betaine *Curr. Genomics* **14** 157–65
[78] Waditee R, Bhuiyan M N H, Rai V, Aoki K, Tanaka Y, Hibino T, Suzuki S, Takano J, Jagendorf A T and Takabe T 2005 Genes for direct methylation of glycine provide high levels of glycinebetaine and abiotic-stress tolerance in *Synechococcus* and *Arabidopsis Proc. Natl. Acad. Sci.* **102** 1318–23
[79] Shirasawa K, Takabe T, Takabe T and Kishitani S 2006 Accumulation of glycinebetaine in rice plants that overexpress choline monooxygenase from spinach and evaluation of their tolerance to abiotic stress *Annal. Bot.* **98** 565–71
[80] Ahmad R, Kim M D, Back K-H, Kim H-S, Lee H-S, Kwon S-Y, Murata N, Chung W-I and Kwak S-S 2008 Stress-induced expression of choline oxidase in potato plant chloroplasts confers enhanced tolerance to oxidative, salt, and drought stresses *Plant Cell Rep.* **27** 687–98
[81] Yu X, Kikuchi A, Matsunaga E, Morishita Y, Nanto K, Sakurai N, Suzuki H, Shibata D, Shimada T and Watanabe K N 2009 Establishment of the evaluation system of salt tolerance on transgenic woody plants in the special netted-house *Plant Biotechnol.* **26** 135–41

[82] He C, Yang A, Zhang W, Gao Q and Zhang J 2010 Improved salt tolerance of transgenic wheat by introducing betA gene for glycine betaine synthesis *Plant Cell, Tissue Organ Cult.* **101** 65–78

[83] Goel D, Singh A K, Yadav V, Babbar S B, Murata N and Bansal K C 2011 Transformation of tomato with a bacterial codA gene enhances tolerance to salt and water stresses *J. Plant Physiol.* **168** 1286–94

[84] Jingjiang L H, He G, Yin Y, Liu M, Liu B, Qiao G, Lin S, Xie L and Zhuo R 2013 Over-expression of the codA gene by RD29a promoter improves salt tolerance in *Nicotiana tabacum Pak. J. Bot.* **45** 821–7

[85] Huang J, Hirji R, Adam L, Rozwadowski K L, Hammerlindl J K, Keller W A and Selvaraj G 2000 Genetic engineering of glycinebetaine production toward enhancing stress tolerance in plants: metabolic limitations *Plant Physiol.* **122** 747–56

[86] Hibino T, Waditee R, Araki E, Ishikawa H, Aoki K, Tanaka Y and Takabe T 2002 Functional characterization of choline monooxygenase, an enzyme for betaine synthesis in plants *J. Biol. Chem.* **277** 41352–60

[87] Su J, Hirji R, Zhang L, He C, Selvaraj G and Wu R 2006 Evaluation of the stress-inducible production of choline oxidase in transgenic rice as a strategy for producing the stress-protectant glycine betaine *J. Exp. Bot.* **57** 1129–35

[88] Khan M S, Yu X, Kikuchi A, Asahina M and Watanabe K N 2009 Genetic engineering of glycine betaine biosynthesis to enhance abiotic stress tolerance in plants *Plant Biotechnol.* **26** 125–34

[89] Park E J, Jeknić Z, Sakamoto A, DeNoma J, Yuwansiri R, Murata N and Chen T H 2004 Genetic engineering of glycinebetaine synthesis in tomato protects seeds, plants, and flowers from chilling damage *Plant J.* **40** 474–87

[90] Park E J, Jeknić Z, Pino M T, Murata N and Chen T H H 2007 Glycinebetaine accumulation is more effective in chloroplasts than in the cytosol for protecting transgenic tomato plants against abiotic stress *Plant Cell Environ.* **30** 994–1005

[91] Leeuwenhoek A V 1978 Observationes D. Anthonii Lewenhoeck, de natis'e semine genitali animalculis *Philos. Trans. R. Soc.* **12** 1040–6

[92] Gupta K, Dey A and Gupta B 2013 Plant polyamines in abiotic stress responses *Acta Physiol. Plant.* **35** 2015–36

[93] Bano C, Amist N and Singh N 2020 *Plant Life Under Changing Environment* (Amsterdam: Elsevier) pp 481–96

[94] Minocha R, Majumdar R and Minocha S C 2014 Polyamines and abiotic stress in plants: a complex relationship *Front. Plant Sci.* **5** 175

[95] Lefevre I, Gratia E and Lutts S 2001 Discrimination between the ionic and osmotic components of salt stress in relation to free polyamine level in rice (*Oryza sativa*) *Plant Sci.* **161** 943–52

[96] Alcázar R, Altabella T, Marco F, Bortolotti C, Reymond M, Koncz C, Carrasco P and Tiburcio A F 2010 Polyamines: molecules with regulatory functions in plant abiotic stress tolerance *Planta* **231** 1237–49

[97] Liu K, Fu H, Bei Q and Luan S 2000 Inward potassium channel in guard cells as a target for polyamine regulation of stomatal movements *Plant Physiol.* **124** 1315–26

[98] Liu J-H, Kitashiba H, Wang J, Ban Y and Moriguchi T 2007 Polyamines and their ability to provide environmental stress tolerance to plants *Plant Biotechnol.* **24** 117–26

[99] Urano K, Yoshiba Y, Nanjo T, Igarashi Y, Seki M, Sekiguchi F, Yamaguchi-Shinozaki K and Shinozaki K 2003 Characterization of *Arabidopsis* genes involved in biosynthesis of polyamines in abiotic stress responses and developmental stages *Plant Cell Environ.* **26** 1917–26

[100] Cuevas J C, López-Cobollo R, Alcázar R, Zarza X, Koncz C, Altabella T, Salinas J, Tiburcio A F and Ferrando A 2008 Putrescine is involved in *Arabidopsis* freezing tolerance and cold acclimation by regulating abscisic acid levels in response to low temperature *Plant Physiol.* **148** 1094–105

[101] Cuevas J C, López-Cobollo R, Alcázar R, Zarza X, Koncz C, Altabella T, Salinas J, Tiburcio A F and Ferrando A 2009 Putrescine as a signal to modulate the indispensable ABA increase under cold stress *Plant Signal. Behav.* **4** 219–20

[102] Alcázar R, Cuevas J C, Patron M, Altabella T and Tiburcio A F 2006 Abscisic acid modulates polyamine metabolism under water stress in *Arabidopsis thaliana Physiol. Plant.* **128** 448–55

[103] Yamaguchi K, Takahashi Y, Berberich T, Imai A, Miyazaki A, Takahashi T, Michael A and Kusano T 2006 The polyamine spermine protects against high salt stress in *Arabidopsis thaliana FEBS Lett.* **580** 6783–8

[104] Evans P T and Malmberg R L 1989 Do polyamines have roles in plant development? *Ann. Rev. Plant Biol.* **40** 235–69

[105] Alcázar R, Marco F, Cuevas J C, Patron M, Ferrando A, Carrasco P, Tiburcio A F and Altabella T 2006 Involvement of polyamines in plant response to abiotic stress *Biotechnol. Lett* **28** 1867–76

[106] Duan J, Li J, Guo S and Kang Y 2008 Exogenous spermidine affects polyamine metabolism in salinity-stressed *Cucumis sativus* roots and enhances short-term salinity tolerance *J. Plant Physiol.* **165** 1620–35

[107] Mapelli S, Brambilla I, Radyukina N, Ivanov Y V, Kartashov A, Reggiani R and Kuznetsov V V 2008 Free and bound polyamines changes in different plants as a consequence of UV-B light irradiation *Gen. Appl. Plant Physiol.* **34** 55–66

[108] Sung M S, Chow T J and Lee T M 2011 Polyamine acclimation alleviates hypersalinity-induced oxidative stress in a marine green macroalga, ulva fasciata, by modulation of antioxidative enzyme gene expression 1 *J. Phycol.* **47** 538–47

[109] Rider J, Hacker A, Mackintosh C, Pegg A, Woster P and Casero R 2007 Spermine and spermidine mediate protection against oxidative damage caused by hydrogen peroxide *Amino Acids* **33** 231–40

[110] Groppa M D, Benavides M P and Tomaro M L 2003 Polyamine metabolism in sunflower and wheat leaf discs under cadmium or copper stress *Plant Sci.* **164** 293–9

[111] Shevyakova N, Cheremisina A and Kuznetsov V V 2011 Phytoremediation potential of *Amaranthus* hybrids: antagonism between nickel and iron and chelating role of polyamines *Russ. J. Plant Physiol.* **58** 634–42

[112] Brieger L 1885 *Ueber ptomaine* Hirschwald

[113] Ladenburg A and Abel J 1888 Ueber das aethylenimin (Spermin?) *Ber. Dtsch. Chem. Ges.* **21** 758–66

[114] Dudley H W, Rosenheim O and Starling W W 1927 The constitution and synthesis of spermidine, a newly discovered base isolated from animal tissues *Biochem. J* **21** 97

[115] Kusano T, Berberich T, Tateda C and Takahashi Y 2008 Polyamines: essential factors for growth and survival *Planta* **228** 367–81

[116] Slocum R D 1991 Polyamine biosynthesis in plants *Biochemistry and Physiology of Polyamines in Plants* (Boca Raton, FL: CRC Press) pp 23–40

[117] Knott J M, Römer P and Sumper M 2007 Putative spermine synthases from *Thalassiosira pseudonana* and *Arabidopsis thaliana* synthesize thermospermine rather than spermine *FEBS Lett.* **581** 3081–6

[118] Bouchereau A, Aziz A, Larher F and Martin-Tanguy J 1999 Polyamines and environmental challenges: recent development *Plant Sci.* **140** 103–25

[119] Kuznetsov V V and Shevyakova N I 2007 Polyamines and stress tolerance of plants *Plant Stress* **1** 50–71

[120] Richards F and Coleman R 1952 Occurrence of putrescine in potassium-deficient barley *Nature* **170** 460

[121] Alcázar R, Cuevas J C, Planas J, Zarza X, Bortolotti C, Carrasco P, Salinas J, Tiburcio A F and Altabella T 2011 Integration of polyamines in the cold acclimation response *Plant Sci.* **180** 31–8

[122] Flores H E and Galston A W 1982 Polyamines and plant stress: activation of putrescine biosynthesis by osmotic shock *Science* **217** 1259–61

[123] Zapata P J, Serrano M a, Pretel M T, Amorós A and Botella M Á 2004 Polyamines and ethylene changes during germination of different plant species under salinity *Plant Sci.* **167** 781–8

[124] Groppa M and Benavides M 2008 Polyamines and abiotic stress: recent advances *Amino Acids* **34** 35

[125] Moschou P N, Paschalidis K A, Delis I D, Andriopoulou A H, Lagiotis G D, Yakoumakis D I and Roubelakis-Angelakis K A 2008 Spermidine exodus and oxidation in the apoplast induced by abiotic stress is responsible for H_2O_2 signatures that direct tolerance responses in tobacco *Plant Cell* **20** 1708–24

[126] Prakash L and Prathapasenan G 1988 Effect of NaCl salinity and putrescine on shoot growth, tissue ion concentration and yield of rice (*Oryza sativa* L. var. GR-3) *J. Agron. Crop Sci.* **160** 325–34

[127] Basu R, Maitra N and Ghosh B 1988 Salinity results in polyamine accumulation in early rice (*Oryza sativa* L.). seedlings *Funct. Plant Biol.* **15** 777–86

[128] Soyka S and Heyer A G 1999 *Arabidopsis* knockout mutation of ADC2 gene reveals inducibility by osmotic stress *FEBS Lett.* **458** 219–23

[129] Roy M and Wu R 2001 Arginine decarboxylase transgene expression and analysis of environmental stress tolerance in transgenic rice *Plant Sci.* **160** 869–75

[130] Roy M and Wu R 2002 Overexpression of S-adenosylmethionine decarboxylase gene in rice increases polyamine level and enhances sodium chloride-stress tolerance *Plant Sci.* **163** 987–92

[131] Kumria R and Rajam M V 2002 Ornithine decarboxylase transgene in tobacco affects polyamines, *in vitro*-morphogenesis and response to salt stress *J. Plant Physiol.* **159** 983–90

[132] Legocka J and Kluk A 2005 Effect of salt and osmotic stress on changes in polyamine content and arginine decarboxylase activity in *Lupinus luteus* seedlings *J. Plant Physiol.* **162** 662–8

[133] Liu J-H, Nada K, Honda C, Kitashiba H, Wen X-P, Pang X-M and Moriguchi T 2006 Polyamine biosynthesis of apple callus under salt stress: importance of the arginine decarboxylase pathway in stress response *J. Exp. Bot.* **57** 2589–99

[134] Jiménez-Bremont J F, Ruiz O A and Rodríguez-Kessler M 2007 Modulation of spermidine and spermine levels in maize seedlings subjected to long-term salt stress *Plant Physiol. Biochem.* **45** 812–21

[135] Tassoni A, Franceschetti M and Bagni N 2008 Polyamines and salt stress response and tolerance in *Arabidopsis thaliana* flowers *Plant Physiol. Biochem.* **46** 607–13

[136] Roychoudhury A, Basu S and Sengupta D N 2011 Amelioration of salinity stress by exogenously applied spermidine or spermine in three varieties of indica rice differing in their level of salt tolerance *J. Plant Physiol.* **168** 317–28

[137] Borrell A, Besford R, Altabella T, Masgrau C and Tiburcio A 1996 Regulation of arginine decarboxylase by spermine in osmotically-stressed oat leaves *Physiol. Plant.* **98** 105–10

[138] An Z, Jing W, Liu Y and Zhang W 2008 Hydrogen peroxide generated by copper amine oxidase is involved in abscisic acid-induced stomatal closure in *Vicia faba J. Exp. Bot.* **59** 815–25

[139] Mo H and Pua E C 2002 Up-regulation of arginine decarboxylase gene expression and accumulation of polyamines in mustard (*Brassica juncea*) in response to stress *Physiol. Plant.* **114** 439–49

[140] Capell T, Bassie L and Christou P 2004 Modulation of the polyamine biosynthetic pathway in transgenic rice confers tolerance to drought stress *Proc. Natl. Acad. Sci.* **101** 9909–14

[141] Hao Y-J, Zhang Z, Kitashiba H, Honda C, Ubi B, Kita M and Moriguchi T 2005 Molecular cloning and functional characterization of two apple S-adenosylmethionine decarboxylase genes and their different expression in fruit development, cell growth and stress responses *Gene* **350** 41–50

[142] Akiyama T and Jin S 2007 Molecular cloning and characterization of an arginine decarboxylase gene up-regulated by chilling stress in rice seedlings *J. Plant Physiol.* **164** 645–54

[143] Urano K, Yoshiba Y, Nanjo T, Ito T, Yamaguchi-Shinozaki K and Shinozaki K 2004 *Arabidopsis* stress-inducible gene for arginine decarboxylase AtADC2 is required for accumulation of putrescine in salt tolerance *Biochem. Biophys. Res. Commun.* **313** 369–75

[144] Prabhavathi V R and Rajam M V 2007 Polyamine accumulation in transgenic eggplant enhances tolerance to multiple abiotic stresses and fungal resistance *Plant Biotechnol.* **24** 273–82

[145] Hummel I, Gouesbet G, El Amrani A, Aïnouche A and Couée I 2004 Characterization of the two arginine decarboxylase (polyamine biosynthesis) paralogues of the endemic subantarctic cruciferous species *Pringlea antiscorbutica* and analysis of their differential expression during development and response to environmental stress *Gene* **342** 199–209

[146] Bhatnagar P 2001 Transgenic manipulation of polyamine metabolism in poplar (*Populus nigra* X maximoviczii) cells *Plant Physiol.* **125** 2139–53

[147] He L, Ban Y, Inoue H, Matsuda N, Liu J and Moriguchi T 2008 Enhancement of spermidine content and antioxidant capacity in transgenic pear shoots overexpressing apple spermidine synthase in response to salinity and hyperosmosis *Phytochemistry* **69** 2133–41

[148] Kasukabe Y, He L, Nada K, Misawa S, Ihara I and Tachibana S 2004 Overexpression of spermidine synthase enhances tolerance to multiple environmental stresses and up-regulates the expression of various stress-regulated genes in transgenic *Arabidopsis thaliana Plant Cell Physiol.* **45** 712–22

[149] Kasukabe Y, He L, Watakabe Y, Otani M, Shimada T and Tachibana S 2006 Improvement of environmental stress tolerance of sweet potato by introduction of genes for spermidine synthase *Plant Biotechnol.* **23** 75–83

[150] Wen X-P, Ban Y, Inoue H, Matsuda N and Moriguchi T 2009 Aluminum tolerance in a spermidine synthase-overexpressing transgenic European pear is correlated with the enhanced level of spermidine via alleviating oxidative status *Environ. Exp. Bot.* **66** 471–8

[151] Wi S J and Park K Y 2002 Antisense expression of carnation cDNA encoding ACC synthase or ACC oxidase enhances polyamine content and abiotic stress tolerance in transgenic tobacco plants *Mol. Cell* **13** 209–20

[152] Waie B and Rajam M V 2003 Effect of increased polyamine biosynthesis on stress responses in transgenic tobacco by introduction of human S-adenosylmethionine gene *Plant Sci.* **164** 727–34

[153] Wi S J, Kim W T and Park K Y 2006 Overexpression of carnation S-adenosylmethionine decarboxylase gene generates a broad-spectrum tolerance to abiotic stresses in transgenic tobacco plants *Plant Cell Rep.* **25** 1111–21

[154] Altabella T, Tiburcio A and Ferrando A 2009 Plant with resistance to low temperature and method of production thereof *Spanish patent application WO2010l004070*

[155] Cheng L, Zou Y, Ding S, Zhang J, Yu X, Cao J and Lu G 2009 Polyamine accumulation in transgenic tomato enhances the tolerance to high temperature stress *J. Integr. Plant Biol.* **51** 489–99

[156] Cook D, Fowler S, Fiehn O and Thomashow M F 2004 A prominent role for the CBF cold response pathway in configuring the low-temperature metabolome of *Arabidopsis Proc. Natl. Acad. Sci. USA* **101** 15243–8

[157] Usadel B, Blaesing O E, Gibon Y, Poree F, Hoehne M, Guenter M, Trethewey R, Kamlage B, Poorter H and Stitt M 2008 Multilevel genomic analysis of the response of transcripts, enzyme activities and metabolites in *Arabidopsis rosettes* to a progressive decrease of temperature in the non-freezing range *Plant Cell Environ.* **31** 518–47

[158] Perez-Amador M A, Leon J, Green P J and Carbonell J 2002 Induction of the arginine decarboxylase ADC2 gene provides evidence for the involvement of polyamines in the wound response in *Arabidopsis Plant Physiol.* **130** 1454–63

[159] Armengaud P, Breitling R and Amtmann A 2004 The potassium-dependent transcriptome of *Arabidopsis* reveals a prominent role of jasmonic acid in nutrient signaling *Plant Physiol.* **136** 2556–76

[160] Bhattacharya S and Kundu A 2020 Sugars and sugar polyols in overcoming environmental stresses *Protective Chemical Agents in the Amelioration of Plant Abiotic Stress: Biochemical and Molecular Perspectives* (New York: Wiley) pp 71–101

[161] Ahmad F, Singh A and Kamal A 2020 Osmoprotective role of sugar in mitigating abiotic stress in plants *Protective Chemical Agents in the Amelioration of Plant Abiotic Stress: Biochemical and Molecular Perspectives* (New York: Wiley) pp 53–70

[162] Romero C, Bellés J M, Vayá J L, Serrano R and Culiáñez-Macià F A 1997 Expression of the yeast trehalose-6-phosphate synthase gene in transgenic tobacco plants: pleiotropic phenotypes include drought tolerance *Planta* **201** 293–7

[163] Keunen E, Peshev D, Vangronsveld J, Van Den Ende W and Cuypers A 2013 Plant sugars are crucial players in the oxidative challenge during abiotic stress: extending the traditional concept *Plant Cell Environ.* **36** 1242–55

[164] Li H-J, Yang A-F, Zhang X-C, Gao F and Zhang J-R 2007 Improving freezing tolerance of transgenic tobacco expressing sucrose: sucrose 1-fructosyltransferase gene from *Lactuca sativa Plant Cell, Tissue Organ Cult.* **89** 37–48

[165] Nishizawa A, Yabuta Y and Shigeoka S 2008 Galactinol and raffinose constitute a novel function to protect plants from oxidative damage *Plant Physiol.* **147** 1251–63

[166] Tarczynski M C, Jensen R G and Bohnert H J 1992 Expression of a bacterial mtlD gene in transgenic tobacco leads to production and accumulation of mannitol *Proc. Natl. Acad. Sci. USA* **89** 2600–4

[167] Sheveleva E, Chmara W, Bohnert H J and Jensen R G 1997 Increased salt and drought tolerance by D-ononitol production in transgenic *Nicotiana tabacum* L. *Plant Physiol.* **115** 1211–9

[168] Flores S 1997 Overexpression of D-myo-inositol-3-phosphate synthase leads to elevated levels of inositol in *Arabidopsis Plant Mol. Biol.* **33** 811–20

[169] Sheveleva E V, Marquez S, Chmara W, Zegeer A, Jensen R G and Bohnert H J 1998 Sorbitol-6-phosphate dehydrogenase expression in transgenic tobacco: high amounts of sorbitol lead to necrotic lesions *Plant Physiol.* **117** 831–9

[170] Yeo E-T, Kwon H-B, Han S-E, Lee J-T, Ryu J-C and Byu M 2000 Genetic engineering of drought resistant potato plants by introduction of the trehalose-6-phosphate synthase (TPS1) gene from *Saccharomyces cerevisiae Mol. Cell* **10** 263–8

[171] Cortina C and Culiáñez-Macià F A 2005 Tomato abiotic stress enhanced tolerance by trehalose biosynthesis *Plant Sci.* **169** 75–82

[172] Garg A K, Kim J-K, Owens T G, Ranwala A P, Do Choi Y, Kochian L V and Wu R J 2002 Trehalose accumulation in rice plants confers high tolerance levels to different abiotic stresses *Proc. Natl. Acad. Sci. USA* **99** 15898–903

[173] Guo W-J, Nagy R, Chen H-Y, Pfrunder S, Yu Y-C, Santelia D, Frommer W B and Martinoia E 2014 SWEET17, a facilitative transporter, mediates fructose transport across the tonoplast of *Arabidopsis* roots and leaves *Plant Physiol.* **164** 777–89

[174] Miranda J A, Avonce N, Suárez R, Thevelein J M, Van Dijck P and Iturriaga G 2007 A bifunctional TPS–TPP enzyme from yeast confers tolerance to multiple and extreme abiotic-stress conditions in transgenic *Arabidopsis Planta* **226** 1411–21

[175] Han S-E, Park S-R, Kwon H-B, Yi B-Y, Lee G-B and Byun M-O 2005 Genetic engineering of drought-resistant tobacco plants by introducing the trehalose phosphorylase (TP) gene from *Pleurotus sajor-caju Plant Cell, Tissue Organ Cult.* **82** 151–8

[176] Pilon-Smits E A, Ebskamp M J, Paul M J, Jeuken M J, Weisbeek P J and Smeekens S C 1995 Improved performance of transgenic fructan-accumulating tobacco under drought stress *Plant Physiol.* **107** 125–30

[177] Pilon-Smits E A, Terry N, Sears T and van Dun K 1999 Enhanced drought resistance in fructan-producing sugar beet *Plant Physiol. Biochem.* **37** 313–7

[178] Hisano H, Kanazawa A, Kawakami A, Yoshida M, Shimamoto Y and Yamada T 2004 Transgenic perennial ryegrass plants expressing wheat fructosyltransferase genes accumulate increased amounts of fructan and acquire increased tolerance on a cellular level to freezing *Plant Sci.* **167** 861–8

[179] Kawakami A, Sato Y and Yoshida M 2008 Genetic engineering of rice capable of synthesizing fructans and enhancing chilling tolerance *J. Exp. Bot.* **59** 793–802

[180] Parvanova D, Popova A, Zaharieva I, Lambrev P, Konstantinova T, Taneva S, Atanassov A, Goltsev V and Djilianov D 2004 Low temperature tolerance of tobacco plants

transformed to accumulate proline, fructans, or glycine betaine. Variable chlorophyll fluorescence evidence *Photosynthetica* **42** 179–85

[181] Zhuo C, Wang T, Lu S, Zhao Y, Li X and Guo Z 2013 A cold responsive galactinol synthase gene from *Medicago falcata* (MfGolS1) is induced by myo-inositol and confers multiple tolerances to abiotic stresses *Physiol. Plant.* **149** 67–78

[182] Wang Z, Zhu Y, Wang L, Liu X, Liu Y, Phillips J and Deng X 2009 A WRKY transcription factor participates in dehydration tolerance in *Boea hygrometrica* by binding to the W-box elements of the galactinol synthase (BhGolS1) promoter *Planta* **230** 1155–66

[183] Taji T, Ohsumi C, Iuchi S, Seki M, Kasuga M, Kobayashi M, Yamaguchi-Shinozaki K and Shinozaki K 2002 Important roles of drought-and cold-inducible genes for galactinol synthase in stress tolerance in *Arabidopsis thaliana Plant J.* **29** 417–26

[184] Pennycooke J C, Jones M L and Stushnoff C 2003 Down-regulating α-galactosidase enhances freezing tolerance in transgenic petunia *Plant Physiol.* **133** 901–9

[185] H-l L, X-y D, Y-y X and Chong K 2007 Over-expression of OsUGE-1 altered raffinose level and tolerance to abiotic stress but not morphology in *Arabidopsis J. Plant Physiol.* **164** 1384–90

[186] Maheswari M, Varalaxmi Y, Vijayalakshmi A, Yadav S, Sharmila P, Venkateswarlu B, Vanaja M and Saradhi P P 2010 Metabolic engineering using mtlD gene enhances tolerance to water deficit and salinity in sorghum *Biol. Plant.* **54** 647–52

[187] Rahnama A, Munns R, Poustini K and Watt M 2011 A screening method to identify genetic variation in root growth response to a salinity gradient *J. Exp. Bot.* **62** 69–77

[188] Abebe T, Guenzi A C, Martin B and Cushman J C 2003 Tolerance of mannitol-accumulating transgenic wheat to water stress and salinity *Plant Physiol.* **131** 1748–55

[189] Nguyen V L, Ribot S A, Dolstra O, Niks R E, Visser R G and van der Linden C G 2013 Identification of quantitative trait loci for ion homeostasis and salt tolerance in barley (*Hordeum vulgare* L.) *Mol. Breed.* **31** 137–52

[190] Shen B, Jensen R G and Bohnert H J 1997 Increased resistance to oxidative stress in transgenic plants by targeting mannitol biosynthesis to chloroplasts *Plant Physiol.* **113** 1177–83

[191] Pujni D, Chaudhary A and Rajam M 2007 Increased tolerance to salinity and drought in transgenic indica rice by mannitol accumulation *J. Plant Biochem. Biotechnol.* **16** 1–7

[192] Tang W, Peng X and Newton R J 2005 Enhanced tolerance to salt stress in transgenic loblolly pine simultaneously expressing two genes encoding mannitol-1-phosphate dehydrogenase and glucitol-6-phosphate dehydrogenase *Plant Physiol. Biochem.* **43** 139–46

[193] Zhifang G and Loescher W 2003 Expression of a celery mannose 6-phosphate reductase in *Arabidopsis thaliana* enhances salt tolerance and induces biosynthesis of both mannitol and a glucosyl-mannitol dimer *Plant Cell Environ.* **26** 275–83

[194] Gao M, Tao R, Miura K, Dandekar A M and Sugiura A 2001 Transformation of Japanese persimmon (*Diospyros kaki* Thunb.) with apple cDNA encoding NADP-dependent sorbitol-6-phosphate dehydrogenase *Plant Sci.* **160** 837–45

[195] Deguchi M, Koshita Y, Gao M, Tao R, Tetsumura T, Yamaki S and Kanayama Y 2004 Engineered sorbitol accumulation induces dwarfism in Japanese persimmon *J. Plant Physiol.* **161** 1177–84

[196] Majee M, Maitra S, Dastidar K G, Pattnaik S, Chatterjee A, Hait N C, Das K P and Majumder A L 2004 A novel salt-tolerant L-myo-inositol-1-phosphate synthase from *Porteresia coarctata* (Roxb.) Tateoka, a halophytic wild rice: molecular cloning, bacterial

overexpression, characterization, and functional introgression into tobacco-conferring salt tolerance phenotype *J. Biol. Chem.* **279** 28539–52

[197] Patra B, Ray S, Richter A and Majumder A L 2010 Enhanced salt tolerance of transgenic tobacco plants by co-expression of PcINO1 and McIMT1 is accompanied by increased level of myo-inositol and methylated inositol *Protoplasma* **245** 143–52

[198] Fukushima E, Arata Y, Endo T, Sonnewald U and Sato F 2001 Improved salt tolerance of transgenic tobacco expressing apoplastic yeast-derived invertase *Plant Cell Physiol.* **42** 245–9

[199] Oberschall A, Deák M, Török K, Sass L, Vass I, Kovács I, Fehér A, Dudits D and Horváth G V 2000 A novel aldose/aldehyde reductase protects transgenic plants against lipid peroxidation under chemical and drought stresses *Plant J.* **24** 437–46

[200] Martínez-Noël G M and Tognetti J A 2018 *Plant Metabolites and Regulation under Environmental Stress* (Amsterdam: Elsevier) pp 397–406

[201] Parida A K and Das A B 2005 Salt tolerance and salinity effects on plants: a review *Ecotoxicol. Environ. Saf.* **60** 324–49

[202] Ismail A M and Horie T 2017 Genomics, physiology, and molecular breeding approaches for improving salt tolerance *Ann. Rev. Plant Biol.* **68** 405–34

[203] Paul S and Roychoudhury A 2016 Seed priming with spermine ameliorates salinity stress in the germinated seedlings of two rice cultivars differing in their level of salt tolerance *Trop. Plant Res.* **3** 616–33

[204] Peng J, Liu J, Zhang L, Luo J, Dong H, Ma Y, Zhao X, Chen B, Sui N and Zhou Z 2016 Effects of soil salinity on sucrose metabolism in cotton leaves *PLoS One* **11** e0156241

[205] He C, Yu Z, da Silva J A T, Zhang J, Liu X, Wang X, Zhang X, Zeng S, Wu K and Tan J 2017 DoGMP1 from *Dendrobium officinale* contributes to mannose content of water-soluble polysaccharides and plays a role in salt stress response *Sci. Rep.* **7** 1–13

[206] Ashraf M and Harris P J 2013 Photosynthesis under stressful environments: an overview *Photosynthetica* **51** 163–90

[207] Nounjan N and Theerakulpisut P 2012 Effects of exogenous proline and trehalose on physiological responses in rice seedlings during salt-stress and after recovery *Plant, Soil Environ.* **58** 309–15

[208] Bohnert H J, Su H and Shen B 1999 Molecular mechanisms of salinity tolerance *Molecular Responses to Cold, Drought, Heat and Salt Stress in Higher Plants* (Austin, TX: R G Landes) pp 29–60

[209] Xu S-M, Liu L-X, Woo K and Wang D-L 2007 Changes in photosynthesis, xanthophyll cycle, and sugar accumulation in two North Australia tropical species differing in leaf angles *Photosynthetica* **45** 348

[210] Arabzadeh N 2012 The effect of drought stress on soluble carbohydrates (sugars) in two species of *Haloxylon persicum* and *Haloxylon aphyllum Asian J. Plant Sci.* **11** 44–51

[211] Osakabe Y, Yamaguchi-Shinozaki K, Shinozaki K and Tran L S P 2014 ABA control of plant macroelement membrane transport systems in response to water deficit and high salinity *New Phytol.* **202** 35–49

[212] Saeidi M, Moradi F and Abdoli M 2017 Impact of drought stress on yield, photosynthesis rate, and sugar alcohols contents in wheat after anthesis in semiarid region of Iran *Arid. Land Res. Manag.* **31** 204–18

[213] Crowe J H, Hoekstra F A and Crowe L M 1992 Anhydrobiosis *Annu. Rev. Physiol.* **54** 579–99

[214] Mohammadkhani N and Heidari R 2008 Drought-induced accumulation of soluble sugars and proline in two maize varieties *World Appl. Sci. J.* **3** 448–53
[215] Yuanyuan M, Yali Z, Jiang L and Hongbo S 2009 Roles of plant soluble sugars and their responses to plant cold stress *Afr. J. Biotechnol.* **8** 2004–10
[216] Li Z-G, Gong M, Xie H, Yang L and Li J 2012 Hydrogen sulfide donor sodium hydrosulfide-induced heat tolerance in tobacco (*Nicotiana tabacum* L.) suspension cultured cells and involvement of Ca^{2+} and calmodulin *Plant Sci.* **185** 185–9
[217] Jain M, Prasad P V, Boote K J, Hartwell A L and Chourey P S 2007 Effects of season-long high temperature growth conditions on sugar-to-starch metabolism in developing microspores of grain sorghum (*Sorghum bicolor* L. Moench) *Planta* **227** 67–79

IOP Publishing

Molecular Response and Genetic Engineering for Stress in Plants, Volume 1

Abiotic stress

Pawan Shukla, Anirudh Kumar, Rakesh Kumar and Manish K Pandey

Chapter 13

Secondary metabolites and plant abiotic stress responses

Himabindu Vasuki Kilambi[1], Shweta[2], Deepanker Yadav[3], Kamal Tyagi[4]*

[1]*School of Biological Sciences, NISER, Jatni, Bhubaneswar, India*

[2]*Department of Chemistry, Institute of Science, Banaras Hindu University, Banaras, India*

[3]*Department of Botany, Guru Ghasidas Vishwavidyalaya, Bilaspur, India*

[4]*Department of Postharvest Science, Agricultural Research Organization, The Volcani Center, Rishon LeZion, Israel*

World agriculture is facing severe challenges in maintaining a sufficient food supply and quality standards. Adverse environmental conditions like increased salt concentration and metal contamination significantly decrease the productivity and quality of crop species globally. Plants use two types of antioxidant systems to reduce the levels of reactive oxygen species: different antioxidative enzymes and different metabolites with antioxidative properties such as ascorbate, polyamine, reduced glutathione, and tocopherol. These metabolites can be primary metabolites or secondary metabolites. Secondary metabolic pathways play a vital role in the plant's interaction with its environment. In particular, secondary metabolites have different physiological roles than basic physiology, such as pollination or seed dispersal, as well as in protection against biotic and abiotic stresses. This chapter summarizes the influence of different abiotic stresses on the production of secondary metabolites and engineering for their production for abiotic stress tolerance.

13.1 Introduction

A plant faces several changes in its surroundings during its life cycle. These chemical, physical, and living factors surrounding the plants keep changing on temporal and spatial scales. The plant must optimize its physiology to survive and complete its life cycle with these changes. Plants have developed specific strategies to survive under

* Author to whom any correspondence should be addressed Email: tyagi.kamal6672@gmail.com.

different environmental conditions. These strategies are the outcome of evolutionary adaptation in plants to grow in various conditions. In addition to this, specific mechanisms help the plant respond to day-to-day changes in its surrounding conditions. The signal for any change in the environment eventually leads to the expression of different genes involved in expressing a range of stress-responsive proteins. These proteins can be categorized based on their function during the stress, including redox activity, ion transporter proteins, chaperones, proteins involved in synthesizing osmolytes, hormone biosynthesis, etc. The plant can produce a diverse range of small molecules during stress responses known as metabolites. This diverse range of metabolites present in plants results from evolutionary processes. Metabolites can be grouped into primary metabolites, common to all plants and essential for their cellular growth and development, and secondary metabolites (SMs), referring to compounds that are typically not required for cell survival and growth but which are of great importance to the plant's continued existence, in particular with regards to environmental conditions. Primary metabolites derived from glycolysis, tricarboxylic acid (TCA) cycle, pentose phosphate, or the shikimate pathway often serve as precursors for synthesizing the tens of thousands of SMs. SMs also have essential roles in pollination, seed dispersal, and interaction of plants with mutualist species of fungi and bacteria. In addition, SMs are also involved in chemical interactions with other plants. Similarly, herbivory-associated plant defense responses often involve SMs [1, 2].

From the human perspective, SMs also have applications in pharmacology, novel materials, agriculture, chemical industry, and forestry, impacting human life in every aspect [2]. Plant SMs such as morphine, codeine, cocaine, quinine, taxol, and perilla alcohol can have medicinal properties. Fragrance and flavoring properties are present in fruits due to SMs (e.g., menthol, nootkatone, sclareol, linalool, geraniol, benzaldehyde, and benzyl alcohol) in flowers, making them commercially important for cosmetics and agricultural products. Different stress conditions cause a change in various phenotypic characters, leading to reducing biomass [3]. In general, SMs' synthesis increases when plants encounter any stresses. Accumulation of phenylpropanoids and phenolic compounds was found to be increased during stress conditions in plants [4, 5].

13.2 Abiotic stresses and their consequences

Plants have evolved strategies to cope with various abiotic stresses. Under stress conditions, a plant's metabolism get disturbed, with a decrease in growth and productivity, and it triggers a series of morphological, molecular, biochemical, and physiological changes within the plant. Plants use certain physiological barricades as a first line of defense against any stress. The second line of defense is chemical modification through synthesis of various SMs and production of reactive oxygen species (ROS)-scavenging enzymes and nonenzymatic antioxidants like ascorbate and glutathione. Also, by increasing the production of certain organic compounds like proline, glycine betaine, raffinose, and mannitol, which help maintain cellular osmotic pressure without disturbing homeostasis [6, 7]. A few proteins, such as late embryogenesis-related proteins, help to maintain protein structure and prevent denaturation when the seed goes through desiccation. Heat shock proteins are also induced during drought stress conditions. Under salt stress conditions, different ion transporters also become activated to maintain the cellular ion homeostasis [8]. Compatible solutes accumulation, such as proline, sugar alcohols, and betaines, is a prevalent response to protect plants against climatic stresses [9]. Proline has been used as a prevalent marker under drought and salt stress conditions and its level was found to increase in many different species under drought and salinity stress.

13.3 Secondary metabolites and their biosynthesis in plants

Plants produce 200,000 specialized metabolites from different secondary metabolic pathways through modifications, such as methylation, hydroxylation, and glycosylation. The growth conditions and temperature can significantly influence the quality and quantity of SMs [10]. A schematic for the SMs' biosynthesis and their interrelations is shown in figure 13.1. Secondary metabolites' synthesis commences from primary metabolic pathways like glycolysis or shikimic acid pathways and successively diversifies, mainly depending on developmental stage, cell type, and environmental cues. Based on their biosynthetic origin, secondary plant metabolites can be divided into three main classes: (1) phenolics, derived from the malonic acid pathway or in some cases the shikimic acid pathway; (2) terpenoids/isoprenoids, derived from the mevalonic acid (MVA) pathway and methylerythritol phosphate (MEP) pathway; and (3) nitrogen/sulfur-containing compounds via the TCA or sometimes via the shikimic acid pathway.

Phenolics includes a vast group of SMs derived from phenylpropanoid. These are produced from primary metabolites like tyrosine or phenylalanine, the products of

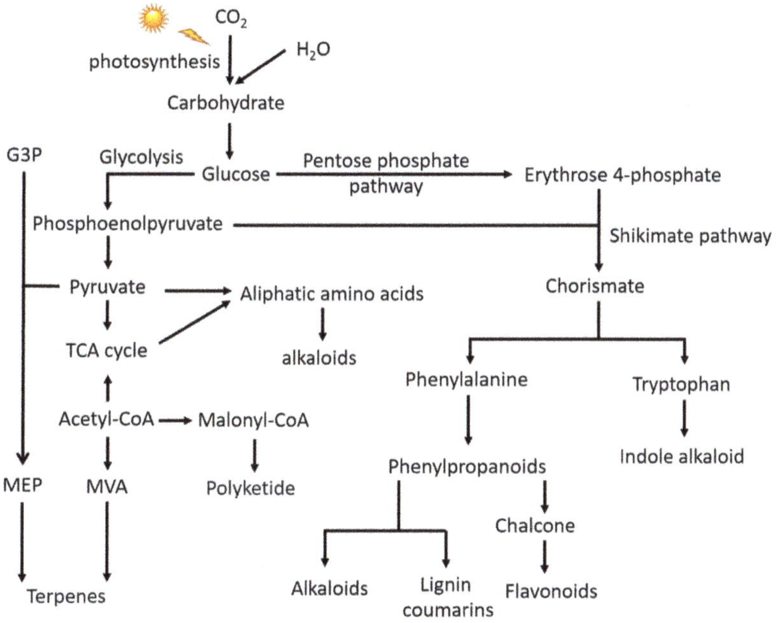

Figure 13.1. Biosynthesis of secondary metabolites and their interrelationships with the primary metabolic pathway. Abbreviations: methylerythritol phosphate, MEP; mevalonic acid pathway, MVA; tricarboxylic acid cycle, TCA; glyceraldehyde-3-phosphate, G3P.

the shikimic acid pathway through a series of enzymatic reactions via the phenylpropanoid, flavonoid, and stilbenoid pathways. On a structural basis, phenylpropanoids can be grouped into five categories: coumarins, flavonoids, phenolic acids, monolignols, and stilbenes [11–14]. Figure 13.2 shows the different classes of

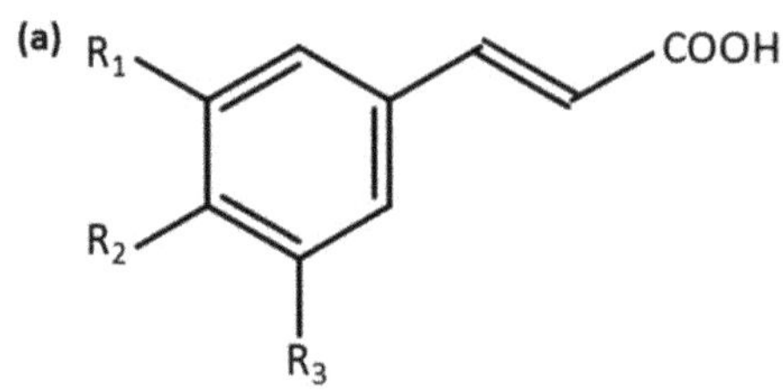

Name	R_1	R_2	R_3
p-Coumaric acid	-H	-OH	-H
Caffeic acid	-OH	-OH	-H
Ferulic acid	$-OCH_3$	-OH	-H

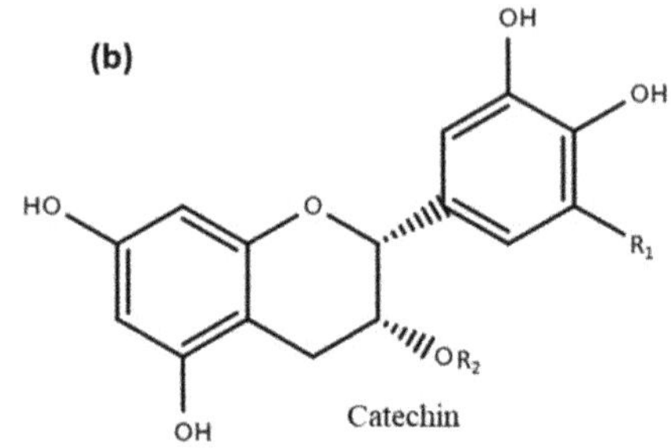

Name	R_1	R_2
Epicatechin	-H	-H
Epicatechin gallate (ECG)	-H	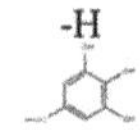
Epigallocatechin (EGC)	-OH	-H

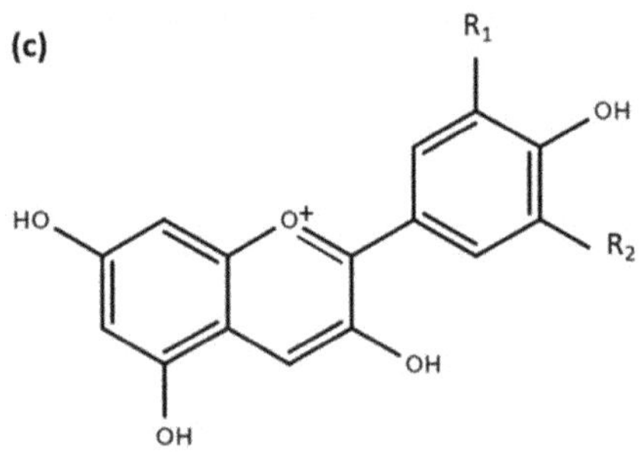

Name	R_1	R_2
Cyanidin	-OH	-H
Delphinidin	-OH	-OH
Peonidin	$-OCH_3$	-H
Petunidin	-OH	$-OCH_3$
Malvidin	$-OCH_3$	$-OCH_3$

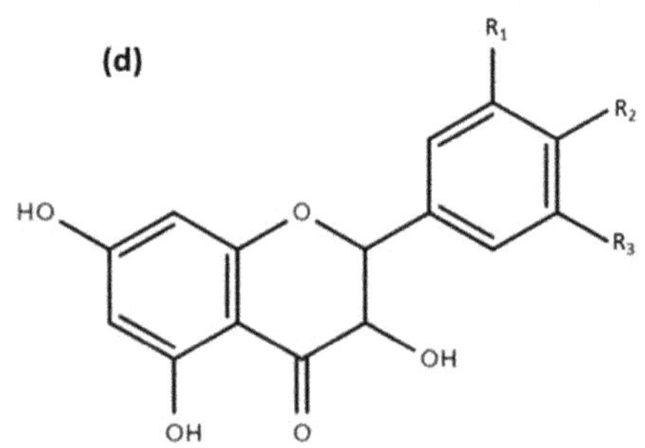

Name	R_1	R_2	R_3
Kaempferol	-H	-OH	-H
Myricetin	-OH	-OH	-OH
Quercetin	-OH	-OH	-H

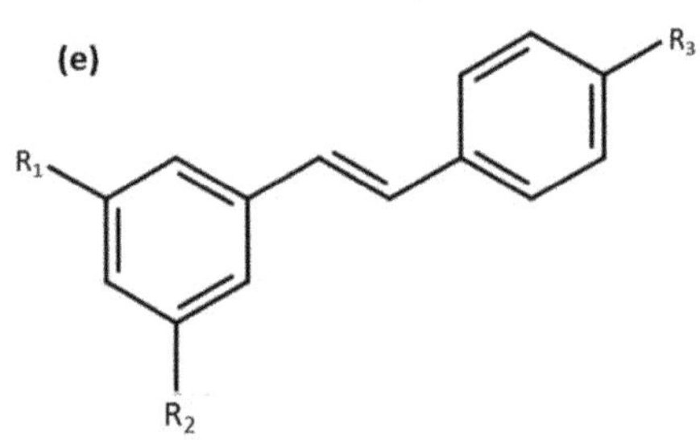

Name	R_1	R_2	R_3
Resveratrol	-OH	-OH	-OH
Pterostilbene	-OCH3	-OCH3	-OH
Pinosylvin	-OH	-OH	-H

Figure 13.2. Structure of different classes of phenolic compounds. (a) Hydroxycinnamic acids, (b) flavan-3-ols, (c) anthocyanins, (d) flavonols, and (e) stilbenoid. The picture was adopted from Tyagi *et al*, 2022 [101].

phenolic compounds. In the initial steps, phenylalanine ammonia-lyase, cinnamate-4-hydroxylase, and 4-coumarate-CoA ligase sequentially catalyzes the production of p-coumaryl-CoA from phenylalanine. These three basic steps provide the basis for further steps of SM biosynthesis. *p*-coumaryl-CoA and malonyl-CoA are the substrates of both chalcone synthase and stilbene synthase, which catalyze the first steps of the flavonoid and stilbenoid pathways, respectively. One of the well-understood downstream branches following the general phenylpropanoids pathway is the flavonoid pathway, including proanthocyanins, anthocyanins, flavones, flavanonols, flavonols, and isoflavones. Most of the enzymatic steps have been very well studied in *Arabidopsis thaliana* and *Vitis vinifera* [15–18].

Terpenes/isoprenoids are derived from the MVA pathway, which is active in cytosol or/and from the plastidial 2-C-MEP pathways. Isoprenoids are the largest such class, structurally and functionally diverse, and encompass ≈55,000 known SMs [19, 20]. Terpenes are simple hydrocarbons, made of dimethylallyl pyrophosphate (DMAPP) and isoprenyl diphosphate/pyrophosphate (ID), whereas terpenoids have an oxygen moiety and additional structural rearrangements in terpene. Both pathways lead to the synthesis of isopentenyl diphosphate (IDP) and its allylic isomer dimethylallyl diphosphate (DMADP), the basic terpenoid biosynthesis building blocks being localized in various cellular compartments. Few terpenes stated as hormones, e.g., brassinosteroids-triterpenes and gibberellins-diterpenes, have different physiological roles within the plants. Isopentenyl pyrophosphate (IPP) and DMAPP are synthesized from pyruvate and glyceraldehyde-3-phosphate (figure 13.1). The enzyme 1-deoxy-D-xylulose-5-phosphate synthase condenses pyruvate and glyceraldehyde-3-phosphate. Its rate-limiting step can increase the end-products' biosynthetic production rate. In the presence of prenyltransferase, DMADP and IDP get condensed and produce bigger molecules of prenyl diphosphates, like the sesquiterpene precursor farnesyl diphosphate (FDP), the monoterpene precursor geranyl diphosphate (GDP), and the diterpene and C40 carotenoid precursor geranyl geranyl diphosphate (GGDP). This is followed by the formation of acyclic precursors like FDP, GDP, and GGDP.

Nitrogen-containing SMs are synthesized from lysine, tyrosine, and tryptophan. This group has mainly four categories of SMs: cyanogenic glycosides, alkaloids, nonprotein amino acids, and glucosinolates. The sulfur-containing SMs include about 200 compounds. Glucosinolates and their breakdown products such as epithionitriles, thiocyanates, oxazolidinethiones and isothiocyanates are some examples of this category [21].

13.4 Role of secondary metabolites in response to abiotic stress

13.4.1 Regulation of secondary metabolites

The SMs' synthesis and their homeostasis are tightly regulated processes affected by various environmental factors. Different transcription factors (TFs) are involved in the transcriptional regulation of SMs' pathways [14, 22]. Among the various studies of TFs, the MYB family has been found to be the largest one in the plant kingdom. It performs significant functions in regulating different plant developmental processes and responses to biotic and abiotic stresses. Moreover, they play a very significant role in regulating primary and secondary metabolisms, especially in phenylpropanoids biosynthesis, as in monolignols and flavonoids [23, 24]. MYB-mediated regulation of SMs is primarily conserved in plants [13, 25–27]. Regulation of flavonoid biosynthesis can be mediated by independent MYB TFs and transcriptional complexes (MYB-bHLH-WDR complex (MBW) or MYB-bHLH complex).

13.4.2 Secondary metabolites in abiotic stress

13.4.2.1 Terpene in abiotic stress

The sessile nature of plants makes them vulnerable to various stresses in the environment. To overcome this, plants have evolved very diverse secondary metabolic pathways that offer an additional tool to respond to various abiotic stressors [28]. SMs are regularly produced in plants, and significantly contribute in many physiological processes essential for the adaptation to a certain condition and completion of their life cycle. SMs can be regulated under different growth conditions and through different signaling and elicitor molecules. A few low-molecular-weight compounds like volatile terpenes (VTs) have been reported in the abiotic stress response [29, 30]. VTs can modulate plants' redox level and create a volatile emission signature profile in stress. Terpenoid and ROS interaction generally happens at the leaf–atmosphere interface; it can also modulate signaling [31, 32]. Monoterpenes' role in mitigating the ozone level by reducing the oxidative damage has been also reported [33]. Heat stress vastly damages the membranes and associated proteins of the various cellular compartments. Isoprene's amphipathic properties are helpful through transient integration with membranes and improving hydrophobic interactions of large protein complexes between themselves or with membrane lipids [34]. Plants can enhance VT production by redirecting resources to the MEP pathway, promoting plant senescence and organ abscission per hormonal presence [35]. VT production is an irreparable investment for plants, yet the production is maintained even during photoinhibition, showing its importance in stress response [31]. On the other hand, carotenoids (such as lutein, neoxanthin, and zeaxanthin) and tocopherols are involved in direct removal of ROS in response to photoinhibition [36].

It has been reported in *Salvia mirzayanni* that an increase in the production of different monoterpenoids causes a loss in photosynthetic productivity [37]; this might be due to the diversion of carbon resources toward the synthesis of SMs [38]. Monoterpene priming enhances photoresiratory and photosynthetic activity, which suggests its importance in enhancing thermotolerance in *Quercus ilex* leaves [39]. ROS-scavenging activity by a 2,2-diphenyl-1-picrylhydrazyl assay of a mixture of citrus oil showed terpenes' role as an antioxidative agent [40]. This was also supported by other findings, where monoterpene vapor caused a decrease in photodamage, which is usually associated with an oxidative burst in plants [41]. Studies concerning H_2O_2's and γ-radiation's effect on the volatile compounds' abundances showed that limonene is a significant inducible compound in rice seedlings [38]. Different toxic metals also induce accumulation of terpenoid phytoalexins momilactone A and B in rice [42]. Different terpenoids like kauralexins, zealexins, and phytoalexins also increase during stress in tissue like roots; *an2* mutant plants with a deficiency of kauralexin showed a drought-sensitive phenotype [43]. Different volatile isoprenoids and monoterpenes are stimulated under different stress environments. Some important monoterpenes, like limonene and linalool, are involved in protecting against heat-induced stress, suggesting their quenching role [44]. Drought stress leads to a remarkable change in the chlorophyll 'a' to 'b' ratio and carotenoids level [45]. Reports on *Chenopodium quinoa* showed a decrease of saponin in plants under water stress. Thermal treatments cause a slight decrease in the β-carotene level in *Brassicaceae*. *Papaver somniferum* and *Cola acuminata* seedlings also show an increase of indole alkaloids such as morphine, narcotine, and codeine under water-limited conditions [46].

Some monoterpenes like α-pinene, carene, and terpinolene are increased under high ultraviolet B (UV-B) radiation [47]. White varieties of grapevines, with higher floral and fruity nodes than red grapevines, could be due to a lack of anthocyanins in the former, which might have adapted to different abiotic stresses by enhancing terpenoid and carotenoid (norisoprenoid precursors) biosynthesis [48]. Bindon *et al* [49] found an increase in the C_{13}-norisoprenoids, β-damascenone, and β-ionone carotenoid cleavage products under low irrigation in Cabernet Sauvignon. Another report on grape C_{13}-norisoprenoids showed a link between environmental conditions and carotenoid metabolism. Primarily, increased UV exposure leads to a decrease in the berries' skins' carotenoid content at maturity due to their accelerated decline post-veraison rather than increased synthesis in green berries pre-veraison [50].

13.4.2.2 Phenolics in abiotic stress

Plant phenolics, namely flavonoids, stilbenes, tannins, coumarins, and lignins, for a major family of SMs that are involved in defense against stress. Plant tissue containing anthocyanin is usually relatively resistant to drought [51]. For instance, a purple cultivar of chili is more resistant to water deficit than a green cultivar [52]. Likewise, several abiotic stressors, such as UV, high-intensity light, wounding, drought, and nutrient deficiency, stimulate anthocyanin accumulation [53]. Under drought stress conditions, the level of phenolic compounds such as salicylic acid (SA), ferulic acid, and 4-hydroxycinnamic acid was upregulated in drought-tolerant *Glycine max* compared to control wild-type [54]. Rutin was upregulated in response to hypoxia and drought stress, whereas quercetin was upregulated under water-deficit conditions [55]. Similarly, drought stress elevates the levels of quercetin and rutin, and betulinic acid was increased in *Hypericum brasiliense* and *Artemisinin artemisia* [56]. An enhanced levels of total phenolics, flavonoids, and anthocyanins was also observed in *Triticum aestivum* under drought stress [57].

Light increases the level of flavonoids such as flavonols and anthocyanins, which are known to act as sunscreens. Flavonols, in particular, are sensitive to light; therefore, there was a reduction in the level of total flavonols in Shiraz bunches kept inside light-proof containers than control bunches [58]. At large, water deficiency results in an increase in the level of anthocyanins, stilbenoids, and sugar contents compared with normal or full irrigation in berries [59–61]. Water stress either does not affect [61] or decreases the level of flavonol content in grapes [62]. However,

UV-B radiation was shown to enhance the level of numerous flavonols (e.g., myricetin, quercetin, and kaempferol 3-glucoside) [61]. In *V. vinifera* cv. Cabernet Sauvignon, high temperatures negatively impact anthocyanin accumulation [63, 64]. Stilbene synthesis in grapes is enhanced by a vast range of abiotic stress factors including UV radiation, water deficit, wounding, ozone, anoxia, and metal ions [18, 60]. Navarro *et al* [65] showed that moderate salinity increases the level of total phenolics content in red peppers. Christie *et al* [66] reported the accumulation of anthocyanin during cold stress and under drought stress at low temperature.

13.5 Secondary metabolites as plant signaling molecules during abiotic stress responses

Abscisic acid (ABA), gibberellins (GAs), brassinosteroids (BRs), and strigolactones are plant hormones which all belong to a large family of terpenes metabolites. They play a critical role in conferring abiotic stress tolerance under unfavorable environmental conditions [67]. CPPU, synthetic cytokinin (forchlorfenuron (*N*-(2-chloro-4-pyridyl)-*N′*-phenylurea)), application altered the level of phenolics metabolites and gene expression in grape berries [68]. Here we describe a few of the important hormones derived from terpenoids and phenolics compounds.

Abscisic acid. A carotenoid derivative phytohormone, usually known as the 'stress hormone', which plays an essential role in different types of climatic stresses [69]. Among abiotic stressors, drought stress is considered as a major stress that negatively impacts plant growth and yield [70]. Accumulation of ABA during drought conditions induces many ABA-dependent drought-responsive physiological changes in the plant; one of the critical changes is the closing of stomata. ABA buildup in response to low temperatures and water deficit have positive effects on grape berry ripening [61, 62, 71]. In actively dividing tissues, the amount of ABA is almost 40 times higher when passing through drought and salt stress [72]. Several researchers have observed that nonfunctional mutants for the ABA biosynthesis pathway display sensitivity toward abiotic stress, while transgenic plants with more ABA synthesis exhibited tolerance [73, 74]. ABA performs the crucial function in desiccation tolerance by increasing the antioxidant capacity in seaweed [75]. A study in maize showed an enhanced production of ABA, H_2O_2, and NO in their leaves after exposure to UV-B radiation [76]. Numerous studies have attempted to establish the link between UV-B and ABA interaction; fascinatingly, a few concluded that the occurrence of ABA enhanced grapevines' performance against UV-B [77, 78].

Gibberellins. A tetracyclic diterpenoid carboxylic acid, associated with enhanced tolerance under abiotic stress [79, 80]. Decrease of GA content and signaling results in restricted plant growth during several stresses, including salinity, cold, and osmotic stress [80]. Drought avoidance is primarily regulated by ABA [81]. However, numerous reports have indicated that changes in the levels of GAs may also be associated [80]. During water-deficit conditions, GA levels are reduced [80], which, in turn, activates various stress-related genes [82], including antioxidant enzymes [83, 84].

Salicylic acid. A phenolic phytohormone which act as a signaling molecule during different climatic stresses [85]. In maize, priming cereal seeds with SA promoted plant growth and development. SA performs a significant function in minimizing salt-induced damage in diverse crop plants. Alhaithloul *et al* [67] observed that the application of SA to salt-stress-grown plants resulted in increased antioxidant enzymes, osmolytes, ABA, IAA (Indole -3- acetic acid), root length, fresh roots, and dry weight. Similarly, the application of SA under drought stress improves growth, photosynthetic pigment contents, fatty acid level, compatible solutes, and SMs, whereas concurrently reducing drought-induced oxidative stress biomarkers.

Brassinosteroids. A plant-steroid-based phytohormone involved in regulating plant growth, development, and responses to different environmental stimuli [67]. BRs also regulate plant responses to various abiotic stresses [67], and their deficiency leads to impairment of important physiological activities in plants that can cause phenotypic deformities. Studies have shown that exogenous BR treatment can boost the ABA level in plants and mitigate the harmful effects of drought [86]. Epibrassinolide application enhanced photosynthetic production, antioxidant defense, and leaf water status, resulting in improved tolerance to drought in tomato [87]. Interestingly, BRs are known to improve thermotolerance in both heat-sensitive and heat-tolerant plant genotypes [88].

13.6 Engineering abiotic stress tolerance

Genetic engineering has performed a significant role in the improvement and generation of various stress-resistant crops. Abiotic stresses adversely affect the plant primary production and cause severe crop losses worldwide. Hence, dealing with these harmful impacts is an urgent need in order to address subsequent issues concerning food quality and security. Considering important roles that SMs play in abiotic stress responses in plants, metabolic engineering toward a stable and proficient production of volatile and essential terpenoids may provide a long-term solution. Consequently, metabolic engineering in plants for terpenoids has been reviewed [89]; however, though increasing the production of terpenoids is the final goal, it is expensive, especially because of the possibility of disturbing cellular homeostasis. Therefore, it is important to take different paradigms into consideration before moving further. A favorable increase in precursor flux in one pathway might inadvertently deplete other pathways with shared precursors needed for the production of equally essential metabolites. Transgenic *Arabidopsis* plants overexpressing *Fragaria vesca* nerolidol synthase 1 (*FaNES1*) displayed an increased amount of linalool as well as its glycosylated and hydroxylated derivatives production, but the transgenic-type was slow in growth and development in comparison to control-type [90] due to low carbon availability for crucial metabolites or phytotoxicity of the newly introduced compounds [90–92]. In retrospect, the role of phytohormones in plant stress responses is well known. Phytohormones initiate stress signaling response in a global manner, climaxing with an integrated stress response including increased production of volatile terpenes [93]. Hence, by changing the level of stress hormones, and thus their potential effect on the stress response, may provide a better substitute than increasing volatile terpene production.

Heterologous expression of 9-cis-epoxycarotenoid dioxygenase (*NCED*) under a stress-inducible promoter (*rd29A*) in petunia plant enhanced the level of endogenous ABA, causing increased drought tolerance and decreased stomatal conductance, with some adverse effects on plant growth and development [94]. The enzyme zeaxanthin epoxidase (ZEP) plays a critical role in ABA biosynthesis and photo-protection xanthophyll cycle. In transgenic tobacco, overexpression of a novel *MsZEP* from alfalfa (*Medicago sativa*) provided drought and salt tolerance [95]. Similarly, ectopic expression of specific the GA 2-oxidase (*GA2ox*) gene showed that GA-deficient rice is slightly more tolerant to high salinity, heat and cold, and more tolerant to dehydration than control [96]. Transgenic plants overproducing glycine betaine showed increased tolerance to abiotic stress in different plant species [9].

13.7 Conclusions

Plants are not free-moving and are fixed in their growth origins, continuously exposed to the surrounding environment and different kinds of biotic and abiotic stresses [97, 98]. SMs are produced by plants to assist them against several stressors imposed by the environment [99, 100]. This chapter consolidates the precise role of SMs in abating abiotic stress and encompass the directions of their production and engineering in plants. A number of different classes of compounds, such as terpenes, phenolics, and phytohormones, are produced by plants to mitigate the negative effects caused due to damage of nucleic acids and proteins by oxidative stress. In the present era, it is possible to combine various genetic, developmental, biochemical, and physiological tools to decipher novel compounds and to assess their role in different stresses. Hence, exploring the regulation of SM synthesis and signaling could facilitate the enhancement of a plant's tolerance to abiotic stress, which can be achieved by metabolic engineering of the known, as well as novel, critical compounds involved in abiotic stress response processes.

Multiple choice questions

1. The metabolic intermediates found in living systems which are essential for growth and life are called …
   (a) Primary metabolites;
   (b) Tannins;
   (c) Secondary metabolites;
   (d) Saponins.

2. Which of the following is NOT a class of secondary metabolite?
   (a) Phenolics;
   (b) Terpenes;
   (c) Amino acids;
   (d) Alkaloids.

3. Which of the following does NOT take part in the biosynthesis of terpenes?
   (a) Acetyl-COA;
   (b) Methylerythritol phosphate;
   (c) Mevalonic acid;
   (d) Phenol.

4. Name the class of secondary metabolites which is characterized by the presence of the hydroxyl group with an aromatic ring.
   (a) Alkaloids;
   (a) Phenolics;
   (c) Glycosides;
   (d) Terpenes.

5. Which of the following classes does NOT belong to phenolic compounds?
   (a) Gossypol;
   (b) Lignans;
   (c) Xanthone;
   (d) Flavonoids.

Answers: 1 (a); 2 (c); 3 (d); 4 (b); 5 (a).

Descriptive-type questions

1. A plant's secondary metabolism is often dynamic and allows the plant to adapt to variable, heterogeneous, and stressful environments. Discuss, and provide evidence, to support this statement. Try to use specific examples from different types of stresses to illustrate your answer.
2. A number of secondary plant metabolites are known to be used for communication or signaling between plants and other organisms (insects, soil bacteria, fungi, etc), as well as between different plants. Provide examples of such signaling chemicals and describe their biological or ecological functions, using three examples from diverse interactions.
3. Primary metabolic pathways tend to be tightly controlled via key regulatory enzymes. Describe the different mechanisms by which such enzymes are regulated at the protein level. Give specific examples of enzymes studied to illustrate your answer.
4. Biotechnology and genetic engineering have found practical applications in the agricultural and food industries. Illustrate how knowledge of plant secondary metabolites has contributed to improved genetically engineered crop plants.
5. Amino acids are used for the biosynthesis of proteins as the building blocks and precursors of various bioactive secondary metabolites. Describe two examples of such pathways in general terms. Include the names of any key enzymes for such pathways that you have learned about.

References

[1] Hartmann T 2007 From waste products to ecochemicals: fifty years research of plant secondary metabolism *Phytochemistry.* **68** 2831–46

[2] do Nascimento N C and Fett-Neto A G 2010 Plant secondary metabolism and challenges in modifying its operation: an overview *Methods Mol. Biol.* **643** 1–13

[3] Pradhan S, Kant C, Verma S and Bhatia S 2017 Genome-wide analysis of the CCCH zinc finger family identifies tissue specific and stress responsive candidates in chickpea (*Cicer arietinum* L.) *PLoS One* **12** e0180469

[4] Dixon R A and Paiva N L 1995 Stress-induced phenylpropanoid metabolism *Plant Cell.* **7** 1085

[5] Chalker-Scott L and Fuchigami L H 2018 The role of phenolic compounds in plant stress responses *Low Temperature Stress Physiology in Crops* (Boca Raton, FL: CRC Press) pp 67–80

[6] Choudhury F K, Rivero R M, Blumwald E and Mittler R 2017 Reactive oxygen species, abiotic stress and stress combination *Plant J.* **90** 856–67

[7] Zhu J-K 2016 Abiotic stress signaling and responses in plants *Cell.* **167** 313–24

[8] Xu Y, Zhou Y, Hong S, Xia Z, Cui D and Guo J *et al* 2013 Functional characterization of a wheat NHX antiporter gene TaNHX2 that encodes a K^+/H^+ exchanger *PLoS One* **8** e78098

[9] Chen T H and Murata N 2002 Enhancement of tolerance of abiotic stress by metabolic engineering of betaines and other compatible solutes *Curr. Opin. Plant Biol.* **5** 250–7

[10] Meena K K, Sorty A M, Bitla U M, Choudhary K, Gupta P and Pareek A *et al* 2017 Abiotic stress responses and microbe-mediated mitigation in plants: the omics strategies *Front. Plant Sci.* **8** 172

[11] Vogt T 2010 Phenylpropanoid biosynthesis *Mol. Plant.* **3** 2–20

[12] Noel J P, Austin M B and Bomati E K 2005 Structure–function relationships in plant phenylpropanoid biosynthesis *Curr. Opin. Plant Biol.* **8** 249–53

[13] Liu J, Osbourn A and Ma P 2015 MYB transcription factors as regulators of phenylpropanoid metabolism in plants *Mol. Plant.* **8** 689–708

[14] Tyagi K, Shukla P, Rohela G K, Shabnam A A and Gautam R 2020 Plant phenolics: their biosynthesis, regulation, evolutionary significance, and role in senescence *Plant Phenolics in Sustainable Agriculture* (Berlin: Springer) pp 431–49

[15] Appelhagen I, Thiedig K, Nordholt N, Schmidt N, Huep G and Sagasser M *et al* 2014 Update on transparent testa mutants from *Arabidopsis thaliana*: characterisation of new alleles from an isogenic collection *Planta.* **240** 955–70

[16] Ichino T, Fuji K, Ueda H, Takahashi H, Koumoto Y and Takagi J *et al* 2014 GFS 9/TT 9 contributes to intracellular membrane trafficking and flavonoid accumulation in *Arabidopsis thaliana Plant J.* **80** 410–23

[17] Petrussa E, Braidot E, Zancani M, Peresson C, Bertolini A and Patui S *et al* 2013 Plant flavonoids—biosynthesis, transport and involvement in stress responses *Int. J. Mol. Sci.* **14** 14950–73

[18] Kuhn N, Guan L, Dai Z W, Wu B-H, Lauvergeat V and Gomès E *et al* 2013 Berry ripening: recently heard through the grapevine *J. Exp. Bot.* **65** 4543–59

[19] Vickers C E, Williams T C, Peng B and Cherry J 2017 Recent advances in synthetic biology for engineering isoprenoid production in yeast *Curr. Opin. Chem. Biol.* **40** 47–56

[20] Christianson D W 2008 Unearthing the roots of the terpenome *Curr. Opin. Chem. Biol.* **12** 141–50

[21] Venditti A and Bianco A 2020 Sulfur-containing secondary metabolites as neuroprotective agents *Curr. Med. Chem.* **27** 4421–36

[22] Patra B, Schluttenhofer C, Wu Y, Pattanaik S and Yuan L 2013 Transcriptional regulation of secondary metabolite biosynthesis in plants *Biochim. Biophys. Acta (BBA)-Gene Regul. Mech.* **1829** 1236–47

[23] Dubos C, Stracke R, Grotewold E, Weisshaar B, Martin C and Lepiniec L 2010 MYB transcription factors in *Arabidopsis Trends Plant Sci.* **15** 573–81

[24] Jin H and Martin C 1999 Multifunctionality and diversity within the plant MYB-gene family *Plant Mol. Biol.* **41** 577–85

[25] Hichri I, Barrieu F, Bogs J, Kappel C, Delrot S and Lauvergeat V 2011 Recent advances in the transcriptional regulation of the flavonoid biosynthetic pathway *J. Exp. Bot.* **62** 2465–83

[26] Zhou M, Sun Z, Wang C, Zhang X, Tang Y and Zhu X *et al* 2015 Changing a conserved amino acid R2R3-MYB transcription repressors results in cytoplasmic accumulation and abolishes their repressive activity in *Arabidopsis Plant J.* **84** 395–403

[27] Xu W, Dubos C and Lepiniec L 2015 Transcriptional control of flavonoid biosynthesis by MYB–bHLH–WDR complexes *Trends Plant Sci.* **20** 176–85

[28] Cramer G R, Urano K, Delrot S, Pezzotti M and Shinozaki K 2011 Effects of abiotic stress on plants: a systems biology perspective *BMC Plant Biol.* **11** 1–14

[29] Dudareva N, Negre F, Nagegowda D A and Orlova I 2006 Plant volatiles: recent advances and future perspectives *Crit. Rev. Plant Sci.* **25** 417–40

[30] Dudareva N, Pichersky E and Gershenzon J 2004 Biochemistry of plant volatiles *Plant Physiol.* **135** 1893–902

[31] Vickers C E, Gershenzon J, Lerdau M T and Loreto F 2009 A unified mechanism of action for volatile isoprenoids in plant abiotic stress *Nat. Chem. Biol.* **5** 283–91

[32] Blande J D, Holopainen J K and Niinemets Ü 2014 Plant volatiles in polluted atmospheres: stress responses and signal degradation *Plant Cell Environ.* **37** 1892–904

[33] Fares S, Loreto F, Kleist E and Wildt J 2007 Stomatal uptake and stomatal deposition of ozone in isoprene and monoterpene emitting plants *Plant Biol.* **9** e69–78

[34] Sharkey T D and Yeh S 2001 Isoprene emission from plants *Annu. Rev. Plant Biol.* **52** 407–36

[35] Korankye E A, Lada R, Asiedu S and Caldwell C 2017 Plant senescence: the role of volatile terpene compounds (VTCs) *Am. J. Plant Sci.* **8** 3120–39

[36] Boncan D A T, Tsang S S, Li C, Lee I H, Lam H-M and Chan T-F *et al* 2020 Terpenes and terpenoids in plants: interactions with environment and insects *Int. J. Mol. Sci.* **21** 7382

[37] Valifard M, Mohsenzadeh S, Kholdebarin B, Rowshan V, Niazi A and Moghadam A 2019 Effect of salt stress on terpenoid biosynthesis in *Salvia mirzayanii*: from gene to metabolite *J. Horticul. Sci. Biotechnol.* **94** 389–99

[38] Lee G W, Lee S, Chung M-S, Jeong Y S and Chung B Y 2015 Rice terpene synthase 20 (OsTPS20) plays an important role in producing terpene volatiles in response to abiotic stresses *Protoplasma.* **252** 997–1007

[39] Loreto F, Förster A, Dürr M, Csiky O and Seufert G 1998 On the monoterpene emission under heat stress and on the increased thermotolerance of leaves of *Quercus ilex* L. fumigated with selected monoterpenes *Plant Cell Environ.* **21** 101–7

[40] Singh P, Shukla R, Prakash B, Kumar A, Singh S and Mishra P K *et al* 2010 Chemical profile, antifungal, antiaflatoxigenic and antioxidant activity of *Citrus maxima Burm.* and *Citrus sinensis* (L.) Osbeck essential oils and their cyclic monoterpene, DL-limonene *Food Chem. Toxicol.* **48** 1734–40
[41] Peñuelas J and Llusià J 2002 Linking photorespiration, monoterpenes and thermotolerance in *Quercus New Phytol.* **155** 227–37
[42] Kato-Noguchi H and Kobayashi K 2009 Jasmonic acid, protein phosphatase inhibitor, metals and UV-irradiation increased momilactone A and B concentrations in the moss *Hypnum plumaeforme J. Plant Physiol.* **166** 1118–22
[43] Vaughan M M, Christensen S, Schmelz E A, Huffaker A, Mcauslane H J and Alborn H T *et al* 2015 Accumulation of terpenoid phytoalexins in maize roots is associated with drought tolerance *Plant Cell Environ.* **38** 2195–207
[44] Khakdan F, Govahi M, Mohebi Z and Ranjbar M 2021 Water deficit stress responses of monoterpenes and sesquiterpenes in different Iranian cultivars of basil *Physiol. Plant.* **173** 896–910
[45] Akula R and Ravishankar G A 2011 Influence of abiotic stress signals on secondary metabolites in plants *Plant Signal. Behav.* **6** 1720–31
[46] Yang L, Wen K-S, Ruan X, Zhao Y-X, Wei F and Wang Q 2018 Response of plant secondary metabolites to environmental factors *Molecules.* **23** 762
[47] Gil M, Pontin M, Berli F, Bottini R and Piccoli P 2012 Metabolism of terpenes in the response of grape (*Vitis vinifera* L.) leaf tissues to UV-B radiation *Phytochemistry.* **77** 89–98
[48] Ferrandino A and Lovisolo C 2014 Abiotic stress effects on grapevine (*Vitis vinifera* L.): Focus on abscisic acid-mediated consequences on secondary metabolism and berry quality *Environ. Exp. Bot.* **103** 138–47
[49] Bindon K A, Dry P R and Loveys B R 2007 Influence of plant water status on the production of C13-norisoprenoid precursors in *Vitis vinifera* L. cv. Cabernet Sauvignon grape berries *J. Agric. Food Chem.* **55** 4493–500
[50] Schultz H R, Löhnertz O, Bettner W, Bálo B and Linsenmeier A 1998 Is grape composition affected by current levels of UV-B radiation? *Vitis.* **37** 191–2
[51] Chalker-Scott L 1999 Environmental significance of anthocyanins in plant stress responses *Photochem. Photobiol.* **70** 1–9
[52] Bahler B D, Steffen K L and Orzolek M D 1991 Morphological and biochemical comparison of a purple-leafed and a green-leafed pepper cultivar *HortScience.* **26** 736
[53] Winkel-Shirley B 2001 Flavonoid biosynthesis. A colorful model for genetics, biochemistry, cell biology, and biotechnology *Plant Physiol.* **126** 485–93
[54] Wang X, Guo R, Li M, Liu Y, Zhao M and Fu H *et al* 2019 Metabolomics reveals the drought-tolerance mechanism in wild soybean (*Glycine soja*) Acta Physiol. Plant. **41** 1–11
[55] de Abreu I N and Mazzafera P 2005 Effect of water and temperature stress on the content of active constituents of *Hypericum brasiliense Choisy Plant Physiol. Biochem.* **43** 241–8
[56] Verma N and Shukla S 2015 Impact of various factors responsible for fluctuation in plant secondary metabolites *J. Appl. Res. Med. Aromat. Plants.* **2** 105–13
[57] Ma D, Sun D, Wang C, Li Y and Guo T 2014 Expression of flavonoid biosynthesis genes and accumulation of flavonoid in wheat leaves in response to drought stress *Plant Physiol. Biochem.* **80** 60–6

[58] Downey M O, Harvey J S and Robinson S P 2004 The effect of bunch shading on berry development and flavonoid accumulation in Shiraz grapes *Aust. J. Grape Wine Res.* **10** 55–73
[59] Castellarin S D, Matthews M A, Di Gaspero G and Gambetta G A 2007 Water deficits accelerate ripening and induce changes in gene expression regulating flavonoid biosynthesis in grape berries *Planta.* **227** 101–12
[60] Deluc L G, Decendit A, Papastamoulis Y, Mérillon J-M, Cushman J C and Cramer G R 2011 Water deficit increases stilbene metabolism in Cabernet Sauvignon berries *J. Agric. Food Chem.* **59** 289–97
[61] Deluc L G, Quilici D R, Decendit A, Grimplet J, Wheatley M D and Schlauch K A *et al* 2009 Water deficit alters differentially metabolic pathways affecting important flavor and quality traits in grape berries of Cabernet Sauvignon and Chardonnay *BMC Genomics.* **10** 1–33
[62] Zarrouk O, Francisco R, Pinto-Marijuan M, Brossa R, Santos R R and Pinheiro C *et al* 2012 Impact of irrigation regime on berry development and flavonoids composition in Aragonez (Syn. Tempranillo) grapevine *Agric. Water Manage.* **114** 18–29
[63] Mori K, Goto-Yamamoto N, Kitayama M and Hashizume K 2007 Loss of anthocyanins in red-wine grape under high temperature *J. Exp. Bot.* **58** 1935–45
[64] Azuma A, Yakushiji H, Koshita Y and Kobayashi S 2012 Flavonoid biosynthesis-related genes in grape skin are differentially regulated by temperature and light conditions *Planta.* **236** 1067–80
[65] Navarro J M, Flores P, Garrido C and Martinez V 2006 Changes in the contents of antioxidant compounds in pepper fruits at different ripening stages, as affected by salinity *Food Chem.* **96** 66–73
[66] Christie P J, Alfenito M R and Walbot V 1994 Impact of low-temperature stress on general phenylpropanoid and anthocyanin pathways: enhancement of transcript abundance and anthocyanin pigmentation in maize seedlings *Planta.* **194** 541–9
[67] Alhaithloul H A S, Abu-Elsaoud A M and Soliman M H 2020 Abiotic stress tolerance in crop plants: role of phytohormones *Abiotic Stress in Plants* (London: IntechOpen)
[68] Tyagi K, Maoz I, Kochanek B, Sela N, Lerno L, Ebeler S E and Lichter A 2021 Cytokinin but not gibberellin application had major impact on the phenylpropanoid pathway in grape *Horticult. Res.* **8** 1–5
[69] Vishwakarma K, Upadhyay N, Kumar N, Yadav G, Singh J and Mishra R K *et al* 2017 Abscisic acid signaling and abiotic stress tolerance in plants: a review on current knowledge and future prospects *Front. Plant Sci.* **8** 161
[70] Tripathi P, Rabara R C, Reese R N, Miller M A, Rohila J S and Subramanian S *et al* 2016 A toolbox of genes, proteins, metabolites and promoters for improving drought tolerance in soybean includes the metabolite coumestrol and stomatal development genes *BMC Genomics.* **17** 1–22
[71] Yamane T, Jeong S T, Goto-Yamamoto N, Koshita Y and Kobayashi S 2006 Effects of temperature on anthocyanin biosynthesis in grape berry skins *Am. J. Enol. Vitic.* **57** 54–9
[72] Zeevaart J A and Creelman R A 1988 Metabolism and physiology of abscisic acid *Annu. Rev. Plant Physiol. Plant Mol. Biol.* **39** 439–73
[73] Iuchi S, Kobayashi M, Taji T, Naramoto M, Seki M and Kato T *et al* 2001 Regulation of drought tolerance by gene manipulation of *9-cis-epoxycarotenoid dioxygenase*, a key enzyme in abscisic acid biosynthesis in *Arabidopsis Plant J.* **27** 325–33

[74] Qin X and Zeevaart J A 2002 Overexpression of a 9-*cis*-epoxycarotenoid dioxygenase gene in *Nicotiana plumbaginifolia* increases abscisic acid and phaseic acid levels and enhances drought tolerance *Plant Physiol.* **128** 544–51
[75] Guajardo E, Correa J A and Contreras-Porcia L 2016 Role of abscisic acid (ABA) in activating antioxidant tolerance responses to desiccation stress in intertidal seaweed species *Planta.* **243** 767–81
[76] Tossi V, Lamattina L and Cassia R 2009 An increase in the concentration of abscisic acid is critical for nitric oxide-mediated plant adaptive responses to UV-B irradiation *New Phytol.* **181** 871–9
[77] Berli F J, Fanzone M, Piccoli P and Bottini R 2011 Solar UV-B and ABA are involved in phenol metabolism of *Vitis vinifera* L. increasing biosynthesis of berry skin polyphenols *J. Agric. Food Chem.* **59** 4874–84
[78] Berli F J, Moreno D, Piccoli P, Hespanhol-Viana L, Silva M F and Bressan-Smith R *et al* 2010 Abscisic acid is involved in the response of grape (*Vitis vinifera* L.) cv. Malbec leaf tissues to ultraviolet-B radiation by enhancing ultraviolet-absorbing compounds, antioxidant enzymes and membrane sterols *Plant Cell Environ.* **33** 1–10
[79] Yamaguchi S 2008 Gibberellin metabolism and its regulation *Annu. Rev. Plant Biol.* **59** 225–51
[80] Colebrook E H, Thomas S G, Phillips A L and Hedden P 2014 The role of gibberellin signalling in plant responses to abiotic stress *J. Exp. Biol.* **217** 67–75
[81] Cutler S R, Rodriguez P L, Finkelstein R R and Abrams S 2010 Abscisic acid: emergence of a core signaling network *Annu. Rev. Plant Biol.* **61** 651–79
[82] Tuna A L, Kaya C, Dikilitas M and Higgs D 2008 The combined effects of gibberellic acid and salinity on some antioxidant enzyme activities, plant growth parameters and nutritional status in maize plants *Environ. Exp. Bot.* **62** 1–9
[83] Achard P, Gong F, Cheminant S, Alioua M, Hedden P and Genschik P 2008 The cold-inducible CBF1 factor–dependent signaling pathway modulates the accumulation of the growth-repressing DELLA proteins via its effect on gibberellin metabolism *Plant Cell.* **20** 2117–29
[84] Omena-Garcia R P, Martins A O, Medeiros D B, Vallarino J G, Ribeiro D M and Fernie A R *et al* 2019 Growth and metabolic adjustments in response to gibberellin deficiency in drought stressed tomato plants *Environ. Exp. Bot.* **159** 95–107
[85] Emamverdian A, Ding Y and Mokhberdoran F 2020 The role of salicylic acid and gibberellin signaling in plant responses to abiotic stress with an emphasis on heavy metals *Plant Signal. Behav.* **15** 1777372
[86] Wang Y-T, Chen Z-Y, Jiang Y, Duan B-B and Xi Z-M 2019 Involvement of ABA and antioxidant system in brassinosteroid-induced water stress tolerance of grapevine (*Vitis vinifera* L.) *Sci. Hortic.* **256** 108596
[87] Yuan L, Yuan Y, Du J, Sun J and Guo S 2012 Effects of 24-epibrassinolide on nitrogen metabolism in cucumber seedlings under $Ca(NO_3)_2$ stress *Plant Physiol. Biochem.* **61** 29–35
[88] Zhang Y P, Zhu X H, Ding H D, Yang S J and Chen Y Y 2013 Foliar application of 24-epibrassinolide alleviates high-temperature-induced inhibition of photosynthesis in seedlings of two melon cultivars *Photosynthetica.* **51** 341–9
[89] Aharoni A, Jongsma M A and Bouwmeester H J 2005 Volatile science? Metabolic engineering of terpenoids in plants *Trends Plant Sci.* **10** 594–602

[90] Aharoni A, Giri A P, Deuerlein S, Griepink F, de Kogel W-J and Verstappen F W *et al* 2003 Terpenoid metabolism in wild-type and transgenic *Arabidopsis* plants *Plant Cell.* **15** 2866–84

[91] Aharoni A, Jongsma M A, Kim T-Y, Ri M-B, Giri A P and Verstappen F W *et al* 2006 Metabolic engineering of terpenoid biosynthesis in plants *Phytochem. Rev.* **5** 49–58

[92] Orlova I, Nagegowda D A, Kish C M, Gutensohn M, Maeda H and Varbanova M *et al* 2009 The small subunit of snapdragon geranyl diphosphate synthase modifies the chain length specificity of tobacco geranylgeranyl diphosphate synthase in planta *Plant Cell.* **21** 4002–17

[93] Wani S H, Kumar V, Shriram V and Sah S K 2016 Phytohormones and their metabolic engineering for abiotic stress tolerance in crop plants *Crop J.* **4** 162–76

[94] Estrada-Melo A C, Reid M S and Jiang C-Z 2015 Overexpression of an ABA biosynthesis gene using a stress-inducible promoter enhances drought resistance in petunia *Horticult. Res.* **2** 1–9

[95] Zhang Z, Wang Y, Chang L, Zhang T, An J and Liu Y *et al* 2016 MsZEP, a novel zeaxanthin epoxidase gene from alfalfa (*Medicago sativa*), confers drought and salt tolerance in transgenic tobacco *Plant Cell Rep.* **35** 439–53

[96] Lo S-F, Ho T-H D, Liu Y-L, Jiang M-J, Hsieh K-T and Chen K-T *et al* 2017 Ectopic expression of specific GA2 oxidase mutants promotes yield and stress tolerance in rice *Plant Biotechnol. J.* **15** 850–64

[97] Kumar A, Gul M Z, Zeeshan A, Bimolata W, Qureshi I A and Ghazi I A 2013 Differential antioxidative responses of three different rice genotypes during bacterial blight infection *Aust. J. Crop Sci.* **7** 1893–900

[98] Kumar A *et al* 2020 Deployment of genetic and genomic tools toward gaining a better understanding of rice-*Xanthomonas oryzae* pv. oryzae interactions for development of durable bacterial blight resistant rice *Front. Plant Sci.* **11** 1152

[99] Kumar R, Bohra A, Pandey A K, Pandey M K and Kumar A 2017 Metabolomics for plant improvement: status and prospects *Front Plant Sci.* **8** 1302

[100] Sharma K, Sarma S, Bohra A, Mitra A, Sharma N K and Kumar A 2018 Plant metabolomics: an emerging technology for crop improvement *New Visions in Plant Science* ed O Çelik 1st edn (London: IntechOpen) pp 65–79

[101] Tyagi K, Lerno L, Rosso M D, Maoz I, Lichter A, Ebeler S E and Flamini R 2022 Extraction and Analysis of phenolic compounds from grape berries *Plant Secondary Metabolism Engineering* ed A G Fett-Neto (New York: Humana) pp 1–17

www.ingramcontent.com/pod-product-compliance
Ingram Content Group UK Ltd.
Pitfield, Milton Keynes, MK11 3LW, UK
UKHW051904100926
8556IPUK00005B/758

9 780750 349192